Hydrometallurgical Recycling of Lithium-Ion Battery Materials

The expanding market share of lithium-ion batteries (LIBs), driven by the secondary battery and electric vehicle markets, has consequently led to the accumulation of spent LIBs. This presents a unique business opportunity for recovering and recycling valuable metals from the spent lithium-ion cathode materials. *Hydrometallurgical Recycling of Lithium-Ion Battery Materials* provides a comprehensive review of the available hydrometallurgical technologies for recycling spent lithium-ion cathode active materials. The aim of this book is to raise awareness of LIB recycling, provide comprehensive knowledge of hydrometallurgical recycling of lithium cathode active materials, and promote an environmentally friendlier hydrometallurgical recycling process.

Key Features

- Summarizes current recycling processes, challenges, and perspectives
- Offers a comprehensive review of current commercialized LIB recycling companies
- Showcases an innovative closed-loop hydrometallurgical recycling process to recycle lithium cathode materials
- Provides detailed modeling and economic analyses of several hydrometallurgical recycling processes
- Features practical cases and data developed by the authors

Offering the most up-to-date information on LIB material recycling, this book is aimed at researchers and professionals in materials, chemical, electrical, and mechanical engineering, as well as chemists working on battery technologies.

Electrochemical Energy Storage and Conversion

Series Editor Jiujun Zhang, National Research Council Institute for Fuel Cell Innovation, Vancouver, British Columbia, Canada

Electrochemical Reduction of Carbon Dioxide: Fundamentals and Technologies
Jinli Qiao, Yuyu Liu, and Jiujun Zhang

Metal–Air and Metal–Sulfur Batteries: Fundamentals and Applications
Vladimir Neburchilov and Jiujun Zhang

Photochemical Water Splitting: Materials and Applications
Neelu Chouhan, Ru-Shi Liu, and Jiujun Zhang

Electrochemical Supercapacitors for Energy Storage and Delivery: Fundamentals and Applications
Aiping Yu, Victor Chabot, Jiujun Zhang

Electrochemical Polymer Electrolyte Membranes
Jianhua Fang, Jinli Qiao, David P. Wilkinson, and Jiujun Zhang

Electrochemical Energy: Advanced Materials and Technologies
Pei Kang Shen, Chao-Yang Wang, San Ping Jiang, Xueliang Sun, and Jiujun Zhang

High-Temperature Electrochemical Energy Conversion and Storage: Fundamentals and Applications
Yixiang Shi, Ningsheng Cai, Tianyu Cao, and Jiujun Zhang

Redox Flow Batteries: Fundamentals and Applications
Huamin Zhang, Xianfeng Li, and Jiujun Zhang

Carbon Nanomaterials for Electrochemical Energy Technologies: Fundamentals and Applications
Shuhui Sun, Xueliang Sun, Zhongwei Chen, Yuyu Liu, David P. Wilkinson, and Jiujun Zhang

Proton Exchange Membrane Fuel Cells
Zhigang Qi

Hydrothermal Reduction of Carbon Dioxide to Low-Carbon Fuels
Fangming Jin

Lithium-Ion Supercapacitors: Fundamentals and Energy Applications
Lei Zhang, David P. Wilkinson, Zhongwei Chen, and Jiujun Zhang

Solid Oxide Fuel Cells: From Fundamental Principles to Complete Systems
Radenka Maric and Gholamreza Mirshekari

Hydrometallurgical Recycling of Lithium-Ion Battery Materials
Joey Jung, Pang-Chieh Sui, and Jiujun Zhang

For more information about this series, please visit: https://www.routledge.com/Electrochemical-Energy-Storage-and-Conversion/book-series/CRCELEENESTO

Hydrometallurgical Recycling of Lithium-Ion Battery Materials

Joey Jung, Pang-Chieh Sui, and Jiujun Zhang

CRC Press
Taylor & Francis Group
Boca Raton London New York

CRC Press is an imprint of the
Taylor & Francis Group, an **Informa** business

First edition published 2023
by CRC Press
6000 Broken Sound Parkway NW, Suite 300, Boca Raton, FL 33487-2742

and by CRC Press
4 Park Square, Milton Park, Abingdon, Oxon, OX14 4RN

CRC Press is an imprint of Taylor & Francis Group, LLC

ISBN: 978-1-032-21602-7 (hbk)
ISBN: 978-1-032-21609-6 (pbk)
ISBN: 978-1-003-26920-5 (ebk)

DOI: 10.1201/9781003269205

Typeset in Times
by KnowledgeWorks Global Ltd.

Contents

Preface

The mass adoption of lithium-ion batteries (LIBs), e.g., in electric vehicles, portable electronics, and stationary energy storage stations, has sparked an enormous demand for battery materials, including lithium, cobalt, manganese, nickel, etc. This trend is causing a risk of LIB materials shortage in sustainable commercialization. With the rapidly increasing volume of production scrap and decommissioned LIBs, the current raw materials for manufacturing LIB materials are gradually becoming scarce, and their prices have risen sharply. In addition to the sustainability issue of available LIBs materials, the environmental impact of the spent and discarded LIBs has also become a serious concern. For these reasons, recycling LIB materials using environmentally friendly methods is essential to regenerate new lithium materials from the spent LIBs for reuse.

This book aims three-fold: to raise awareness of spent LIBs recycling, to provide comprehensive knowledge of hydrometallurgical recycling, and to promote an environmentally friendly recycling process. This book gives an overview of the hydrometallurgical recycling of LIB materials with scientific and technological fundamentals, the current state of technology of recycling processes, technical challenges, and perspectives. Furthermore, this book also showcases an innovative and environmentally friendly closed-loop hydrometallurgical recycling process to recycle lithium cathode materials with detailed modeling, economic analysis, and life cycle assessment.

We believe this book will be a resource benefiting researchers, students, and industrial professionals by providing an overview of the materials, system design, and related issues for the development of the LIB recycling process. The information in this book will be beneficial for readers in selecting existing materials/technologies and developing new materials/technologies.

We would like to express our appreciation to Dr. Marcus Tomlinson from Turnstone Metallurgical Services for providing METSIM simulation data on the processes, Ms. Phoebe Whattoff and Mr. Jordan Lindsay from Minviro Ltd. for providing a life cycle assessment of the recycling process, and Allison Shatkin from Taylor & Francis for her guidance and support in smoothing this book preparation process.

If there are any technical errors in this book, we would sincerely appreciate the reader's constructive comments for further improvement.

Joey Jung
Vancouver, British Columbia, Canada

Pang-Chieh Sui
Victoria, British Columbia, Canada

Jiujun Zhang
Richmond, British Columbia, Canada
September 13, 2022

Authors

Dr. Joey Jung is the Vice President Operations of Kemetco Research Inc. and co-founder and treasurer of the International Academy of Electrochemical Energy Science (IAOEES). Dr. Jung was appointed as the Director of Hydrogen & Fuel Cell Research Center at the Institute for Sustainable Energy, Shanghai University from 2018 to 2022. Dr. Jung has more than 20 years of R&D experience in the subject areas of fuel cell catalysts, fuel cell gas diffusion layers, high surface area carbon foam and carbon paper, lead-acid batteries, lithium-ion batteries, metal-air batteries, electroplating/electrowinning, electroreduction of carbon dioxide, lithium extraction and processing, and lithium-ion battery recycling. Dr. Jung has edited and co-authored two books, three book chapters, over 20 publications, and more than 20 US/EU/JP/CA/CN patents, and he has been involved in over 130 industrial projects. Dr. Jung coedited the book *Lead-Acid Battery Technologies: Fundamentals, Materials, and Applications*, published in 2015 by CRC Press.

Professor Pang-Chieh Sui currently holds dual appointments at the Wuhan University of Technology (WUT, from 2016) and Tsinghua-Sichuan Energy Internet Research Institute (EIRI, 2018). Before joining the WUT/EIRI, he was a Senior Researcher and Tech Lab Manager of the Institute for Integrated Energy Systems, University of Victoria (Canada) during 2003–2015, and a Research Scientist at the National Advanced Driving Simulator (USA) during 1997–2003. Prof. Sui received a Bachelor's degree from National Tsing Hua University (Taiwan) in 1986 and an MS and a Ph.D. from the University of Iowa, USA, in 1992 and 1997, respectively. He is a recipient of the Hanse-Wissenschaftskolleg Fellowship of Germany (2015), the 6th Hubei National Telent Plan Award (2016), the Sichuan-1000 Plan (2019), and the Overseas, Hong Kong & Macao Scholars Collaborated Researching Fund of China (2014).

Professor Jiujun Zhang is the Dean of Fuzhou University College of Materials Sciences and Engineering and Dean of Shanghai University Institute for Sustainable Energy/College of Sciences. He is a former Principal Research Officer at the National Research Council of Canada (NRC), Fellow of the Academy of Science of the Royal Society of Canada (FRSC-CA), Fellow of the International Society of Electrochemistry (FISE), Fellow of the Engineering Institute of Canada (FEIC), Fellow of the Canadian Academy of

Engineering (FCAE), Fellow of the Royal Society of Chemistry (FRSC-UK), and the Founder/Chairman of The International Academy of Electrochemical Energy Science (IAOEES). From 2014 to 2022, Prof. Zhang was ranked in the top 1% of Highly Cited Researchers in the world and has also been listed as one of the "3000 World's Most Influential Scientific Minds" by Thomson Reuters in 2014–2016. Dr. Zhang has more than 750 publications with approximately 63000 citations with an h-index of 103, including 550 refereed journal papers, 28 edited /co-authored books, 43 book chapters, 220 conference keynotes, and invited oral presentations, as well as over 16 US/EU/WO/JP/CA patents, and produced more than 90 industrial technical reports. Dr. Zhang serves as the Editor-in-Chief of *Electrochemical Energy Reviews* (Springer Nature), Associate Editor of *Green Energy & Environment* (KeAi), and the Editor/Editorial Board member for several international journals as well as Editor for the book series *Electrochemical Energy Storage and Conversion* (CRC Press).

1 Hydrometallurgical Recycling of Lithium-Ion Battery Cathode Material

Joey Jung and Jiujun Zhang

CONTENTS

1.1 OBJECTIVES AND SIGNIFICANCE OF BATTERY RECYCLE

For sustainable survival and development of human society, electric vehicles play a critical role in reducing local air pollution and climate change caused by fossil fuels fed vehicles. It is estimated that more than 245 million electric vehicles will be on

DOI: 10.1201/9781003269205-1

1

the road by 2030, which are more than 30 times above today's 7.2 million level [1]. Electric vehicles can be powered by batteries such as lithium-ion batteries (LIBs), which receive and store electricity by plugging into the grid. LIBs have been recognized as one type of the most practical and commercially feasible batteries for electric vehicles [2]. With the booming growth of electric vehicles, the production of LIBs has increased drastically in the past decade.

Advances in the commercial development of LIBs have spawned significant demand for lithium (Li), cobalt (Co), manganese (Mn), and nickel (Ni), which are the compositing elements in LIB electrode materials. For example, in 2000, the mined production of cobalt in the world was only 33,300 metric tons per year, with the majority of products used for super-alloys in high-temperature services. In 2019, the world mine production of cobalt was approximately 140,000 metric tons per year (as metal), of which 80% were destined for the LIB market [3]. As 98% of the cobalt production worldwide is a by-product of copper (Cu) and nickel mining, the increase in cobalt demand leads to a shortage in supply and a dramatic increase in price, evidenced by the 311% price hike between 2016 (USD 12.01/lb) and 2018 (USD 37.43/lb) [4].

In 2019, LIBs for electric vehicle production consumed about 19-kilo metric tons (kmt) of cobalt, 17 kmt of lithium, 22 kmt of manganese, and 65 kmt of nickel. With the projection of more than 245 million electric vehicles by 2030, the required cobalt demand would expand to about 180 kmt/year, lithium to around 185 kmt/year, manganese to 177 kmt/year, and nickel to 925 kmt/year [5]. As shown in Figure 1.1, with the data published in Mineral Commodity Summaries from 2017 and 2021 by U.S. Geological Survey, lithium reserves have increased from 15,566 kmt in 2017 to 22,425 kmt in 2021, thanks to the continuing exploration [3, 5]. Identified lithium resources include continental brines, geothermal brines, hectorite, oilfield brines, pegmatites, and searlesite.

Although the total lithium reserve increased, the cost to recover lithium also increased. Lithium supply security has become a top priority for technology companies in Asia, Europe, and the United States. To secure lithium resources, joint ventures among technology companies and exploration companies continued to be established, and vehicle manufacturers and governments are forging alliances to safeguard their needs for lithium which is treated as the future energy source, for example, Toyota Corporations, Magna International, Mitsubishi, and Tesla have forged partnerships with lithium exploration companies and have invested large sums to develop lithium deposits around the world.

As the upcoming mass adoption of electric vehicles will cause a significant supply crunch of LIB materials, automobile Original Equipment Manufacturers (OEMs) need to start to look at overcoming the possible shortage of LIB materials through the reuse of LIBs retired from electric vehicles for other applications, such as electricity energy storage (battery second use), and through recycling LIB materials to obtain the important metals once they have finished their lifecycle. Industry analysts have indicated that in China, the weight of spent or discarded LIBs is over 500,000 tons by 2020 [6]; and the worldwide number of spent LIBs will hit 2 million metric tons per year by 2030 [7].

Recycling those spent LIBs can provide a source of LIB materials such as lithium, nickel, cobalt, manganese, and aluminum [8]. Spent LIB recycling processes initially

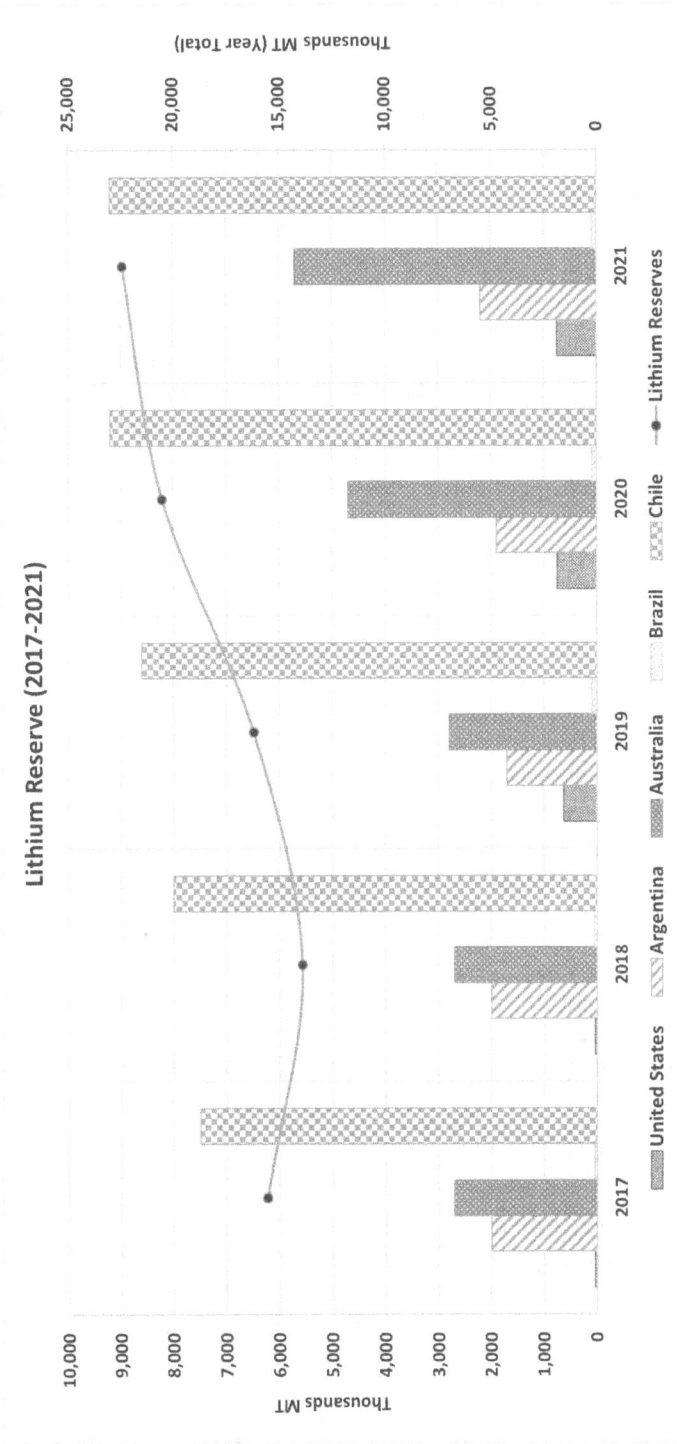

FIGURE 1.1 World Lithium Reserves by U.S. Geological Survey [3, 5].

```
┌──────────────┐   ┌─────────────┐   ┌─────────────┐   ┌─────────────┐   ┌──────────┐   ┌───────────────┐
│Spent Lithium-│→ │Energy Release│→ │ Mechanical  │→ │  Electrode  │→ │ Cathode  │→ │ Spent Cathode │
│ ion Battery  │   │             │   │ Separation  │   │  Assembly   │   │          │   │   Material    │
└──────────────┘   └─────────────┘   └─────────────┘   └─────────────┘   └──────────┘   └───────────────┘
                                      ┌─────────────┐   ┌─────────────┐   ┌─────────────┐   ┌─────────────┐
                                      │  Recycled   │   │             │   │             │   │             │
                                      │Stainless Steel│ │    Anode    │   │ Electrolyte │   │ Aluminum Foil│
                                      │    Case     │   │             │   │             │   │             │
                                      └─────────────┘   └─────────────┘   └─────────────┘   └─────────────┘
                                                        ┌─────────────┐   ┌─────────────┐
                                                        │ Copper Foil │   │  Graphite   │
                                                        └─────────────┘   └─────────────┘
```

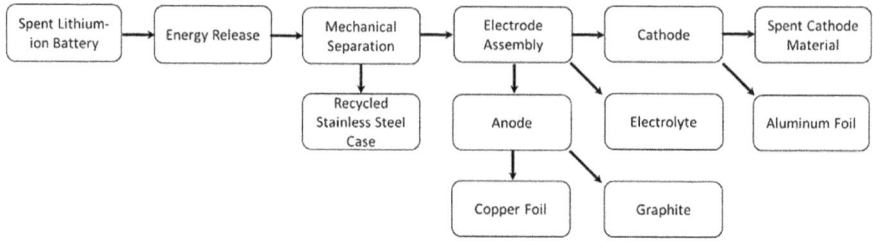

FIGURE 1.2   Lithium-ion battery physical dissemble.

go through sorting the batteries by battery chemistries followed by deep discharge to avoid violent reaction from the charged electrode materials exposed to the air. During discharge, positive charged lithium ions produced move from anode through electrolyte and separator to cathode, making spent cathode active material enriched with lithium. The discharged batteries then go through physical dissemble, as shown in Figure 1.2, before undergoing electrolyte extraction, high-temperature melting-and-extraction (pyrometallurgy), smelting, direct recycling, or chemical/hydro-metallurgy extraction. The spent cathode active materials can be separated from current collectors by using dimethylacetamide, N-methyl-pyrrolidone, N-methyl-2-pyrrolidone (NMP), or acetone at 60°C to dissolve the binder. He et al. [9] reported a patented complex aqueous peeling agent, namely exfoliating and extracting solution (AEES), which could weaken the mechanical interlocking force and Coulombic force between cathode active materials and foils, and separately recovered Al foil, Cu foil, and active materials.

Li et al. [10] investigated replacing organic NMP solvent with water during electrode fabrication in order to simplify the recycling process, and found that water-processed electrodes exhibited a comparable electrochemical performance with NMP-based electrodes and the electrode materials were easier to be recycled.

Other than cathode active material enriched in lithium, anode material like lithium titanate (LTO) and battery electrolyte, $LiPF_6$, $LiBF_4$, or $LiClO_4$, also enrich with lithium. Recycling cathode active material, LTO, and electrolyte can provide a sustainable lithium source.

LTO can be treated with hydrochloric acid (HCl) to extract Li and recover titanium as titanium oxide (TiO_2) [11, 12]. Electrolyte can be extracted by vacuum distillation or extracted by sub- and supercritical carbon dioxide (CO_2) extraction using a flow-through or batch autoclave setup. Sub- and supercritical CO_2 extraction have demonstrated that up to 90% of the electrolyte can be recovered from the spent LIBs [13–16]. The recovered and decomposed salt containing electrolytes can be purified with a weakly basic anion exchange resin [16].

As identified, the spent cathode active material can contain up to 7 weight percent (wt%) lithium, as shown in Figure 1.3; and the spent lithium nickel cobalt aluminum (NCA) battery contains up to 31 wt% of cathode active material [17].

Furthermore, Figure 1.4 shows the main components of lithium cobalt oxide (LCO) battery and lithium nickel manganese cobalt oxide (NMC) battery, where the cathode active materials account for 41 wt% and 26 wt% of the battery, respectively [18].

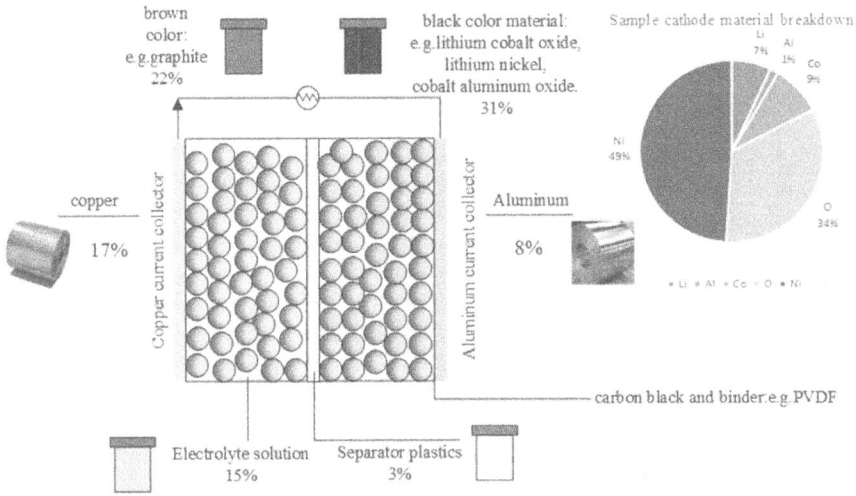

FIGURE 1.3   Components of a lithium-ion battery with NCA cathode [17].

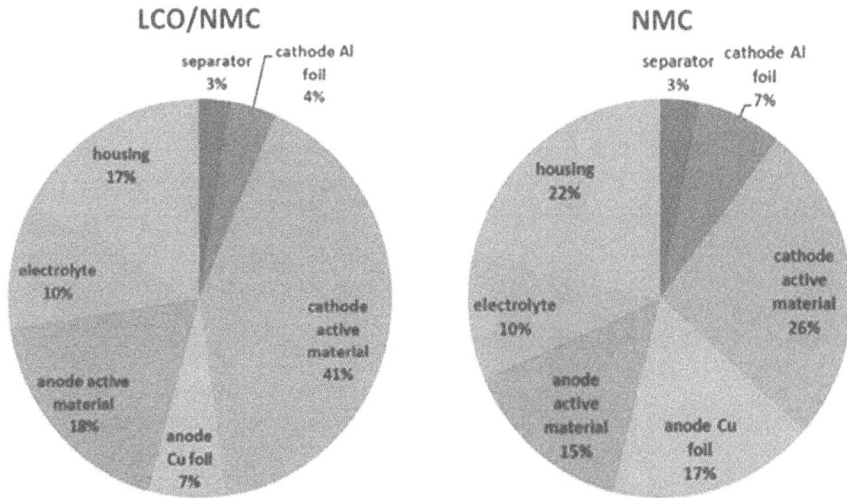

FIGURE 1.4   Components of a lithium-ion battery with LCO and NMC cathode [18].

Recycling these spent cathode active materials can not only recycle lithium but also recover nickel, cobalt, manganese, and aluminum.

1.2 PYROMETALLURGICAL RECYCLING PROCESS

Currently, there are several pyrometallurgy or smelting facilities that commercially recycle spent LIBs. The pyrometallurgical process often runs at near 1500°C to recover cobalt, nickel, and copper but not lithium, aluminum, or any organic compounds. Figure 1.5 shows a schematic of pyrometallurgy recycling process [19].

```
┌─────────────────────────────────────────────────────────────────┐
│              Pyrometallurgical recycling of Li-ion battery         │
└─────────────────────────────────────────────────────────────────┘
```

```
                        ┌──────────────────────────┐
                        │       Mechanical          │
                        │      pre-treatment        │
┌───────────────────┐   │       Crushing,           │   ┌──────────────┐
│                   │   │        Milling,           │   │    Casing     │
│ Spent Li-ion      │──▶│        Sieving,           │──▶│    Plastic    │
│ battery           │   │   Gravity separation,     │   │    Label      │
│                   │   │  Magnetic separation,     │   │              │
└───────────────────┘   │  Pneumatic separation     │   └──────────────┘
                        └──────────────────────────┘
                                    │
                                    ▼
                        ┌──────────────────────────┐
                        │     Anode & Cathode        │
                        └──────────────────────────┘
                                    │
                                    ▼
                        ┌──────────────────────────┐   ┌──────────────┐
                        │     Heat treatment         │──▶│ Electrolyte & │
                        │     150°C – 500°C          │   │ Organic solvent│
┌───────────────────┐   └──────────────────────────┘   └──────────────┘
│ Reducing agents:  │
│ Activated carbon, │   ┌──────────────────────────┐   ┌──────────────┐
│ NaHSO₄, CaCl₂,    │──▶│     Heat treatment         │   │  Black mass  │
│ NH₄Cl             │   │     1400°C – 1700°C        │──▶│  Ni / Co alloy│
└───────────────────┘   │  Graphite & aluminum foil  │   │  Co oxide    │
                        │  can act as reducing agent │   │  Mn oxide    │
┌───────────────────┐   └──────────────────────────┘   │  Ni oxide    │
│ Slag modifiers:   │              │                     └──────────────┘
│ Al₂O₃, CaO,       │──▶           ▼
│ MgO, SiO₂         │   ┌──────────────────────────┐
└───────────────────┘   │         Li₂O,              │
                        │         Li₂CO₃             │
                        │       Slag residue         │
                        └──────────────────────────┘
```

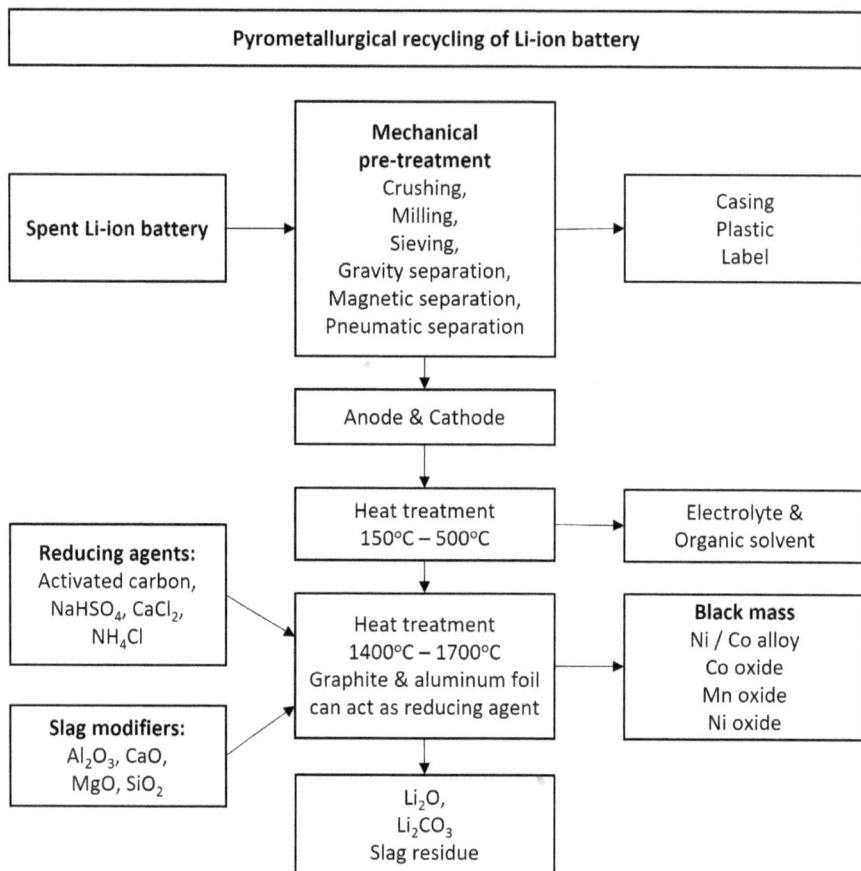

FIGURE 1.5 Schematic of pyrometallurgical recycling of spent LIB [19].

Pyrometallurgical recycling process is capital intensive, in part because of the need to treat the emission of toxic fluorine compounds released during the smelting. After the pyrometallurgy process, valuable metals are reduced and then recovered in the form of alloys. For example, Umicore develops the pyrometallurgical and hydrometallurgical processes to recycle Ni, Co, and Li from spent LIBs [20]. Georgi-Maschler et al. [21] used a reductive smelting process to recover valuable metals from the spent LIBs. The Fe, Co, Ni, and Mn in the batteries were converted to alloys, and Li during the reductive smelting reported to slag, which could be recovered with further leaching using H_2SO_4.

Trager et al. [22] used vacuum evaporation and selective carrier gas evaporation at temperatures above 1400°C to evaporate Li from the spent LIBs. Liang et al. [23] reported a high-temperature pyrolysis to regenerate spent $LiFePO_4$ cathode active materials with smooth material surface and small particle sizes. Tang et al. [24] reported a vacuum pyrolysis approach to recover Li and Co from the spent $LiCoO_2$.

Wang et al. [25] roasted Li_2CoO_2 with $NaHSO_4 \cdot H_2O$ to selectively separate Li and Co. Wang et al. [26] investigated catalytic carbothermic reduction by heating the mixture of $LiCoO_2$, graphite, and NaOH at 520°C for 180 minutes. Water leach of the heated mixture showed that 93% Li could be extracted while Co retained in the solid.

1.3 HYDROMETALLURGICAL RECYCLING PROCESSES

Hydrometallurgical recycling method of the spent lithium cathode active materials, compared to other methods, has many advantages, including high recovery of valuable metals, high purity of the recovered metals, and low energy consumption. However, the disadvantages of hydrometallurgical recycling method are the relatively complicated process and the generation of more wastewater [27].

Hydrometallurgical recycling of spent lithium cathode active material generally runs below 100°C and involves leaching, impurity removal, metal extraction, and lithium recovery. During leaching, metal ions such as Ni^{2+}, Co^{2+}, Mn^{2+}, Al^{3+}, Fe^{2+}, and Li^+ in the spent cathode active material dissolve into the leach solution. The impurity removal process removes the unwanted metal ions such as Al^{3+} and Fe^{2+} as well as unleached solids like carbon and graphite powder. Metal extraction and lithium recovery can recover Ni^{2+}, Co^{2+}, Mn^{2+}, and Li^+ from the purified leach solution.

Leaching is often conducted using acids together with a reducing agent. The most common acid and reducing agent used are sulfuric acid (H_2SO_4) and hydrogen peroxide (H_2O_2), respectively. Impurity removal and metal extraction are often conducted using sodium hydroxide (NaOH) to precipitate metal ions as metal hydroxides or using solvent extraction to extract individual metal ions. Lithium recovery is often conducted by adding sodium carbonate (Na_2CO_3) into the lithium enriched solution to precipitate lithium as lithium carbonate (Li_2CO_3). [27].

Using NaOH for impurity removal and metal extraction together with adding Na_2CO_3 to recover lithium can result in a solution enriched with Na^+ ions. The sodium containing solution needs to be treated before discharge or reuse. The Li_2CO_3 will need to be further refine before it can be reused in battery material fabrication, which further complicates the hydrometallurgical recycling process. Due to the high content of the Na^+ ions, the process is an open, one-through process and generates lots of wastewater for treatment and discharge. Furthermore, despite recent progress in recycling process of spent LIBs, one of the challenges lies in the unsatisfactory purity of cathode active materials; or more specifically, it remains difficult to achieve commercial-grade cathode active materials via a near-complete removal of impurity elements.

It therefore poses a need to develop an environmental friendlier hydrometallurgical process to recycle spent lithium cathode active materials. The objectives of this book is to introduce a closed-loop recycling process for spent lithium cathode active materials; to regenerate reagent consumed in the process for reducing reagent costs; to regenerate commercial-grade cathode precursors and commercial-grade cathode active materials; to reuse all process water back in the process to minimize discharge; and to eliminate the environmental impact of high Na^+ ions. By achieving these objectives, a more economically attractive and an environmentally friendlier process for recycling spent lithium ion battery cathode active materials will be realized.

1.4 CURRENT HYDROMETALLURGY TREATMENTS

Hydrometallurgy extraction process or chemical leaching can offer a less energy-intensive alternative and lower capital cost. These processes employed regents such as hydrochloric acid (HCl), nitric acid (HNO_3), sulfuric acid (H_2SO_4), and hydrogen peroxide (H_2O_2) for extracting and separating cathode metals, generally run below 100°C and can recover lithium in addition to the other transition metals [28].

Various hydrometallurgy methods have been developed in the recent decade for recycling cathode active materials of LIBs from various battery chemistries including lithium cobalt dioxide, $LiCoO_2$ (LCO), lithium manganese dioxide, $LiMn_2O_4$ (LMO), lithium nickel manganese cobalt oxide, $LiNiMnCoO_2$ (NMC), and lithium nickel cobalt aluminum oxide, $LiNiCoAlO_2$ (NCA), and lithium iron phosphate $LiFePO_4$ (LFP), to recover cobalt, nickel, manganese, and lithium. Figure 1.6 shows the hydrometallurgy process for recycling battery cathode active materials. The process can be categorized into four sections, which are leaching, impurity removal, metal like Ni, Co, Mn recovery, and lithium recovery. The valuable metals, mainly cobalt, manganese, nickel, and lithium can be extracted from the spent lithium-ion cathode active materials through leaching treatment, as summarized in Table 1.1.

As shown in Figure 1.6, the spent cathode active material is first slurried with weak acid and then transferred to the leaching tanks. Acid and reducing agents are then added into the leaching tanks to leach out Li^+, Ni^{2+}, Co^{2+}, Mn^{2+}, Fe^{2+}, and Al^{3+}. In the impurity removal section, the unwanted impurity can be removed through pH adjustment. After impurity removal, the solution is transferred to metal recovery section where metal ions can be recovered via sedimentation, chemical precipitation, or solvent extraction. After metal recovery, the remaining lithium enriched solution is transferred to the lithium recovery section where lithium can be recovered by chemical precipitation using sodium carbonate or by crystallization by distillation.

The recovered materials can be reformulated to regenerate lithium cathode active material.

1.4.1 Hydrometallurgical Recycle of $LiCoO_2$, $LiMn_2O_4$, NMC, and NCA Cathode Active Materials

1.4.1.1 Hydrochloric Acid and Sodium Hydroxide as Hydrometallurgy Treatment Reagents

Zheng et al. [28], Contestabile et al. [30], and Takacova et al. [31] investigated the leaching of spent lithium cobalt dioxide ($LiCoO_2$) cathode active material with hydrochloric acid (HCl). Based on the experimental results and literature reports, the extraction of cobalt and lithium could achieve 100% at 80°C with either 2 M or 4 M HCl within 90 minutes and 60 minutes, respectively. Wang et al. [32] achieved a leaching efficiency of more than 99% of cobalt (Co), manganese (Mn), nickel (Ni), and lithium (Li) with 4 M HCl at 80°C and 60 minutes of leaching time on $LiCoO_2$, $LiMn_2O_4$, and $LiNi_{1/3}Mn_{1/3}Co_{1/3}O_2$.

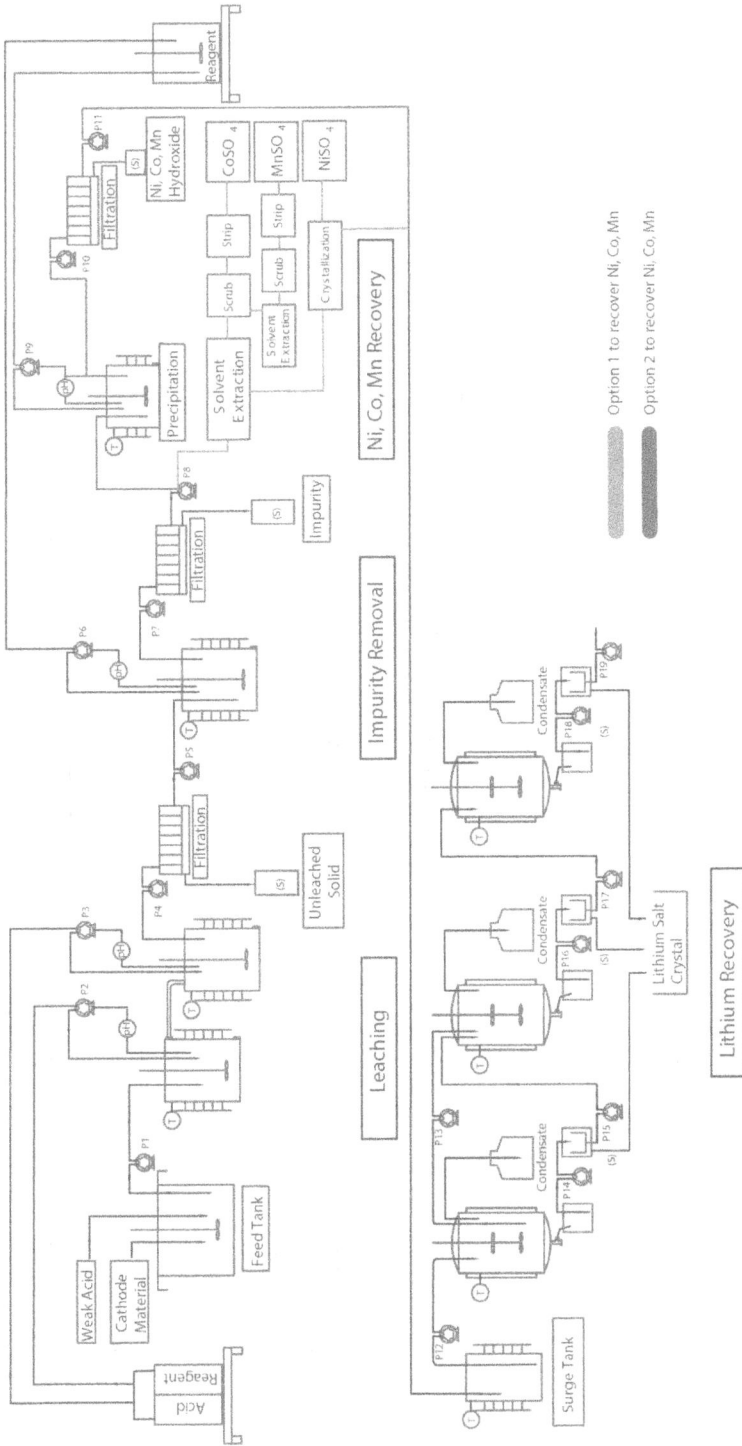

FIGURE 1.6 Schematic of the hydrometallurgy process for recycling battery cathode active materials.

TABLE 1.1

Reported Leaching Processes for Recycling Spent Cathode Active Materials of LIBs with LCO, LMO, NMC, and NCA Chemistries

No	Leaching Treatment		Leach Temp (°C)	Leach Efficiency (%)			Ref [#]
	Inorganic Leaching Agent	Reducing Agent		Li	Co	Mn	
A							
1	HCl (4 mol L^{-1})	N/A	80	100	100		[29, 30]
2	HCl (2 mol L^{-1})	N/A	60-80	100	100		[31]
3	HCl (4 mol L^{-1})	N/A	80	100	100	100	[32]
4	HNO$_3$ (2 mol L^{-1})	N/A	80	100		95	[34]
5	HNO$_3$ (1 mol L^{-1})	H$_2$O$_2$ (1.7 Vol%)	75	85	85		[35]
6	HNO$_3$ (1 mol L^{-1})	H$_2$O$_2$ (1.7 Vol%)	75	85	85		[36]
7	HNO$_3$ (1 mol L^{-1})	H$_2$O$_2$ (1.0 Vol%)	80	100	100		[37]
8	H$_2$SO$_4$ (8.0 Vol%)	N/A	80	95	80		[8]
9	H$_2$SO$_4$ (2 mol L^{-1})	H$_2$O$_2$ (5.0 Vol%)	80	99	99		[39]
10	H$_2$SO$_4$ (3 mol L^{-1})	N/A	70	98	98		[41]
11	H$_2$SO$_4$ (6.0 Vol%)	H$_2$O$_2$ (4.0 Vol%)	65	37	55		[42]
12	H$_2$SO$_4$ (4 mol L^{-1})	H$_2$O$_2$ (10.0 Vol%)	85	96	95		[43]
13	H$_2$SO$_4$ (2 mol L^{-1})	H$_2$O$_2$ (6.0 Vol%)	60	99	99		[44]
14	H$_2$SO$_4$ (2 mol L^{-1})	H$_2$O$_2$ (15.0 Vol%)	75	100	100		[45]
15	H$_2$SO$_4$ (2 mol L^{-1})	H$_2$O$_2$ (2.0 Vol%)	60	88	96		[46]
16	H$_2$SO$_4$ (2 mol L^{-1})	H$_2$O$_2$ (5.0 Vol%)	75	94	93		[47]
17	H$_2$SO$_4$ (3 mol L^{-1})	H$_2$O$_2$ (3.0 Vol%)	70	100	100		[48]
18	H$_2$SO$_4$ (6.0 Vol%)	H$_2$O$_2$ (1.0 Vol%)	60	90	90		[40]

(Continued)

TABLE 1.1 (Continued)

Reported Leaching Processes for Recycling Spent Cathode Active Materials of LIBs with LCO, LMO, NMC, and NCA Chemistries

No	Leaching Treatment		Reducing Agent	Leach Temp (°C)	Leach Efficiency (%)			Ref [#]
					Li	Co	Mn	
A	**Inorganic Leaching Agent**							
19	H_2SO_4	(2 mol L^{-1})	H_2O_2 (10.0 Vol%)	70				[49]
20	H_2SO_4	(3 mol L^{-1})	H_2O_2 (3.0 Vol%)	80	98	97	97	[50]
21	H_2SO_4	(4 mol L^{-1})	$Na_2S_2O_3$ (0.25molL^{-1})	50–80				[51, 52]
22	NH_3		Na_2SO_4	50–80	61	81		[65, 66]
23	H_3PO_4	(0.7 mol L^{-1})	H_2O_2 (4.0 Vol%)	40	99	99		[79]
B	**Inorganic Leaching Agent**		**Organic Reducing Agent**					
24	H_2SO_4		Glucose	90	100	100		[53, 54]
25	H_3PO_4	(0.2 mol L^{-1})	$C_6H_8O_7$ (0.4 mol L^{-1})	90	100	92	92	[78]
C	**Organic Leaching Agent**		**Reducing Agent**					
26	$H_2C_2O_2 2H_2O$	(1.5 mol L^{-1})	N/A	80	98	98		[64]
27	$C_4H_5O_6$ (DL-malic Acid)	(1.5 mol L^{-1})	H_2O_2 (2.0 Vol%)	90	94	93		[69]
28	$C_4H_5O_6$ (DL-malic Acid)	(1.5 mol L^{-1})	H_2O_2 (3.0 Vol%)	80	98	99		[70]
29	$C_4H_5O_6$ (DL-malic Acid)	(1.5 mol L^{-1})	Grape seed (0.6g/g)	80	99	92		[71]
30	$C_4H_5O_6$ (DL-malic Acid)	(1.5 mol L^{-1})	Glucose	55	100	98		[72]

(Continued)

TABLE 1.1 (*Continued*)

Reported Leaching Processes for Recycling Spent Cathode Active Materials of LIBs with LCO, LMO, NMC, and NCA Chemistries

No	Leaching Treatment		Leach Temp (°C)	Leach Efficiency (%)			Ref [#]
				Li	Co	Mn	
C	**Organic Leaching Agent**	**Reducing Agent**					
31	$C_4H_5O_6$ (DL-malic Acid) (1.5 mol L^{-1})	Electrochemical reduction	60	100	99	99	[73, 74]
32	$C_6H_8O_7 \cdot H_2O$ (Citric Acid) (1.25 mol L^{-1})	H_2O_2 (1.0 Vol%)	90	100	90		[75, 76]
33	$C_6H_8O_6$ (Ascorbic Acid) (1.25 mol L^{-1})		70	98	97		[80]
34	$C_6H_6O_3S$ (1.3 mol L^{-1})	CH_2O_2 (1.5 mol L^{-1})	50	99	97		[81]
D	**Bioleaching Agent**	**Additives**					
35	Acidithiobacillus ferrooxidans	S + Fe	30	10	65		[82]
36	Mesophilic potential sulfur-oxidizing bacteria		25				[86]

The hydrochloric acid leaches Co, Mn, Ni, and Li according to the following reactions [29–32]:

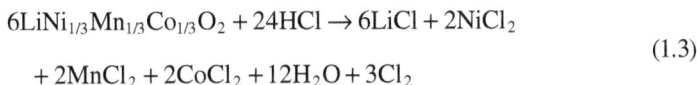

$$2LiCoO_2 + 8HCl \rightarrow 2LiCl + 2CoCl_2 + 2H_2O + Cl_2 \qquad (1.1)$$

$$2LiMn_2O_4 + 16HCl \rightarrow 2LiCl + 4MnCl_2 + 8H_2O + 3Cl_2 \qquad (1.2)$$

$$6LiNi_{1/3}Mn_{1/3}Co_{1/3}O_2 + 24HCl \rightarrow 6LiCl + 2NiCl_2$$
$$+ 2MnCl_2 + 2CoCl_2 + 12H_2O + 3Cl_2 \qquad (1.3)$$

The leached Co^{2+}, Mn^{2+}, and Ni^{2+} could be recovered through selective precipitation as metal hydroxides could slowly increase the leachate pH using sodium hydroxide (NaOH) [31]. The precipitation of Mn^{2+}, Ni^{2+}, and Co^{2+} began at pH of 1, 2, and 3, and was completely precipitated at pH 12, 8, and 10, respectively. The leached Li^+ could be precipitated as lithium carbonate via adding sodium carbonate (Na_2CO_3) after the leachate clear of Co^{2+}, Mn^{2+}, and Ni^{2+}.

Alternatively, the manganese could be selectively recovered from the leachate using potassium permanganate ($KMnO_4$) via the following redox reaction [33]:

$$3Mn^{2+} + 2MnO_4^- + 2H_2O \rightarrow 5MnO_2 + 4H^+ \qquad (1.4)$$

The optimum operating conditions of Mn precipitation via $KMnO_4$ had a molar ratio of Mn^{2+} to $KMnO_4$ at 2, pH at 2, and an operation temperature at 40°C. After removing Mn from the leachate, the Ni could be selectively recovered using dimethylglyoxime ($C_4H_8N_2O_2$) and ammonia (NH_3) solution via adding 28wt% NH_3 solution into the leachate to from $Ni(NH_3)_6^{2+}$ which then reacted with $C_4H_8N_2O_2$ to form a red precipitate. Nickel in the red precipitate could then be re-leached into solution with 4 M HCl and then precipitated with NaOH to form nickel hydroxide ($Ni(OH)_2$). The optimum operating conditions for Ni recovery using $C_4H_8N_2O_2$ and NH_3 had a molar ratio of $C_4H_8N_2O_2$ to $Ni(NH_3)_6^{2+}$ at 2 and pH at 9. Cobalt could then be recovered as cobalt hydroxide by raising the leachate pH to 11 via NaOH, and Li was recovered as Li_2CO_3 via adding Na_2CO_3.

Figure 1.7 illustrates the recover flow sheet with the two alternatives. It should be noted that the employed HCl as a leaching agent possibly generates toxic chlorine gas (Cl_2) as a byproduct as shown in Eqs. (1.1)–(1.3).

1.4.1.2 Nitric Acid as a Hydrometallurgy Treatment Reagent

Castillo et al. [34] studied the leaching of $LiMn_2O_4$ from spent lithium-ion battery using nitric acid (HNO_3). The leaching reactions can be expressed as follows:

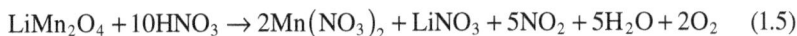

$$LiMn_2O_4 + 10HNO_3 \rightarrow 2Mn(NO_3)_2 + LiNO_3 + 5NO_2 + 5H_2O + 2O_2 \qquad (1.5)$$

They stated that 100% of Li and up to 95% of Mn could be leached with 2M HNO_3 at 80°C in 120 minutes. Lee et al. [35, 36], and Li et al. [37] studied the reductive leaching of $LiCoO_2$ and $LiMn_2O_4$ using a lower concentration of nitric acid (HNO_3)

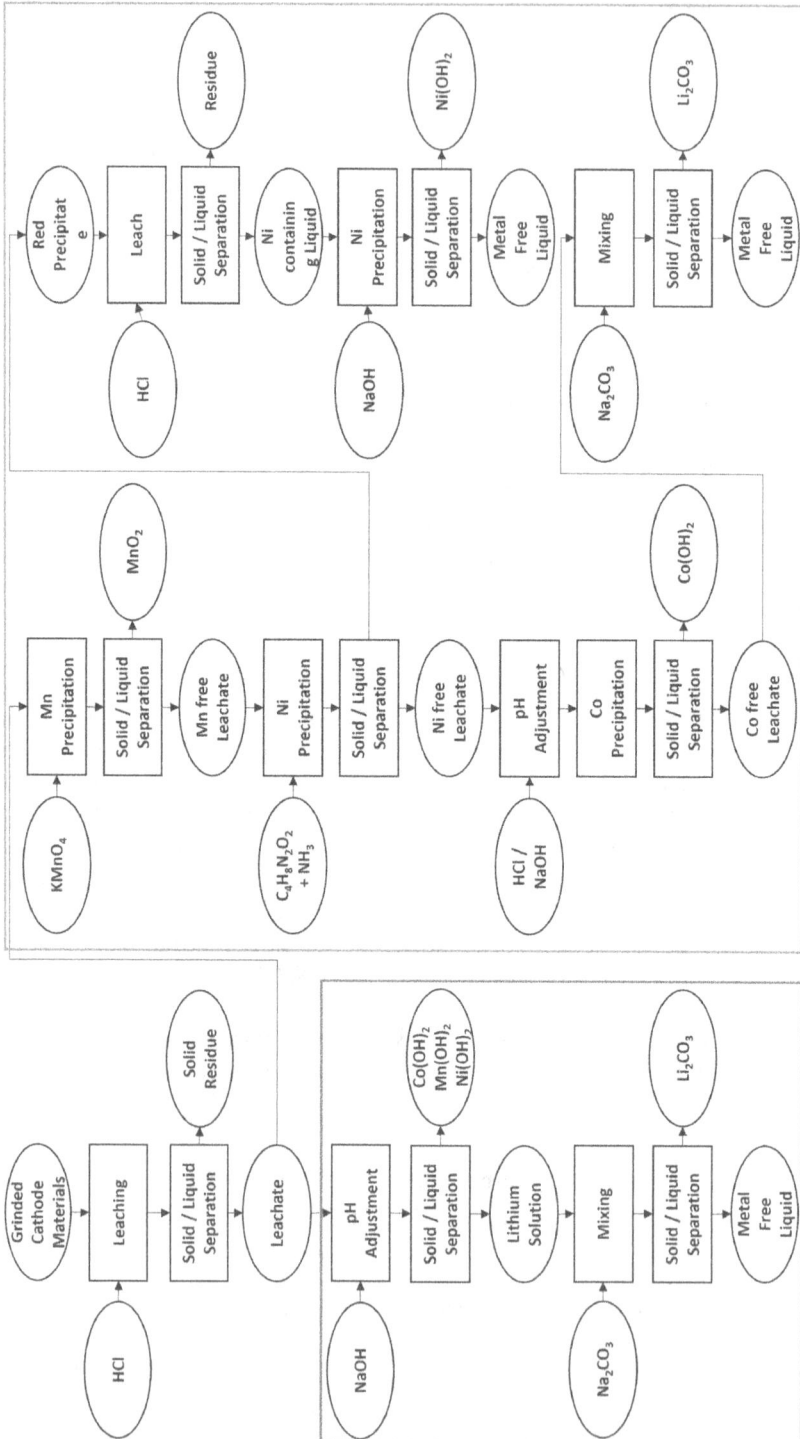

FIGURE 1.7 Two alternative flowsheets (solid and dash) to recover Li, Mn, Co, and Ni from HCl leach [29–31, 33].

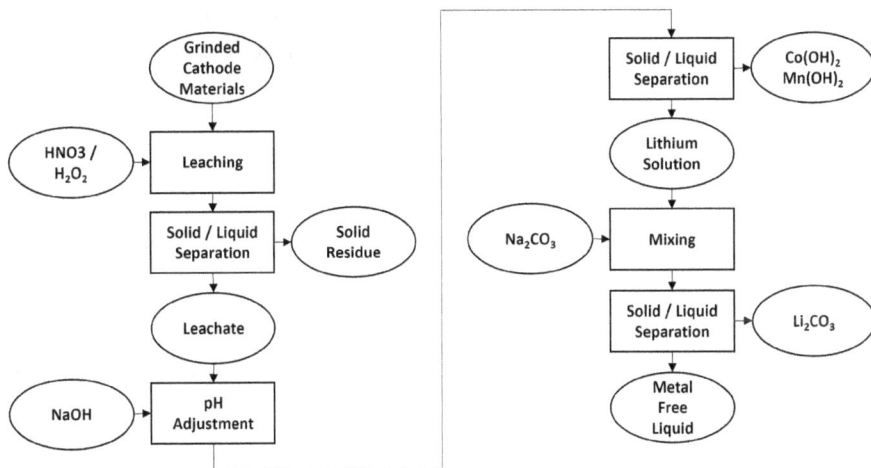

FIGURE 1.8 HNO$_3$ leach flowsheets [34–37].

with hydrogen peroxide (H$_2$O$_2$). The leaching reaction of LiMn$_2$O$_4$ is illustrated as follows:

$$2LiMn_2O_4 + 10HNO_3 + H_2O_2 \rightarrow 2LiNO_3 + 4Mn(NO_3)_2 + 6H_2O + 2O_2 \quad (1.6)$$

Their studies showed that when leaching with 1 M HNO$_3$, the leaching efficiency of Co and Li could reach to 40% and 75%, respectively. The leaching efficiency was improved significantly when adding H$_2$O$_2$ to the leachate.

Lee et al. [36] reported that with 1 M HNO$_3$ and 1.7 vol% H$_2$O$_2$ at 75°C, up to 99% of Co and Li could be leached in 30 minutes. They believed that the improvement of leaching efficiency was due to the addition of H$_2$O$_2$ acting as a reducing agent to reduce Co^{3+} to Co^{2+}, Mn^{3+} to Mn^{2+}, which were readily dissolved in HNO$_3$.

Using lower concentration of HNO$_3$ in the leach is of particular interest to the battery recycle industry as it reduces the reagent consumption in the following material recovery steps. Figure 1.8 shows the flow sheets of HNO$_3$ leach and HNO$_3$ + H$_2$O$_2$ leach. The drawback of using HNO$_3$ to leach is that leaching with HNO$_3$ generates NO$_x$. NO$_x$, if released into the atmosphere, is a prominent air pollutant thus needs to be captured for treatment. On the other hand, NO$_x$ can be collected to react with water to generate HNO$_3$ and feed back to the leach operation, which is worth of further investigation.

1.4.1.3 Sulfuric Acid as a Hydrometallurgy Treatment Reagent

Leaching spent LIB cathode active materials with sulfuric acid (H$_2$SO$_4$) is normally slow. For LiCoO$_2$, the bonding nature between oxygen and cobalt entities (O-Co-O) is chemical and hence comparably very strong. The breaking of these bonds requires significantly more energy [38].

Sun et al. [39] reported using H$_2$SO$_4$ to leach spent LiCoO$_2$. Leaching LiCoO$_2$ with H$_2$SO$_4$ involved the reduction of Co^{3+} in the solid phase to Co^{2+} in the aqueous phase. It was believed that CoO_2 first formed Co_3O_4 solid which required excess of H_2SO_4 to be converted to soluble $CoSO_4$ as showed in the following reactions [40]:

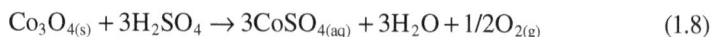

$$4LiCoO_{2(s)} + 3H_2SO_4 \rightarrow Co_3O_{4(s)} + 2Li_2SO_{4(aq)} + CoSO_{4(aq)} + 3H_2O + 1/2O_{2(g)} \quad (1.7)$$

$$Co_3O_{4(s)} + 3H_2SO_4 \rightarrow 3CoSO_{4(aq)} + 3H_2O + 1/2O_{2(g)} \quad (1.8)$$

Total:

$$4LiCoO_2 + 6H_2SO_4 \rightarrow 2Li_2SO_4 + 4CoSO_4 + 6H_2O + O_2 \quad (1.9)$$

Sun et al. [39] discovered that with 2M H_2SO_4, leaching efficiency of 76% could be achieved. With 3 M H_2SO_4 at 70°C, Nan et al. [48] could achieve 99% leaching efficiency. However, higher acid concentration was not favored as it increased the overall reagent consumption in the overall recovery process. To increase leaching efficiency with lower acid concentration, Sun et al. [39] reported that adding H_2O_2 as a reductant during the leaching process to reduce Co^{3+} to Co^{2+} could facilitate the forward reaction as shown in the following reaction:

$$2LiCoO_2 + 3H_2SO_4 + H_2O_2 \rightarrow Li_2SO_4 + 2CoSO_4 + 4H_2O + O_2 \quad (1.10)$$

With the addition of 5 vol% H_2O_2, the leaching efficiency of cobalt could be increased to 99%. This was because H_2O_2 could help break the strong chemical bond between Co and O. In the absence of a reducing agent, O from $LiCoO_2$ was oxidized into O_2, leading to a low Co leaching efficiency. With H_2O_2 acting as a reducing agent, Co^{3+} in the solid species was reduced to Co^{2+} in the aqueous phase. As Co^{2+} was readily dissolved in aqueous solution, the Co leaching efficiency was much higher. On the other hand, Li^+ is readily dissolved in aqueous solution thus the addition of H_2O_2 showed a negligible impact on the leaching efficiency of lithium, which was also reported by Ferreira et al. [40]. This behavior could be expended to other spent lithium cathode active materials. Nayl et al. [41] reported that in the leaching $LiCoO_2$ and a $LiMnO_2$ with 2 M H_2SO_4, only 51.6% of Mn and about 42.7% of Co were obtained. With the addition of 4% H_2O_2, 97.8% of Mn and 99.6% of Co were leached. The equations for $LiMnO_2$ leach can be expressed as follows:

$$4LiMnO_2 + 6H_2SO_4 \rightarrow 2Li_2SO_4 + 4MnSO_4 + 6H_2O + O_2 \quad (1.11)$$

$$2LiMnO_2 + 3H_2SO_4 + H_2O_2 \rightarrow Li_2SO_4 + 2MnSO_4 + 4H_2O + O_2 \quad (1.12)$$

Dorella et al. [42], Chen et al. [43], Kang et al. [44], Shin et al. [45], Zhu et al. [46], Swain et al. [47], Nan et al. [48], and Chen et al. [49] reported the leaching of spent lithium cathode active materials with various H_2SO_4 and H_2O_2 concentrations. Based on their studies, the optimum conditions could be concluded as follows: 2M H_2SO_4, 5vol% H_2O_2, leaching time of 60–90 minutes at temperature between 60°C to 80°C, and a solid/liquid ratio of 50 gL^{-1}.

Yang et al. [50] reported a leaching and direct precipitation process to regenerate $LiNi_{0.6}Co_{0.2}Mn_{0.2}O_2$ cathode active material with carbonate co-precipitation. Leaching efficiencies of 97.8% for Li, 98.1% for Ni, 96.5% for Co, and 97.0% for Mn were obtained with 3M H_2SO_4, 3vol% H_2O_2, and leaching time 60 minutes at 80°C. The regenerated $LiNi_{0.6}Co_{0.2}Mn_{0.2}O_2$ cathode could produce a great initial discharge capacity of 173.4 mAh/g at 0.1C discharge under a calcination temperature of 850°C and 93.6% capacity retention after 100 cycles at 1C.

Wang et al. [51] and Vieceli et al. [52] reported the leaching of spent lithium-ion cathode active materials using H_2SO_4 and sodium thiosulfate ($Na_2S_2O_3$) as the reducing agents. It concluded that the leaching efficiencies of Co and Li could be increased at higher acid concentration. The leaching rates of Co, Cu, and Li were increased first and then decreased afterwards with increasing $Na_2S_2O_3$ concentration. The optimum conditions, at which 99.95% Co and 99.71% Li were obtained, were 3 M H_2SO_4 and 0.25 M $Na_2S_2O_3$ at 90°C for 3 hours with liquid/solid ratio at 15:1. The following equation illustrates the reaction process:

$$8LiCoO_2 + Na_2S_2O_3 + 11H_2SO_4 \rightarrow 4Li_2SO_4 + 8CoSO_4 + Na_2SO_4 + 11H_2O \quad (1.13)$$

Figure 1.9 shows the flowsheets using sulfuric acid with various reducing agents to leach waste/spent cathode material.

Granata et al. [53] and Furlani et al. [54] reported using glucose ($C_6H_{12}O_6$) as reducing agent in H_2SO_4 leaching according to the following reaction:

$$24LiCoO_2 + C_6H_{12}O_6 + 36H_2SO_4 \rightarrow 12Li_2SO_4 + 24CoSO_4 + 6CO_2 + 42H_2O \quad (1.14)$$

Glucose is a low cost and non-hazardous chemical and was first proposed for Mn reducing leaching of pyrolusite ores [55, 56] and zinc manganese dioxide alkaline battery [57]. It was discovered that the leaching solution contained high concentration of formic acid along with mono-carboxylic poly-hydroxyl acids like glyceric acid and glycolic acid, suggesting that glucose was oxidized. The use of glucose as the reducing agent should be an innovative and environmentally-friendly option in H_2SO_4 leaching of spent LIB cathode active materials. Furlani et al. [54] reported that 88% extraction of Co could be obtained with glucose as a reducing agent. Therefore, the leached lithium, cobalt, and other metals can be recovered through many methods such as chemical precipitation, solvent extraction, and crystallization.

Chemical precipitation can selectively precipitate the metals. For example, for leachate containing Mn^{2+}, Ni^{2+}, and Co^{2+}, precipitation can be achieved using sodium hydroxide (NaOH) and sodium sulfide (Na_2S) based on each metal's hydroxide solubility or sulfide solubility at certain pH values. For example, in order to separate Mn from Ni and Co in the solution, NaOH and Na_2S solutions were used based on different hydroxide solubility and sulfide solubility of metals at a certain pH [51, 52]. For Mn precipitation, the concentrated Na_2CO_3 solution was used. For the final stage, the Na_2CO_3 solid was used. Cobalt, nickel, and manganese could be totally precipitated from the solution at pH above 10 and leaving on lithium in the solution.

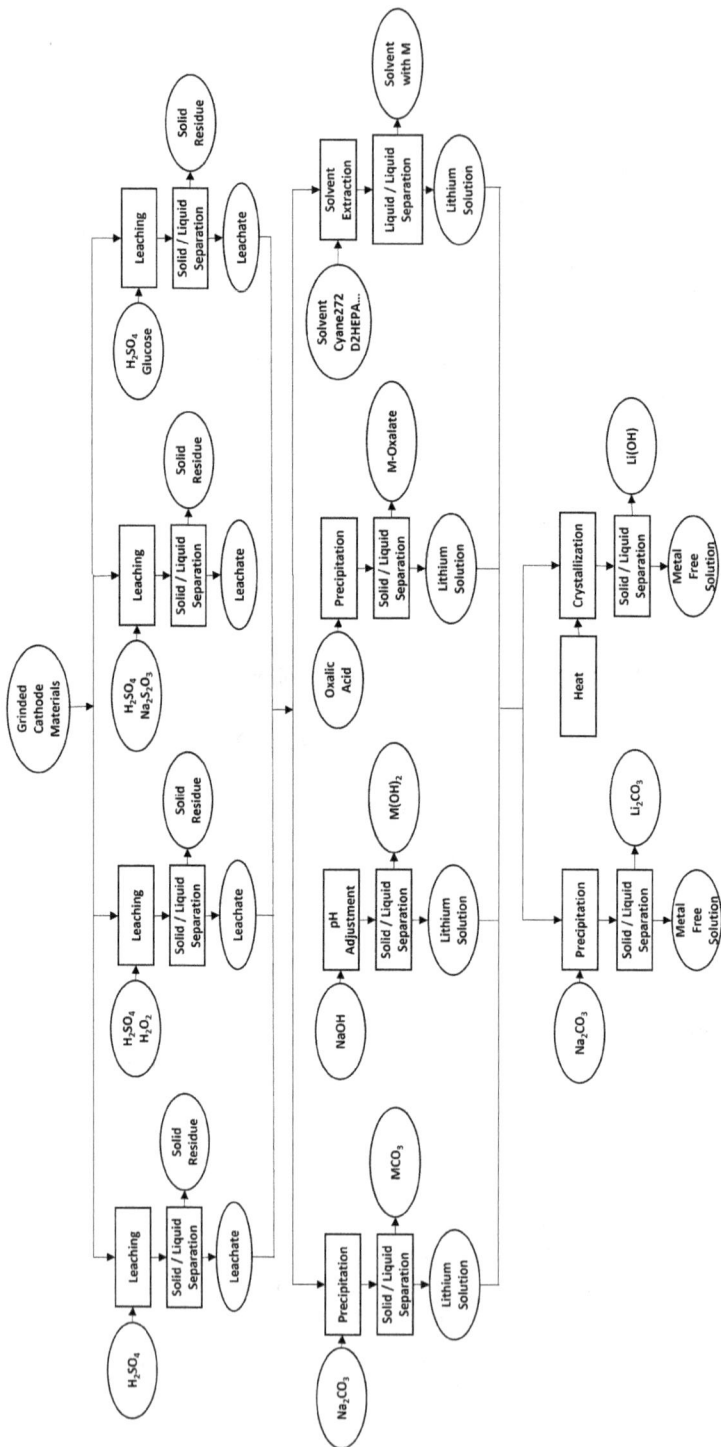

FIGURE 1.9 H₂SO₄ leach flowsheet [39–57].

Solvent extraction could be employed to selectively extract cobalt [58]. The extraction of metal was based on the following mechanism [59]:

$$M_{Aq}^{2+} + A_{Org}^{-} + 2(HA)_{2Org} \rightarrow MA_2 3HA_{Org} + H_{Aq}^{+} \qquad (1.15)$$

where $A_{Org}^{-} + 2(HA)_{2Org}$ represents the extractant saponified by the following reaction:

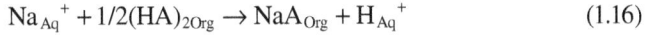

$$Na_{Aq}^{+} + 1/2(HA)_{2Org} \rightarrow NaA_{Org} + H_{Aq}^{+} \qquad (1.16)$$

The efficiency of metal extraction was calculated by the distribution coefficients of metals in aqueous and organic phases. When the solution containing Co^{2+} and Ni^{2+}, the distribution coefficients (D) can be illustrated as follows:

$$D_{Co} = [Co]_{Org}/[Co]_{aq} \qquad (1.17)$$

$$D_{Ni} = [Ni]_{Org}/[Ni]_{aq} \qquad (1.18)$$

where metal concentration in organic phase could be determined by a mass balance considering the residual concentration in aqueous phase. The index of extraction selectivity (β) of Co over Ni can then be calculated by the following equation:

$$\beta = D_{Co}/D_{Ni} \qquad (1.19)$$

Lupi et al. [60] reported that Co could be extract from nickel with Cyanex 272 [di-(2,4,4-trimethylpentyl) phosphoric acid] in Kerosene. Granata et al. [48] tested two extractants, D2HEPA and Cyanex 272, and found that Cyanex 272 could selectively extract Co completely before extracting Ni. Dorella et al. [42] used Cyanex 272 to extract cobalt from a cobalt lithium enriched leachate. Nan et al. [61] used Acorga M5640 and Cyanex 272 to effectively extract 98% of the copper and 97% of the cobalt from sulfate solution. Their work also demonstrated that Acorga M5640 and Cyanex 272 could be recycled after striping with H_2SO_4. Zhang et al. [29] conducted a batch solvent extraction using 0.90 M of PC-88A with Kerosene at an O:A ratio of 0.85:1 and pH 6.7 to extract cobalt from lithium enriched solution and found that 99.99% purity of Co could be obtained after H_2SO_4 stripping.

Zante et al. [62] used solvent extraction to individually extract $Mn^{2+}, Co^{2+}, Ni^{2+}$, and Li^{+}. Mn^{2+} was first removed using N,N,N',N'-tetra(noctyl) diglycolamide (TODGA) extractant diluted in an ionic liquid and then stripped with water. Second, Co^{2+} was extracted using tri-hexyl tetradecylphosphonium chloride ([P66614] [Cl]) and then stripped with water. Third, Ni^{2+} was recovered using deep eutectic solvents (DES) made of carboxylic acids and lidocaine and stripped with sodium oxalate. Finally, Li^{+} was precipitated by chemical precipitation. Sun et al. [63] selectively extracted lithium using benzo-15-crown-5 ether (B15C5). At pH of 6.0, temperature of 30°C, and 120 minutes of extraction time, the extraction rate of Li^{+} was 37%, and the extraction rate of Ni^{2+} was 5.18%, and about zero for Co^{2+} and Mn^{2+}.

Crystallization to recover Co^{2+} was studied by Ferreira et al. [40], who reported that by evaporating 85% H_2O from the leachate containing 11.3 g/L Co, 0.6 g/L Al, and 1.2 g/L Li, a purified monohydrated $CoSO_4 \cdot H_2O$ was produced with only 0.4% Al, and 0.6% Li contamination. Granata et al. [53] recovered 80% lithium as carbonate by evaporating 80% of H_2O to obtain a purity higher than 98% Li_2CO_3 crystal after removing Co^{2+}, Ni^{2+}, and Mn^{2+} through precipitation using carbonate.

1.4.1.4   Oxalate as a Hydrometallurgy Treatment Reagent

Oxalate is a common organic acid and can be dissolved in warm H_2O. Sun et al. [64] used oxalate to leach cobalt and lithium into leachate, where cobalt was then precipitate out as cobalt oxalate. The reaction between $LiCoO_2$ and oxalate is a multiphase reaction. The following shows the leaching reactions using oxalate:

$$7H_2C_2O_4 + 2LiCoO_{2(s)} \rightarrow 2LiHC_2O_4 + 2Co(HC_2O_4)_2 + 4H_2O + 2CO_{2(g)} \quad (1.20)$$

$$4H_2C_2O_4 + 2LiCoO_{2(s)} \rightarrow Li_2C_2O_4 + 2CoC_2O_{4(s)} + 4H_2O + 2CO_{2(g)} \quad (1.21)$$

During the oxalate leaching process, carbon dioxide was evolved from oxalate that converted Co^{3+} to Co^{2+} then promoted the dissolution of Co. Sun et al. [64] also investigated the addition of H_2O_2 on promoting the Co leaching efficiency through the following reactions:

$$6H_2C_2O_4 + 2LiCoO_{2(s)} + 3H_2O_2 \rightarrow 2LiHC_2O_4 + 2Co(HC_2O_4)_2 + 6H_2O + 2O_{2(g)}$$
$$(1.22)$$

$$3H_2C_2O_4 + 2LiCoO_{2(s)} + H_2O_2 \rightarrow Li_2C_2O_4 + 2CoC_2O_{4(s)} + 4H_2O + O_{2(g)} \quad (1.23)$$

The experimental data suggested that H_2O_2 had little influence on the leaching of $LiCoO_2$ with oxalate. The oxalate leaching efficiencies with and without H_2O_2 was 96.3% and 96.7%, respectively. They concluded that the optimized leaching conditions using oxalate was 1.0 M oxalate at 80°C for 120 minutes with solid/liquid ration of 50 g/L. Figure 1.10 shows the proposed flowsheet.

1.4.1.5   Ammonia with Sodium Sulfite as a Hydrometallurgy Treatment Reagent

Leaching spent cathode active material such as Ni/Mn/Co containing cathode active material (NMC) with ammonia (NH_3) requires sodium sulfite (Na_2SO_3) as a reductant to reduce Co^{3+} to Co^{2+}. The Ni^{2+} and Co^{2+} will react with ammonia according to the following equations while Mn and ammonia reaction is not favorable.

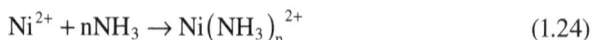

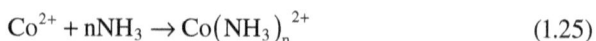

$$Ni^{2+} + nNH_3 \rightarrow Ni(NH_3)_n^{2+} \quad (1.24)$$

$$Co^{2+} + nNH_3 \rightarrow Co(NH_3)_n^{2+} \quad (1.25)$$

$Co(NH_3)_n^{2+}$ is more soluble at pH around 9-11 while $Ni(NH_3)_n^{2+}$ is soluble at pH from 8.5 to 10.5. As such, Ni^{2+}, Co^{2+}, Li^+, and Mn^{2+} can be selectively recovered.

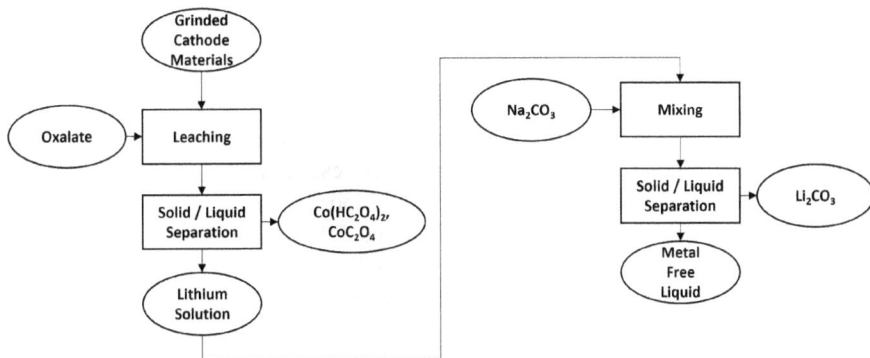

FIGURE 1.10 Flowsheet of leaching $LiCoO_2$ with oxalate [64].

Zheng et al. [65] reported leaching NMC with NH_3, ammonium sulfate $(NH_4)_2SO_4$ and Na_2SO_3. Ni, Co, and Li could be selectively leached out from cathode scrap powder. With 4 M NH_3, 1.5 M $(NH_4)_2SO_4$ and 0.5 M Na_2SO_3 at 353 K, 10 g/L pulp density and 300 minutes of retention time, more than 98% of Ni, Co, and Li could be selectively leached with 6.3% of Mn.

Wu et al. [66] reported that leaching 2% pulp density spent Ni/Co/Al-containing material (NCA) with NH_3 and Na_2SO_3 and using ammonia bicarbonate (NH_4HCO_3) as a pH buffer could obtain 60.5% leaching efficiency for Li and 81% for Co while Al was hardly leached.

1.4.1.6 DL-malic Acid as a Hydrometallurgy Treatment Reagent

DL-malic acid is a natural organic acid and a member of the C4-dicarboxylic acid family [67] with ionization constants of $K_1 = 4.0 \times 10^{-4}$ and $K_2 = 9.0 \times 10^{-6}$ [68]. DL-malic acid degrades easily under aerobic and anaerobic conditions thus its waste can be treated easily. Li et al. [69] investigated leaching $LiCoO_2$ with DL-malic acid. The reaction between DL-malic acid and $LiCoO_2$ is a multiphase reaction. The following shows the leaching reactions:

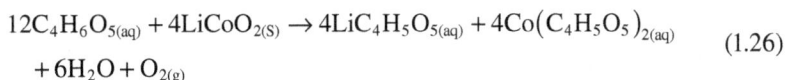

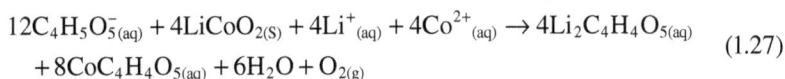

$$12C_4H_6O_{5(aq)} + 4LiCoO_{2(S)} \rightarrow 4LiC_4H_5O_{5(aq)} + 4Co\left(C_4H_5O_5\right)_{2(aq)} \\ + 6H_2O + O_{2(g)} \tag{1.26}$$

$$12C_4H_5O_{5(aq)}^- + 4LiCoO_{2(S)} + 4Li^+_{(aq)} + 4Co^{2+}_{(aq)} \rightarrow 4Li_2C_4H_4O_{5(aq)} \\ + 8CoC_4H_4O_{5(aq)} + 6H_2O + O_{2(g)} \tag{1.27}$$

They also investigated the addition of H_2O_2 on promoting the conversion of Co^{3+} to Co^{2+} to promote Co dissolution and improving Co leaching efficiency through the following equations:

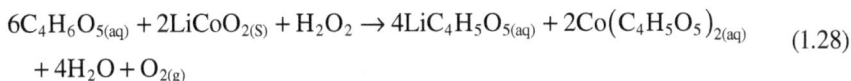

$$6C_4H_6O_{5(aq)} + 2LiCoO_{2(S)} + H_2O_2 \rightarrow 4LiC_4H_5O_{5(aq)} + 2Co\left(C_4H_5O_5\right)_{2(aq)} \\ + 4H_2O + O_{2(g)} \tag{1.28}$$

$$6C_4H_5O_{5(aq)}^- + 2LiCoO_{2(S)} + 2Li^+_{(aq)} + 2Co^{2+}_{(aq)} + H_2O_2 \rightarrow 2Li_2C_4H_4O_{5(aq)}$$
$$+ 4CoC_4H_4O_{5(aq)} + 4H_2O + O_{2(g)} \tag{1.29}$$

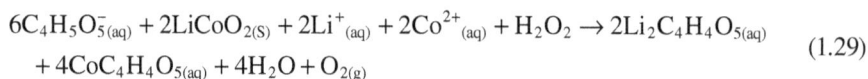

They found that when $LiCoO_2$ leached with only DL-malic acid, 37wt% Co and 54wt% Li were leached. The addition of H_2O_2 was essential as it could significantly increase the leaching efficiency where 93wt% Co and 99wt% Li were obtained. They observed that the best conditions for cobalt and lithium leaching with DL-malic acid were 1.5 M DL-malic acid, 2.0vol% H_2O_2, 90°C for 40 minutes with a solid/liquid ratio of 20 g/L. At these conditions, 100wt% Li and more than 90wt% Co could be leached. Zhou et al. [70] reported that leaching efficiency of 98.13wt% for Li and 98.86wt% for Co could be reached by adopting ultrasonically enhanced leaching using 1.5M DL-malic acid, 3vol% H_2O_2, and 80°C for 25 minutes with spray drying to regenerate $LiCoO_2$. Zhang et al. [71] used grape seed as reductant. The catechin, epicatechin and epigallocatechin gallate (EGCG) contained in grape seed were employed as the efficient reductants during leaching. About 92% Co and 99% Li could be leached with 0.6 g/g of grape seed, 1.5M malic acid, 80°C during 180 minutes, and slurry density 20 g/L. Leaching could also be done with malic acid with glucose as reductant [72], and 98% of Co could be precipitated using ammonium oxalate.

Meng et al. [73, 74] leached spent $LiNi_{1/3}Co_{1/3}Mn_{1/3}O_2$ and $LiCoO_2$ materials with malic acid and electrochemical cathode reduction. For $LiNi_{1/3}Co_{1/3}Mn_{1/3}O_2$, the leaching efficiencies of Li, Ni, Co and Mn reached to 100%, 99.87%, 99.58% and 99.82%, respectively, with 1.5 mol/L malic acid and a working voltage of 8 V at 300 r/min and 60°C for 30 minutes [73]. For $LiCoO_2$, the leaching efficiencies reached about 90% for cobalt and nearly 94% for lithium using 1.25 mol/L of malic acid and a working voltage of 8 V for 180 minutes at 70°C [74]. Figure 1.11 shows the proposed flowsheet.

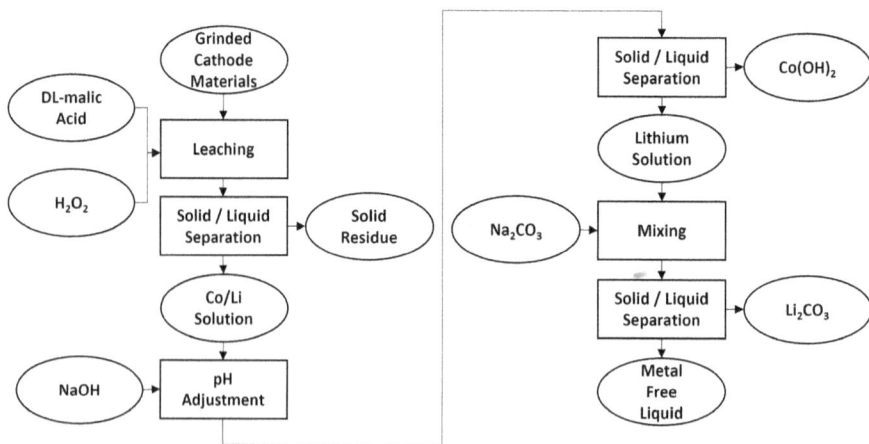

FIGURE 1.11   Proposed flowsheet for DL-malic acid leaching [67, 69, 70].

1.4.1.7 Citric Acid as a Hydrometallurgy Treatment Reagent

Citric acid has three carboxyls in one $C_6H_8O_7 \cdot H_2O$ molecule. 1 mole of citric acid can release up to 3 mol H^+ theoretically. The dissociation reaction and the corresponding reaction constants of citric acid are as follows:

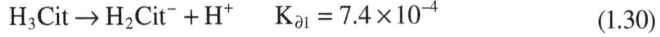

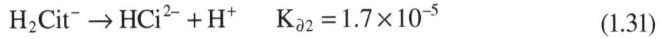

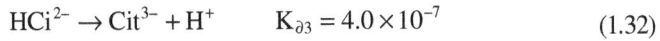

$$H_3Cit \rightarrow H_2Cit^- + H^+ \qquad K_{\partial 1} = 7.4 \times 10^{-4} \qquad (1.30)$$

$$H_2Cit^- \rightarrow HCi^{2-} + H^+ \qquad K_{\partial 2} = 1.7 \times 10^{-5} \qquad (1.31)$$

$$HCi^{2-} \rightarrow Cit^{3-} + H^+ \qquad K_{\partial 3} = 4.0 \times 10^{-7} \qquad (1.32)$$

Li et al. [75], Golmohammadzadeh et al. [76], and Musariri et al. [77] used citirc acid with H_2O_2 to leach spent $LiCoO_2$ with the leaching reactions shown in the following:

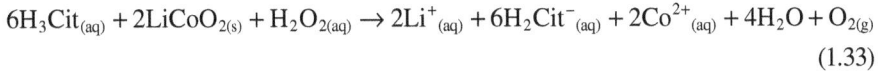

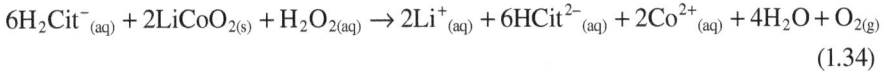

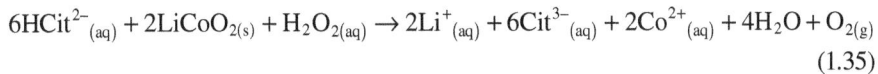

$$6H_3Cit_{(aq)} + 2LiCoO_{2(s)} + H_2O_{2(aq)} \rightarrow 2Li^+_{(aq)} + 6H_2Cit^-_{(aq)} + 2Co^{2+}_{(aq)} + 4H_2O + O_{2(g)} \qquad (1.33)$$

$$6H_2Cit^-_{(aq)} + 2LiCoO_{2(s)} + H_2O_{2(aq)} \rightarrow 2Li^+_{(aq)} + 6HCit^{2-}_{(aq)} + 2Co^{2+}_{(aq)} + 4H_2O + O_{2(g)} \qquad (1.34)$$

$$6HCit^{2-}_{(aq)} + 2LiCoO_{2(s)} + H_2O_{2(aq)} \rightarrow 2Li^+_{(aq)} + 6Cit^{3-}_{(aq)} + 2Co^{2+}_{(aq)} + 4H_2O + O_{2(g)} \qquad (1.35)$$

Based on their investigation, the best conditions for the leaching of spent $LiCoO_2$ using citric acid were 1.25 M citric acid with 1.0 vol% H_2O_2 at 90°C for 30 minutes and a solid/liquid ratio of 20 g/L. At these conditions, up to 100% Li and up to 96% Co could be extracted. It should be noted that citric acid with H_2O_2 was not able to dissolve Co_3O_4 thus not able to fully extract the Co. Part of $LiCoO_2$ could be transferred to Co_3O_4 during the pre-treatment at high temperature for removing $LiPF_6$ electrolyte. Figure 1.12 shows the proposed flowsheet.

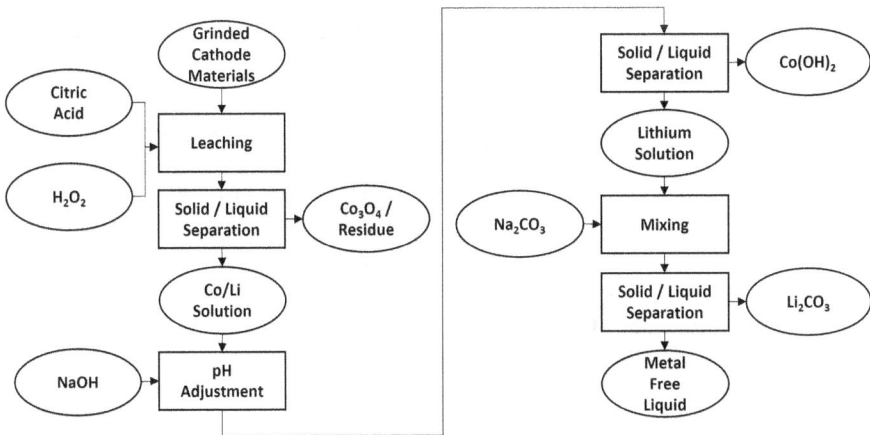

FIGURE 1.12 Proposed flowsheet for citric acid leaching [75–77].

1.4.1.8 Phosphoric Acid as a Hydrometallurgy Treatment Reagent

Zhuang et al. [78] reported using phosphoric acid (H_3PO_4) and citric acid mixture to leach spent NMC cathode active material. With 0.2 M phosphoric acid and 0.4 M citric acid with a solid to liquid (S/L) ratio of 20 g/L at 90°C for 30 minutes of retention time, leaching efficiency of 100% for Li, 93.38% for Ni, 91.63% for Co, and 92.00% for Mn can be achieved, respectively.

Chen et al. [79] reported using H_3PO_4 with H_2O_2 to leach Li_2CoO_2. By using 0.7 M H_3PO_4 and 4vol% H_2O_2, liquid to solid ratio of 20 mL g^{-1} at 40°C for 60 minutes of retention time, Co could be recovered as cobalt phosphate $Co_3(PO_4)_2$ precipitate while Li remained in the leachate with the following reactions:

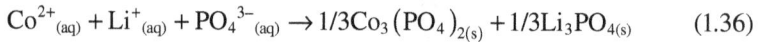

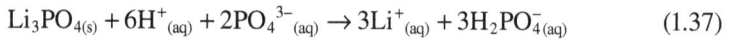

$$Co^{2+}_{(aq)} + Li^+_{(aq)} + PO_4^{3-}_{(aq)} \rightarrow 1/3Co_3(PO_4)_{2(s)} + 1/3Li_3PO_{4(s)} \quad (1.36)$$

$$Li_3PO_{4(s)} + 6H^+_{(aq)} + 2PO_4^{3-}_{(aq)} \rightarrow 3Li^+_{(aq)} + 3H_2PO_4^-_{(aq)} \quad (1.37)$$

On the other hand, excess H_3PO_4 could result in $Co_3(PO_4)_2$ re-dissolving back into the leachate according to the following equation:

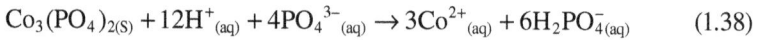

$$Co_3(PO_4)_{2(S)} + 12H^+_{(aq)} + 4PO_4^{3-}_{(aq)} \rightarrow 3Co^{2+}_{(aq)} + 6H_2PO_4^-_{(aq)} \quad (1.38)$$

1.4.1.9 Ascorbic Acid as a Hydrometallurgy Treatment Reagent

Ascorbic acid is a nature occurring organic acid. Ascorbic acid behaves as a vinylogous carboxylic acid where the electrons in the double bond, hydroxyl group lone pair, and the carbonyl double bond form a conjugated system, which results in the hydroxyl group being more acidic than typical hydroxyl groups. The pK_a values of ascorbic acid are $pKa_1 = 4.10$ and $pKa_2 = 11.6$, respectively. Ascorbic acid is known as a reducing agent that can be oxidized by one electron to a radical state or doubly oxidized to dehydroascorbic acid ($C_6H_6O_6$).

Li et al. [80] leached the spent $LiCoO_2$ with ascorbic acid. During leaching, $LiCoO_2$ was first dissolved and formed $C_6H_6O_6Li_2$, where Co^{3+} was reduced to Co^{2+} with ascorbic acid acting as a reducing agent and ascorbic acid ($C_6H_8O_6$) was oxidized to dehydroascorbic acid ($C_6H_6O_6$). Eq. (1.39) shows the leaching reaction, and Figure 1.13 shows the proposed flow sheet. The optimum leaching conditions were found to be 1.25 M ascorbic acid at 70°C for 20 minutes with solid/liquid ratio of 25 g/L, where 94.8% Co and 98.5% Li were leached into the solution.

$$4C_6H_8O_6 + 2LiCoO_2 \rightarrow C_6H_6O_6 + C_6H_6O_6Li_2 + 2C_6H_6O_6Co + 4H_2O \quad (1.39)$$

1.4.1.10 Benzenesulfonic Acid with Formic Acid

Fu et al. [81] reported using a mixture of 1.3 mol/L benzenesulfonic acid ($C_6H_6O_3S$) and 1.5 mol/L formic acid (CH_2O_2) to leach Li_2CoO_2 and achieved 97% leaching efficiency for Co and 99% for Li with a solid to liquid (S/L) ratio of 30 g/L, and 40 minutes of retention time at 50°C. The leached Co^{2+} was recovered from the leachate as pure cobalt benzene sulfonic with a 99% of recovery efficiency, and Li^+ could be entirely precipitated after Co removal with adding phosphoric acid to produce lithium phosphate, Li_3PO_4

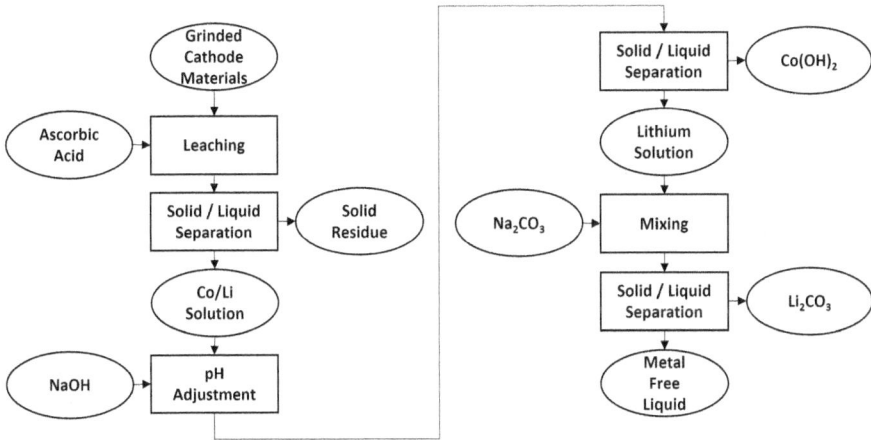

FIGURE 1.13 Proposed flowsheet for ascorbic acid leach [80].

1.4.1.11 Bioleaching

Mishra et al. [82], Xin et al. [83], and Zeng et al. [84] reported bioleaching of spent $LiCoO_2$ using chemolithotrophic and acidophilic bacteria, and acidithiobacillus fer-rooxidans (AF). AF utilized ferrous ion (Fe^{2+}) as the energy source to produce metabolites such as H_2SO_4 and ferric ions (Fe^{3+}), which assisted leaching of spent $LiCoO_2$. The generated Fe^{3+} could induce a series of redox reactions, leading to the formation of the strong reducing agent Fe^{2+}, which then promoted the reduction of Co^{3+} to Co^{2+}. The generated H_2SO_4 enabled direct acid dissolution of Co^{2+}. The following equations illustrate the bioleaching process:

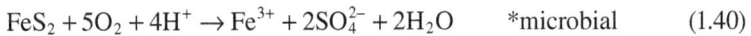

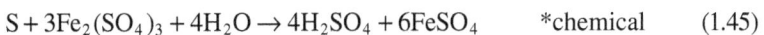

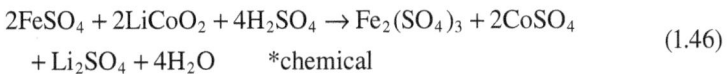

$$FeS_2 + 5O_2 + 4H^+ \rightarrow Fe^{3+} + 2SO_4^{2-} + 2H_2O \qquad \text{*microbial} \qquad (1.40)$$

$$FeS_2 + Fe_2(SO_4)_3 \rightarrow 3FeSO_4 + 2S \qquad \text{*chemical} \qquad (1.41)$$

$$FeS_2 + 7Fe_2(SO_4)_3 + 8H_2O \rightarrow 15FeSO_4 + 8H_2SO_4 \qquad \text{*chemical} \qquad (1.42)$$

$$2S + 3O_2 + 2H_2O \rightarrow 2H_2SO_4 \qquad \text{*microbial} \qquad (1.43)$$

$$4Fe^{2+} + O_2 + 4H^+ \rightarrow 4Fe^{3+} + 2H_2O \qquad \text{*microbial} \qquad (1.44)$$

$$S + 3Fe_2(SO_4)_3 + 4H_2O \rightarrow 4H_2SO_4 + 6FeSO_4 \qquad \text{*chemical} \qquad (1.45)$$

$$2FeSO_4 + 2LiCoO_2 + 4H_2SO_4 \rightarrow Fe_2(SO_4)_3 + 2CoSO_4$$
$$+ Li_2SO_4 + 4H_2O \qquad \text{*chemical} \qquad (1.46)$$

Mishra et al. [82] and Silverman et al. [85] conducted bioleaching with AF and using 1% elemental sulfur and 3 g/L Fe^{2+} as the energy source for AF. The result suggested that bioleaching of Co was faster than lithium.

Zeng et al. [84] reported Cu-catalyzed bioleaching with AF and found that the dissolution of Co was increased from 43.1% in 10 days without Cu to 99.9% in 6 days with 0.75 g/L Cu. The result indicated that the copper ions could enhance the oxidation of $LiCoO_2$. The following equations illustrates the catalytic mechanisms:

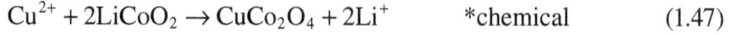

$$Cu^{2+} + 2LiCoO_2 \rightarrow CuCo_2O_4 + 2Li^+ \qquad \text{*chemical} \qquad (1.47)$$

$$CuCo_2O_4 + 6Fe^{3+} \rightarrow 6Fe^{2+} + Cu^{2+} + 2O_2 + 2Co^{2+} \qquad \text{*chemical} \quad (1.48)$$

$$4Fe^{2+} + O_2 + 4H^+ \rightarrow 4Fe^{3+} + 2H_2O \qquad \text{*microbial} \qquad (1.49)$$

The work from Mishra et al. [82], Xin et al. [83], and Zeng et al. [84] indicated that it was possible to recover metal from spent lithium-ion cathode active material by acidophilic bacteria. Huang et al. [86] reported using potential sulfur-oxidizing bacteria with a bio-electro-hydrometallurgical platform to recover Li, Mn, and Co from spent LIBs. The bioleaching process and selective adsorption by PC-88A/TOA-modified granular activated carbon were both incorporated into an electrokinetics approach to achieve maximum recoveries of 91.45%, 93.64% and 87.92% for Co, Li, and Mn, respectively. The proposed chemical reactions are as follows:

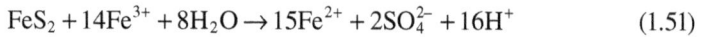

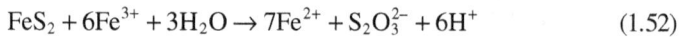

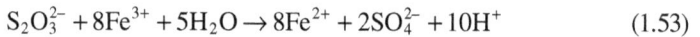

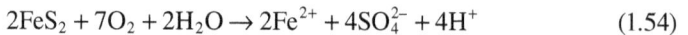

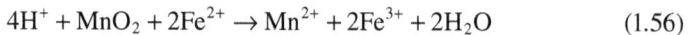

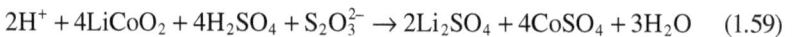

$$Fe_2O_3 + 6H^+ \rightarrow 2Fe^{3+} + 3H_2O \qquad (1.50)$$

$$FeS_2 + 14Fe^{3+} + 8H_2O \rightarrow 15Fe^{2+} + 2SO_4^{2-} + 16H^+ \qquad (1.51)$$

$$FeS_2 + 6Fe^{3+} + 3H_2O \rightarrow 7Fe^{2+} + S_2O_3^{2-} + 6H^+ \qquad (1.52)$$

$$S_2O_3^{2-} + 8Fe^{3+} + 5H_2O \rightarrow 8Fe^{2+} + 2SO_4^{2-} + 10H^+ \qquad (1.53)$$

$$2FeS_2 + 7O_2 + 2H_2O \rightarrow 2Fe^{2+} + 4SO_4^{2-} + 4H^+ \qquad (1.54)$$

$$Li_2CoMn_3O_8 + 2H_2SO_4 \rightarrow Li_2SO_4 + CoSO_4 + 3MnO_2 + 2H_2O \qquad (1.55)$$

$$4H^+ + MnO_2 + 2Fe^{2+} \rightarrow Mn^{2+} + 2Fe^{3+} + 2H_2O \qquad (1.56)$$

$$2H^+ + MnO_2 + S_2O_3^{2-} \rightarrow Mn^{2+} + SO_4^{2-} + S\downarrow + H_2O \qquad (1.57)$$

$$2LiCoO_2 + 3H_2SO_4 + 2Fe^{2+} + 2H^+ \rightarrow Li_2SO_4 + 2CoSO_4 + 4H_2O + 2Fe^{3+} \quad (1.58)$$

$$2H^+ + 4LiCoO_2 + 4H_2SO_4 + S_2O_3^{2-} \rightarrow 2Li_2SO_4 + 4CoSO_4 + 3H_2O \quad (1.59)$$

$$Mn^{2+} + 2OH^- \rightarrow Mn(OH)_{2(S)} \qquad (1.60)$$

$$2Mn(OH)_2 + O_2 \rightarrow 2MnO(OH)_{2(s)} \qquad (1.61)$$

Hydrometallurgical recycling involves dissolving the valuable cathode active materials in acids and separating the constituent metals using chemical precipitation or

TABLE 1.2

Comparison of Hydrometallurgy Treatments

Hydrometallurgy Treatment		Chemical Used	Advantage	Disadvantage
Leaching	1	HCl	Chemical inexpensive	Equipment corrosion, Cl contamination
	2	HNO_3	High extraction rate on Co and Li	Dissolve Al, oxidize Mn and Co, NOx in discharge
	3	H_2SO_4	Chemical inexpensive	Sulfate ion has solubility limitation
	4	H_3PO_4	Low acid consumption,	Phosphate has very limited solubility
	5	Organic acid	Environmental friendlier	Lower extraction rate
	6	Bioleaching	Environmental friendlier	Low kinetic, low extraction rate
Precipitation		NaOH, LiOH, Na_2CO_3, Li_2Co_3	Low energy consumption, low cost	pH adjustment difficult to control, metals often co-precipitated
Solvent extraction		D2EHPA, Cyanex 272, Acorga M5640, PC-88A, TODGA, P66614-Cl, DES	Able to extract individual metal ions, high separation efficiency, low energy consumption	Process very complicated, multiple extraction stages, high cost of solvents, generated wastewater and emission, Toxic solvents needed to be waste treated

solvent extraction. This approach can recover the valuable metals such as Ni, Co, Mn, and Li into individual chemicals, chemical precursor or new cathode active materials. As the trend is to decrease valuable metal content like Co in cathode active material, it is important to ensure that the simple processes, low reagent consumption rate, and high metal recovery can be achieved in order to keep hydrometallurgical process profitable. Table 1.2 shows the comparison of various leaching treatments and metal recover processes.

1.4.2 Hydrometallurgical Recycle of LiFePO₄ Cathode Active Material

Lithium iron phosphate ($LiFePO_4$ or LFP) batteries are used in energy storage and electric vehicles like Tesla Model 3 (China version). Processes to recycle of spent LFP can be categorized to direct recycling and hydrometallurgical recycling.

Direct recycling regenerates $LiFePO_4$ through replenish lithium into spent $LiFePO_4$ and reshape material structure using hydrothermal method [87–95]. Table 1.3 summarizes the operation conditions from literatures.

$LiFePO_4$ can also be recovered through hydrometallurgical recycling. Hydrometallurgical recycling is consisted of leaching and metal recovery. Various chemicals including H_2SO_4, citric acid ($C_6H_8O_7$), $Na_2S_2O_8$, oxalic acid ($C_2H_2O_4$), acetic

TABLE 1.3

Process Summary to Direct Regeneration of LiFePO$_4$

No	Process	Additives	Heat Treated Time (h)	Temp (°C)	Li Source	Heating Inner Atmosphere	Reference
1	Hydrothermal	LiOH, L-ascorbic acid, Dodecyl benzene sulfonic acid sodium (SDBS)	6	160	LiOH		[87]
2	Thermal ball milling	Sucrose, Li$_2$Co$_3$	2	800/650	Li$_2$Co$_3$	H$_2$/Ar	[88]
3	Ball milling	Li$_2$Co$_3$	N/A		Li$_2$Co$_3$		[89]
	Heat treatment			700		N$_2$	
4	Heat treatment	Glucose, Li$_2$Co$_3$	2	900	Li$_2$Co$_3$		[90]
5	Heat treatment	LiFePO$_4$	8	700	LiFePO$_4$		[91]
6	Heat treatment	Li$_2$Co$_3$	1	650	Li$_2$Co$_3$	H$_2$/Ar	[92]
7	Heat treatment		1	650	N/A	H$_2$/Ar	[93]
8	Leaching + reprecipitate	H$_3$PO$_4$	9	85			[94]
	Heat treatment	Li$_2$Co$_3$			Li$_2$Co$_3$	N$_2$	
9	Annealing		10	700			[95]
	Leaching	6 M HCl, NH$_4$OH					
	Ball milling	LiOH, Sucrose			LiOH		
	Heat treatment		8	700		H$_2$/Ar	

acid (CH$_3$COOH), HCl, phosphoric acid (H$_3$PO$_4$), and H$_2$O$_2$ have been reported to be employed for leaching. Metal recovery is conducted through precipitation, solvent extraction (SX), or ion-exchange [96]. Table 1.4 shows the comparison of the leach and metal recover operation conditions of the hydrometallurgy process.

Yang et al. [97] used mechanochemical activation with EDTA-2Nato recover lithium using sodium hydroxide (NaOH) and H$_3$PO$_4$, achieved a 94.3% Li leaching efficiency and an overall of 82.6% Li recovery. He et al. [98] demonstrated that leaching with (NH$_4$)$_2$S$_2$O$_8$ could selectively leach 97.5% Li. Yadav et al. [99] reported that methanesulfonic acid with H$_2$O$_2$ could have a 94% Li leaching efficiency. Zhang et al. [100] demonstrated that sodium persulfate (Na$_2$S$_2$O$_8$) could selectively leach 99% Li within 20 minutes; and with sodium carbonate (Na$_2$CO$_3$), 95% of Li could be recovered as lithium carbonate (Li$_2$CO$_3$). Li et al. [101] used oxalic acid to achieve 98% Li leaching efficiency.

Yang et al. [102] showed that leaching with acetic acid and H$_2$O$_2$ and precipitating with Na$_2$CO$_3$, over 95% of Li could be leached and recovered. Furthermore, Jing et al. [103] showed that with H$_2$O$_2$, up to 95.4% of Li could be leached. Li et al. [104] used citric acid to leach and Na$_2$CO$_3$ to recover Li and achieved 99.35% of leaching efficiency and 95% recovery.

TABLE 1.4

Hydrometallurgy Processes to Leach and Recover Li from Spent LiFePO$_4$

No	Chemical	Leaching				Metal Recovery (Precipitation, SX, Ion exchange)			Ref [#]
		Time (h)	Temp (°C)	Pulp Density	Li Leaching Efficiency (%)	Chemical	Temp (°C)	Yield (%)	
1	0.6M H$_3$PO$_4$ Mechanochemical activation with EDTA-2Na	20 30			94.3	5M NaOH		82.6	[97]
2	(NH$_4$)$_2$ S$_2$O$_8$	1			97.5	Na$_2$S to form NaFeS$_2$			[98]
3	Methanesulfonic acid, p-toluenesulfonic acid H$_2$O$_2$	1.5	25	40g/L	94	Na$_2$CO$_3$	96	85	[99]
4	Na$_2$S$_2$O$_8$	1/3	25	300g/L	99	Na$_2$CO$_3$	95		[100]
5	0.3M Oxalic acid	1	80	60g/L	98				[101]
6	0.8M Acetic acid, 6vol% H$_2$O$_2$	1/2	50	120g/L	95.1	Na$_2$CO$_3$	95	84.8	[102]
7	2.8M H$_2$O$_2$	4	25	200g/L	95.4				[103]
8	Citric acid, H$_2$O$_2$	2		20g LFP/20g acid	99.35	Na$_2$CO$_3$	95	85.4	[104]
9	0.28M H$_2$SO$_4$	4	85		98.46	Na$_2$PO$_4$		84.2	[105]
10	2.5M H$_2$SO$_4$	4	60	100g/L	97	Na$_2$CO$_3$			[106]

(Continued)

TABLE 1.4 (Continued)
Hydrometallurgy Processes to Leach and Recover Li from Spent $LiFePO_4$

		Leaching				Metal Recovery (Precipitation, SX, Ion exchange)			
No	Chemical	Time (h)	Temp (°C)	Pulp Density	Li Leaching Efficiency (%)	Chemical	Temp (°C)	Yield (%)	Ref [#]
11	$2M\ H_2SO_4$		80			H_3PO_4			[107]
12	$0.3M\ H_2SO_4,\ H_2O_2$	2	60			Na_3PO_4	65	92.5	[108]
13	$4M\ H_2SO_4,\ 2vol\%\ H_2O_2$		60			Na_3PO_4	60	90	[109]
14	$H_2SO_4,\ O_2/H_2O_2$	2	80–120	100g/L	92	Na_2CO_3			[110]
15	HCl					SX: 40% D2EHPA with Kerosene	35	90	[111]
						Strip with HCl	35		
						Precipitate with Na_2CO_3	80		
16	4M HCl					Na_3PO_4		96	[112]
17	6.5M HCl, $15vol\%\ H_2O_2$	2	60	200g/L	92.2	Na_3PO_4	90	80.9	[113]

On the other hand, H_2SO_4 and HCl are widely reported to leach Li from spent $LiFePO_4$. Tao et al. [105], Zheng et al. [106], and Tedjar et al. [107] used H_2SO_4 to leach above 97% of Li within 4 hours. Li et al. [108], Cai et al. [109], and Schurmans et al. [110] added H_2O_2 as an oxidant into H_2SO_4 leaching, the leaching time could be reduced.

Song et al. [111] used HCl to leach and using SX to recover up to 90% of Li. Wang et al. [112] illustrated that leaching with HCl and precipitating with Na_3PO_4 could recover up to 96% of Li. Huang et al. [113] combined flotation with HCl leaching of Li. Li et al. [114] used electrolysis cell with anionic membrane to delithiate and oxidize $LiFePO_4$ to $FePO_4$ on the anode compartment. Then Li^+ joined with OH^- ions produced from hydrogen evolution and passed through the anionic membrane to form LiOH. Li^+ could be recovered through adding Na_2CO_3 to form Li_2CO_3.

Due to the low cost of iron phosphate, hydrometallurgical recycling to recover individual materials like Fe, PO_4^{3-} and Li is not economic feasible. Direct recycling to regenerate $LiFePO_4$ or directly generating $LiFePO_4$ during hydrometallurgical recycling should be the focus.

1.4.3 Hydrometallurgical Recycle of Lithium Polymer Battery Cathode Active Material

Lithium metal polymer battery is consisted of an ultra-thin lithium metal foil anode, a solid copolymer electrolyte containing a lithium salt, and a lithium vanadium oxide cathode. To recycle lithium polymer battery, Cardarelli et al. [115] used a freezing method to harden the polymer electrolyte and comminute the cooled and hardened material before the comminuted material subjected to incineration, leaching and precipitation to recover lithium and vanadium.

REFERENCES

1. Global EV Outlook 2020, IEA Publications, June (2020).
2. Ding Y., Cano Z.P., Yu A., Lu J., Chen Z., Automotive Li-ion batteries: Current status and future perspectives, Electrochemical Energy Reviews, 2, 1–28, (2019).
3. Mineral Commodity Summaries 2020, U.S. Geological Survey, U.S. Department of the Interior, January (2020).
4. Garside M., Average cobalt spot price in the United States from 2008 to 2019 (in U.S. dollars per pound), https://www.statista.com/statistics/339743/average-spot-price-of-cobalt-in-the-us/#:~:text=In%202019%2C%20the%20average%20spot,37.4%20 U.S.%20dollars%20per%20pound March (2022)
5. Jewell S., Kimball S.M., *Mineral Commodity Summaries 2017.* (US Department of the Interior, US Geological Survey, Reston, VA 2017).
6. Zeng X., Li J., Singh N., Recycling of spent lithium-ion battery: A critical review, Critical Reviews in Environmental Science and Technology, 44(10), 1129, (2014).
7. Jacoby M., It's time to recycle lithium-ion batteries, Chemical & Engineering News, 97(28), 29–32, (2019).
8. Li L., Zhang X., Li M., Chen R., Wu F., Amine K., Lu J., The recycling of spent lithium-ion batteries: A review of current processes and technologies, Electrochemical Energy Reviews, 1, 461–482, (2018).
9. He K., Zhang Z.Y., Alai L., Zhang F.S., A green process for exfoliating electrode materials and simultaneously extracting electrolyte from spent lithium-ion batteries, Journal of Hazardous Materials, 375, 43–51, (2019).

10. Li J., Lu Y., Yang T., Ge D., Wood D., Li Z., Water-based electrode manufacturing and direct recycling of lithium-ion battery electrodes – A green and sustainable manufacturing system, iScience, 23, 101081, May (2020).

11. Larouche F., Tedjar F., Amouzegar K., Houlachi G., Bouchard P., Demopoulos G.P., Zaghib K., Progress and status of hydrometallurgical and direct recycling of Li-ion batteries and beyond, Materials, 13, 801, (2020).

12. Kasnatscheew J., Wagner R., Winter M., Cekic-Laskovic C., Interfaces and materials in lithium ion batteries: Challenges for theoretical electrochemistry, Topics in Current Chemistry, 376, 16, (2018).

13. Grutzke M., Kraft V., Weber W., Wendt C., Friesen A., Klamor S., Winter M., Nowak S., Supercritical carbon dioxide extraction of lithium-ion battery electrolytes, Journal of Supercritical Fluids, 94, 216–222, (2014).

14. Grutzke M., Monnighoff X., Horsthemke F., Kraft V., Winter M., Nowak S., Extraction of lithium-ion battery electrolytes with liquid and supercritical carbon dioxide and additional solvents, RSC Advances, 5, 43209–43217, May (2015).

15. Monnighoff X., Friesen A., Konersmann B., Horsthemke F., Grutzke M., Winter M., Nowak S., Supercritical carbon dioxide extraction of electrolyte from spent lithium ion batteries and its characterization by gas chromatography with chemical ionization, Journal of Power Sources, 352, 56–63, (2017).

16. Liu Y., Mu D., Li R., Ma Q., Zheng R., Dai C., Purification and characterization of reclaimed electrolytes from spent lithium-ion batteries, The Journal of Physical Chemistry C, 121, 4181–4187, (2017).

17. Jacoby M., It's time to get serious about recycling lithium-ion batteries, Chemical & Engineering News, 97(28), (2019); https://cen.acs.org/materials/energy-storage/time-serious-recycling-lithium/97/i28

18. Golubkov A.W., Fuchs D., Wagner J., Wiltsche H., Stangl C., Fauler G., Voitic G., Thaler A., Hacker V., Thermal-runaway experiments on consumer Li-ion batteries with metal-oxide and olivin-type cathodes, RSC Advances, 4, 3633, (2014).

19. Assefi M., Maroufi S., Yamauchi Y., Sahajwalla V., Pyrometallurgical recycling of Li-ion, Ni-Cd, and Ni-MH batteries: A minireview, Current Opinion in Green and Sustainable Chemistry, 24, 26–31, (2020).

20. Meshram P., Pandey B.D., Mankhd T.R., Extraction of lithium from primary and secondary sources by pre-treatment, leaching, and separation: A comprehensive review, Hydrometallurgy, 150, 192–208, (2014).

21. Georgi-Maschler T., Friedrich B., Weyhe R., Heegn H., Rutz M., Development of a recycling process for Li-ion batteries, Journal of Power Sources, 207, 173–82, (2012).

22. Trager T., Friedrich B., Weyhe R., Recovery concept of value metals from automotive lithium-ion batteries, Chemie Ingenieur Technik, 87(11), 1550–1557, (2015)

23. Liang Q., Yue H., Wang S., Yang S., Lam K.H., Hou X., Recycling and crystal regeneration of commercial used $LiFePO_4$ cathode active materials, Electrochimica Acta, 330(10), 135323, January (2020).

24. Tang Y., Xie H., Zhang B., Chen X., Zhao Z., Qu J., Xing P., Yin H., Recovery and regeneration of $LiCoO_2$-based spent lithium-ion batteries by a carbothermic reduction vacuum pyrolysis approach: Controlling the recovery of CoO or Co, Waste Management, 97, 140–148, September (2019).

25. Wang D., Zhang X., Chen H., Sun J., Separation of Li and Co from the active mass of spent Li-ion batteries by selective sulfating roasting with sodium bisulfate and water leaching, Minerals Engineering, 125, 28–35, September (2018).

26. Wang W., Han Y., Zhang T., Zhang L., Xu S., Alkali metal salt catalyzed carbothermic reduction for sustainable recovery of LiCoO2: Accurately controlled reduction and efficient water leaching, ACS Sustainable Chemistry & Engineering, 7, 16729–16737, (2019).

27. Dalini E.A., Karimi G., Zandevakili S., Goodarzi M., A review on environmental, economic and hydrometallurgical processes of recycling spent lithium-ion batteries, Mineral Processing and Extractive Metallurgy Review. (2020). DOI:10.1080/08827508 .2020.1781628

28. Zheng X., Zhu Z., Lin X., Zhang Y., He Y., Cao H., Sun Z., A mini-review on metal recycling from spent lithium ion batteries, Engineering, 4, 351–379, (2018).

29. Zhang P., Yokoyama T., Itabashi O., Suzuki T., Inoue K., Hydrometallurgical process for recovery of metal values from spent lithium-ion secondary batteries, Hydrometallurgy, 47, 259–271, (1998).

30. Contestabile M., Panero S., Scrosati B., A laboratory-scale lithium-ion battery recycling process, Journal of Power Sources, 92, 65–69, (2001).

31. Takacova Z., Havlik T., Kukurugya F., Orac D., Cobalt and lithium recovery from active mass of spent li-ion batteries: Theoretical and experimental approach, Hydrometallurgy, 163, 9–17, (2016).

32. Wang R., Lin Y.C., Wu S.H., A novel recovery process of metal values from the cathode active materials of the lithium-ion secondary batteries, Hydrometallurgy, 99, 194–201, (2009).

33. Yang Y., Lei S., Song S., Sun W., Wang L., Stepwise recycling of valuable metals from Ni-rich cathode active material of spent lithium-ion batteries, Waste Management, 102, 131–138, October (2019).

34. Castillo S., Ansart F., Labberty R., Portal J., Advances of recovering of spent lithium battery compounds, Journal of Power Sources, 112, 247–254, (2002).

35. Lee C.K., Rhee K.I., Preparation of LiCoO2 from spent lithium-ion batteries, Journal of Power Sources, 109, 17–21, (2002).

36. Lee C.K., Rhee K.I., Reductive leaching of cathodic active materials from lithium ion battery wastes, Hydrometallurgy, 68, 5–10, (2003).

37. Li L., Chen R., Sun F., Wu F., Liu J., Preparation of LiCoO2 films from spent lithium-ion batteries by a combined recycling process, Hydrometallurgy, 108, 220–225, (2011).

38. Vu C., Han K.N., Lawson F., Leaching behaviour of cobaltous and cobalto-cobaltic oxides in ammonia and in acid solutions, Hydrometallurgy, 6, 75–87, (1980).

39. Sun L., Qiu K., Vacuum pyrolysis and hydrometallurgical process for the recovery of valuable metals from spent lithium-ion batteries, Journal of Hazard Mater, 194, 378–384, (2011).

40. Ferreira D.A., Prados L.M., Majuste D., Mansur M.B., Hydrometallurgical separation of aluminum, cobalt, copper, and lithium from spent Li-ion batteries, Journal of Power Sources, 187, 238–246, (2009).

41. Nayl A.A., Elkhashab R.A., Badawy S.M., El-Khateeb M.A., Acid leaching of mixed spent li-ion batteries, Arabian Journal of Chemistry, 10, S3632–S3639, (2017).

42. Dorella G., Mansur M.B., A study of the separation of cobalt from spent Li-ion battery residues, Journal of Power Sources, 170, 210–215, (2007).

43. Chen L., Tang X., Zhang Y., Li L., Zhen Z., Zhang Y., Process for the recovery of cobalt oxalate from lithium ion batteries, Hydrometallurgy, 108, 80–86, (2011).

44. Kang J., Senanayake G., Sohn J.S., Shin S.M., Recovery of cobalt sulfate from spent lithium ion batteries by reductive leaching and solvent extraction with cyanex 272, Hydrometallurgy, 100, 168–171, (2010).

45. Shin S.M., Kim N.H., Sohn J.S., Yang D.H., Kim Y.H., Development of a metal recovery process from Li-ion battery wastes, Hydrometallurgy, 79, 172–181, (2005).

46. Zhu S., He W., Li G., Zhou X., Zhang X., Huang J., Recovery of Co and Li from spent lithium-ion batteries by combination method of acid leaching and chemical precipitation, Transactions of Nonferrous Metals Society of China, 22, 2274–2281, (2012).

47. Swain B., Jeong J., Lee J., Lee G.H., Sohn J., Hydrometallurgical process for recovery of cobalt from waste cathodic active material generated during manufacturing of lithium ion batteries, Journal of Power Sources, 167, 536–544, (2007).
48. Nan J., Han D., Yang M., Cui M., Hou X., Recovery of metal values from a mixture of spent lithium-ion batteries and nickel-metal hydride batteries, Hydrometallurgy, 84, 75–80, (2006).
49. Chen W.S., Ho H.J., Recovery of valuable metals from lithium-ion batteries NMC cathode waste materials by hydrometallurgical methods, Metals, 8, 321, May (2018).
50. Yang X., Dong P., Hao T., Zhang Y., Meng Q., Li Q., Zhou S., A combined method of leaching and co-precipitation for recycling spent $LiNi_{0.6}Mn_{0.2}O_2$ cathode active materials: Process optimization and performance aspects, The Journal of the Minerals, Metals & Materials Society (TMS), 72(11), 3843–3852, (2020).
51. Wang J., Chen M., Chen H., Luo T., Xu Z., Leaching study of spent Li-ion batteries, Procedia Environmental Sciences, 16, 443–450, (2012).
52. Vieceli N., Nogueira C., Guimaraes C., Pereira M., Durao F., Margarido F., Hydrometallurgical recycling of lithium-ion batteries by reductive leaching with sodium metabisulphite, Waste Management, 71, 350–361, (2018).
53. Granata G., Moscardini E., Pagnanelli F., Trabucco F., Toro L., Product recovery from li-ion battery waste coming from an industrial pre-treatment plant: Lab scale tests and process simulations, Journal of Power Sources, 206, 393–401, (2012).
54. Furlani G., Pagnanelli F., Toro L., Reductive acid leaching of manganese dioxide with glucose: Identification of oxidation derivatives of glucose, Hydrometallurgy, 81, 234–240, (2006).
55. Pagnanelli F., Garavini M., Veglio F., Toro L., Preliminary screening of purification processes of liquor leach solutions obtained from reductive leaching of low-grade manganese ores, Hydrometallurgy, 71, 319–327, (2004).
56. Veglioi F., Trifoni M., Toro L., Leaching of manganiferous ores by glucose in a sulfuric acid solution: Kinetic modeling and related statistical analysis, Industrial & Engineering Chemistry Research, 40, 3895–3901, (2001).
57. Sayilgan E., Kukrer T., Civelekoglu G., Ferella F., Akcil A., Veglio F., Kitis M., A review of technologies for the recovery of metals from spent alkaline and zinc-carbon batteries, Hydrometallurgy, 97, 158–166, (2009).
58. Xu J., Thomas H.R., Francis R.W., Lum K., Wang J., Liang B., A review of processes and technologies for the recycling of lithium-ion secondary batteries, Journal of Power Sources, 177, 512–527, (2008).
59. Swain B., Jeong J., Lee J.C., Lee G.H., Development of process flow sheet for recovery of high pure cobalt from sulfate leach liquor of lib industry waste: A mathematical model correlation to predict optimum operational conditions, Separation and Purification Technology, 63, 360–369, (2008).
60. Lupi C., Pasquali M., Electrolytic nickel recovery from lithium-ion batteries, Minerals Engineering, 16, 537–542, (2003).
61. Nan J., Han D., Han D., Zuo X., Recovery of metal values from spent lithium-ion batteries with chemical deposition and solvent extraction, Journal of Power Sources, 152, 278–284, (2005).
62. Zante G., Braun A., Masmoudi A., Barillon R., Trebouet D., Boltoeva M., Solvent extraction fractionation of manganese, cobalt, nickel and lithium using ionic liquids and deep eutectic solvents, Minerals Engineering, 156, 106512, (2020).
63. Sun Y., Zhu M., Yao Y., Wang H., Tong B., Zhao Z., A novel approach for the selective extraction of Li^+ from the leaching solution of spent lithium-ion batteries using benzo-15-crown-5 ether as extractant, Separation and Purification Technology, 237, 116325, (2020).

64. Sun L., Qiu K., Organic oxalate as leachant and precipitant for the recovery of valuable metals from spent lithium-ion batteries, Waste Manage, 32, 1575–1582, (2012).

65. Zheng X., Goa W., Zhang X., He M., Lin X., Cao H., Zhang Y., Sun Z., Spent lithium-ion battery recycling – Reductive ammonia leaching of metals from cathode scrap by sodium sulphite, Waste Management, 60, 680–688, February (2017).

66. Wu C., Li B., Yuan C., Ni S., Li L., Recycling valuable metals from spent lithium-ion batteries by ammonium sulfite-reduction ammonia leaching, Waste Management, 93, 153–161, 15 June (2019).

67. Moon S.Y., Hong S.H., Kim T.Y., Metabolic engineering of escherichia coil for the production of malic acid, Biochemical Engineering Journal, 40, 312–320, (2008).

68. Donald E.P., Gellan G.W., IFT's food technology industrial achievement award, Food Technology, 47–49, 94–96, (1993).

69. Li L., Ge J., Chen R., Wu F., Chen S., Zhang X., Environmental friendly leaching reagent for cobalt and lithium recovery from spent lithium-ion batteries, Waste Management, 30, 2615–2621, (2010).

70. Zhou S., Zhang Y., Meng Q., Dong P., Fei Z., Li Q., Recycling of $LiCoO_2$ cathode active material from spent lithium ion batteries by ultrasonic enhanced leaching and one-step regeneration, Journal of Environmental Management, 277, 111426, (2021).

71. Zhang Y., Meng Q., Dong P., Duan J., Lin Y., Une of grape seed as reductant for leaching of cobalt from spent lithium-ion batteries, Journal of Industrial and Engineering Chemistry, 66, 86–93, (2018).

72. Meng Q., Zhang Y., Dong P., A combined process for cobalt recovering and cathode active material regeneration from spent LiCoO2 batteries: Process optimization and kinetics aspects, Waste Management, 71, 372–380, (2018).

73. Meng Q., Zhang Y., Dong P., Liang F., A novel process for leaching of metals from LNi1/3Co1/3Mn1/3O2 material of spent lithium ion batteries: Process optimization and kinetics aspects, Journal of Industrial and Engineering Chemistry, 61, 131–141, (2018).

74. Meng Q., Zhang Y., Dong P., Use of electrochemical cathode-reduction method for leaching of cobalt from spent lithium-ion batteries, Journal of Cleaner Production, 180, 64–70, (2018).

75. Li L., Ge J., Wu F., Chen R., Chen S., We B., Recovery of cobalt and lithium from spent lithium ion batteries using organic citric acid as leachant, Journal of Hazardous Materials, 176, 288–293, (2010).

76. Golmohammadzadeh R., Rashchi F., Vahidi E., Recovery of lithium and cobalt from spent lithium-ion batteries using organic acids: Process optimization and kinetic aspects, Waste Management, 64, 244–254, June (2017).

77. Musariri B., Akdogan G., Dorfling C., Bradshaw S., Evaluating organic acids as alternative leaching reagents for metal recovery from lithium ion batteries, Minerals Engineering, 137, 108–117, 15 June (2019).

78. Zhuang L., Sun C., Zhou T., Li H., Dai A., Recovery of valuable metals from $LiNi_{0.5}Co_{0.2}Mn_{0.3}O_2$ cathode active materials of spent Li-ion batteries using mild mixed acid as leachant, Waste Management, 85(15), 175–185, February (2019).

79. Chen X., Ma H., Luo C., Zhou T., Recovery of valuable metals from waste cathode active materials of spent lithium-ion batteries using mild phosphoric acid, Journal of Hazardous Materials, 326, 77–86, 15 March (2017).

80. Li L., Lu J., Ren Y., Zhang X.X., Chen R., Wu F., Amine K., Ascorbic acid assisted recovery of cobalt and lithium from spent li-ion batteries, Journal of Power Source, 218, 21–27, (2012).

81. Fu Y., He Y., Chen H., Ye C., Lu Q., Li R., Xie W., Wang J., Effective leaching and extraction of valuable metals from electrode material of spent lithium-ion batteries using mixed organic acids leachant, Journal of Industrial and Engineering Chemistry, 79(25), 154–162, November (2019).

82. Mishra D., Kim D.J., Ralph D.E., Ahn J.G., Rhee Y.H., Bioleaching of metals from spent lithium ion secondary batteries using *Acidithiobacillus ferrooxidans*, Waste Management, 28, 333–338, (2008).

83. Xin B., Zhang D., Zhang X., Bioleaching mechanism of Co and Li from spent lithium-ion battery by the mixed culture of acidophilic sulphur-oxidising and iron-oxidising bacteria, Bioresource Technology, 100–124, 6163–6169, (2009).

84. Zeng G., Deng X., Luo S., Luo X., Zou J., A copper-catalyzed bioleaching process for enhancement of cobalt dissolution from spent lithium-ion batteries, Journal of Hazardous Materials, 199–200, 164–169, (2012).

85. Silverman M., Lundgren D., Studies on the chemoautotrophic iron bacterium ferrobacillus ferrooxidans I. An improved medium and a harvesting procedure for securing high cell yields, Journal of Bacteriology, 77, 642–647, (1959).

86. Huang T., Liu L., Zhang S., Recovery of cobalt, lithium, and manganese from the cathode active materials of spent lithium-ion batteries in a bio-electro-hydrometallurgical process, Hydrometallurgy, 188, 101–111, (2019).

87. Song W., Liu J., You L., Wang S., Zhou Q., Gao Y., Yin R., Xu W., Guo Z., Re-synthesis of nano-structured LiFePO$_4$/graphene composite derived from spent lithium-ion battery for booming electric vehicle application, Journal of Power Sources, 419, 192–202, (2019).

88. Sun Q., Li X., Zhang H., Song D., Shi X., Song J., Li C., Zhang L., Resynthesizing LiFePO$_4$/C materials from the recycled cathode via a green full-solid route, Journal of Alloys and Compounds, 818, 153292, (2020).

89. Li J., Wang Y., Wang L., Liu B., Zhou H., A facile recycling and regeneration process for spent LiFePO$_4$ batteries, Journal of Materials Sciences: Materials in Electronics, 30, 14580–14588, (2019).

90. Jiang X., Wang P., Li L., Yu J., Yin Y., Hou F., Recycling process for spent cathode active materials of LiFePO$_4$ batteries, Materials Science Forum, 943, 141–148, (2019).

91. Song X., Hu T., Liang C., Long H.L., Zhou L., Song W., You L., Wu Z.S., Liu J.W., Direct regeneration of cathode active materials from spent lithium iron phosphate batteries using a solid phase sintering method, RSC Advances, 7(8), 4783–4790, (2017).

92. Li X., Zhang J., Song D., Song J., Zhang L., Direct regeneration of recycled cathode active material mixture from scrapped LiFePO$_4$ batteries, Journal of Power Sources, 345, 78–84, (2017).

93. Chen J., Li Q., Song J., Song D., Zhang L., Shi X., Environmentally friendly recycling and effective repairing of cathode powders from spent LiFePO$_4$ batteries, Green Chemistry, 18(8), 2500–2506, (2016).

94. Bian D., Sun Y., Li S., Tian Y., Yang Z., Fan X., Zhang W., A novel process to recycle spent LiFePO$_4$ for synthesizing LiFePO$_4$/C hierarchical microflowers, Electrochimica Acta, 190, 134–140, (2016).

95. Shin E., Kim S., Noh J.K., Byun D., Chung K.Y., Kim H.S., Cho B.W., A green recycling process designed for LiFePO$_4$ cathode active materials for Li-ion batteries, Journal of Materials Chemistry A, 3(21), 11493–11502, (2015).

96. Forte F., Pietrantonio M., Pucciarmati S., Puzone M., Fontana D., Lithium iron phosphate batteries recycling: An assessment of current status, Critical Reviews in Environmental Science and Technology, 1054–3389, 1547–6537, (2020).

97. Yang Y., Zheng X., Cao H., Zhao C., Lin X., Ning P., Zhang Y., Jin W., Sun Z., A closed-loop process for selective metal recovery from spent lithium iron phosphate batteries through mechanochemical activation, ACS Sustainable Chemistry & Engineering, 5, 9972–9980, (2017).

98. He K., Zhang Z.Y., Zhang F.S., A green process for phosphorus recovery from spent LiFePO$_4$ batteries by transformation of delithiated LiFePO$_4$ crystal into NaFeS$_2$, Journal of Hazardous Materials, 395, 122614, (2020).

99. Yadav P., Jie C.J., Tan S., Srinivasan M., recycling of cathode from spent lithium iron phosphate batteries, recycling of cathode from spent lithium iron phosphate batteries, Journal of Hazardous Materials, 300, 123068, (2020).

100. Zhang J., Hu J., Liu Y., Jing Q., Yang C., Chen Y., Wang C., Sustainable and facile method for the selective recovery of lithium from cathode scrap of spent LiFePO$_4$ batteries, ACS Sustainable Chemistry & Engineering, 7(6), 5626–5631, (2019).

101. Li L., Lu J., Zhai L., Zhang X., Curtiss L., Jin Y., Wu F., Chen R., Amine K., A facile recovery process for cathodes from spent lithium iron phosphate batteries by using oxalic acid, CSEE Journal of Power and Energy Systems, 4(2), 219–255, (2018).

102. Yang Y., Meng X., Cao H., Lin X., Liu C., Sun Y., Zhang Y., Sun Z., Selective recovery of lithium from spent lithium iron phosphate batteries: A sustainable process, Green Chemistry, 20(13), 3121–3133, (2018).

103. Jing Q., Zhang J., Liu Y., Yang C., Ma B., Chen Y., Wang C., E-pH diagrams for the Li-Fe-P-H$_2$O system from 298 to 473 K: Thermodynamic analysis and application to the wet chemical processes of the LiFePO$_4$ cathode active material, The Journal of Physical Chemistry C, 123(23), 14207–14215, (2019).

104. Li L., Bian Y., Zhang X., Yao Y., Xue Y., Fan E., Wu F., Chen R., A green and effective room-temperature recycling process of LiFePO$_4$ cathode active materials for lithium-ion batteries, Waste Management, 85, 437–444, (2019).

105. Tao S., Li J., Wang L., Hu L., Zhou H., A method for recovering Li$_3$PO$_4$ from spent lithium iron phosphate cathode active material through high-temperature activation, Ionics, 25(12), 5643–5653, (2019).

106. Zheng R., Zhao L., Wang W., Liu Y., Ma Q., Mu D., Li R., Dai C., Optimized Li and Fe recovery from spent lithium-ion batteries via a solution-precipitation method, RSC Advances, 6(49), 43613–43625, (2016).

107. Tedjar F., Foudraz J.C., Method for the mixed recycling of lithium-based anode batteries and cells, US Patent No: US7820317B2 (2010).

108. Li H., Xing S., Liu Y., Li F., Guo H., Kuang G., Recovery of lithium, iron, and phosphorus from spent LiFePO$_4$ batteries using stoichiometric sulfuric acid leaching system, ACS Sustainable Chemistry & Engineering, 5(9), 8017–8024, (2017).

109. Cai G., Fung K.Y., Ng K.M., Wibowo C., Process development for the recycle of spent lithium ion batteries by chemical precipitation, Industrial & Engineering Chemistry Research, 53(47), 18245–18259, (2014).

110. Schurmans M., Thijs B., Process for the recovery of lithium and iron from LFP batteries, Patent Application WO2012072619A1, (2012).

111. Song Y., He L., Zhao Z., Liu X., Separation and recovery of lithium from Li$_3$PO$_4$ leaching liquor using solvent extraction with saponified D2EHPA, Separation and Purification Technology, 229, 115823, (2019).

112. Wang X., Wang X., Zhang R., Wang Y., Shu H., Hydrothermal preparation and performance of LiFePO$_4$ by using Li3PO4 recovered from spent cathode scraps as Li source, Waste Management, 78, 208–216, (2018).

113. Huang Y., Han G., Liu J., Chai W., Wang W., Yang S., Su S., A stepwise recovery of metals from hybrid cathodes of spent Li-ion batteries with leaching-flotation-precipitation process, Journal of Power Sources, 325, 555–564, (2016).

114. Li Z., Liu D., Xiong J., He L., Zhao Z., Wang D., Selective recovery of lithium and iron phosphate/carbon from spent lithium iron phosphate cathode active material by anionic membrane slurry electrolysis, Waste Management, 107, 1–8, (2020).

115. Cardarelli F., Dube J., Method for Recycling Spent Lithium Metal Polymer Rechargeable Batteries and Related Materials, US Patent No: US 7,192,564 B2, (2007).

2 Current Commercial Hydrometallurgical Recycling Process

Joey Jung and Jiujun Zhang

CONTENTS

DOI: 10.1201/9781003269205-2

2.1 BRIEF DESCRIPTION OF RECYCLING PROCESSES IN THE MARKET

2.1.1 AKKUSER RECYCLING

AkkuSer, a Finnish company, developed a Dry-Technology method to recycle high-grade cobalt (Co) LIBs. The Dry-Technology method is a mechanical process and does not require any preprocessing steps for discharging the spent LIBs. The mechanical process is based on a two-stage crushing line followed by a magnetic and mechanical separation unit. Process outputs are metal concentrates that are delivered as raw material to metal refineries. The process uses no water, chemicals, or heating thus does not produce any related emissions [1].

2.1.2 DIRECT RECYCLING AND ONTO DIRECT RECYCLING

Direct recycling, as shown in Figure 2.1(a), separates different active battery materials through physical processes, such as gravity separation and flotation, without causing chemical changes to recover cathode active materials that are reusable.

The reusable cathode active materials are replenished with lithium through a solid-state method to top up with the stoichiometric addition of lithium. Also, the process resets transition metal oxidation numbers to the original spinel structures. However, contamination such as that by polyvinyl difluoride polymers that bind cathode-carbon mixtures and transition metal dissolution have limited the development [2].

OnTo Technology LLC, a company in Oregon USA, has developed a direct recycling process with CO_2 deactivation and hydrothermal processing, as shown in Figure 2.1(b), to directly reutilize the parts taken from the used batteries to make new LIBs [2, 3].

2.1.3 LITHOREC PROCESS – DUESENFELD

Duesenfeld GmbH, a company in Weendeburg Germany, has developed a patented mechanical, thermodynamic, and hydrometallurgical process that adopts the LithoRec process, as shown in Figure 2.2. The LithoRec process first deeply discharges the spent LIBs and then shreds the batteries under an inert atmosphere. Second, battery electrolyte, consisting of a mixture of linear and cyclic carbonates and a conducting salt, is extracted by either vacuum distillation, sub- and supercritical CO_2 extraction, dimethyl carbonate (DMC) liquid extraction, or a thermal drying step. Third, the subsequent iron parts are removed via magnetic separation and transferred to scrap metal recycling. Fourth, the residual non-magnetic material is fed to a zig-zag air classifier, where the shredded material is further separated into two fractions containing the current collectors, active materials, and the other fraction consisting of the separator and plastic foils. Fifth, the active material, graphite, and the current collectors are separated and collected by heating up to 400–600°C to remove the binder using air jet sieves. Sixth, the active material is fed to a hydrometallurgical step where the active material is dissolved in an acidic mixture and Li can be leached out [4, 5].

(1) Direct Recycling: Solid State

| Spent Li-ion batteries | → | HAZMAT shipping & HAZMAT storage | → | Cryo or aqua shredding | → | Sieving | → | Binder incineration | → | Dense Fluid separation to separate anode and cathode | → | Exact Li addition and calcination |

(a)

(2) OnTo Direct Recycling with Cathode-Healing™

| Spent Li-ion batteries | → | Deactivation with CO_2 to be safe Non-HAZMAT shipping | → | Disassembly | → | Sieving | → | Hydrothermal | → | Froth flotation to separate anode and cathode | → | Exact Li addition and calcination |

(b)

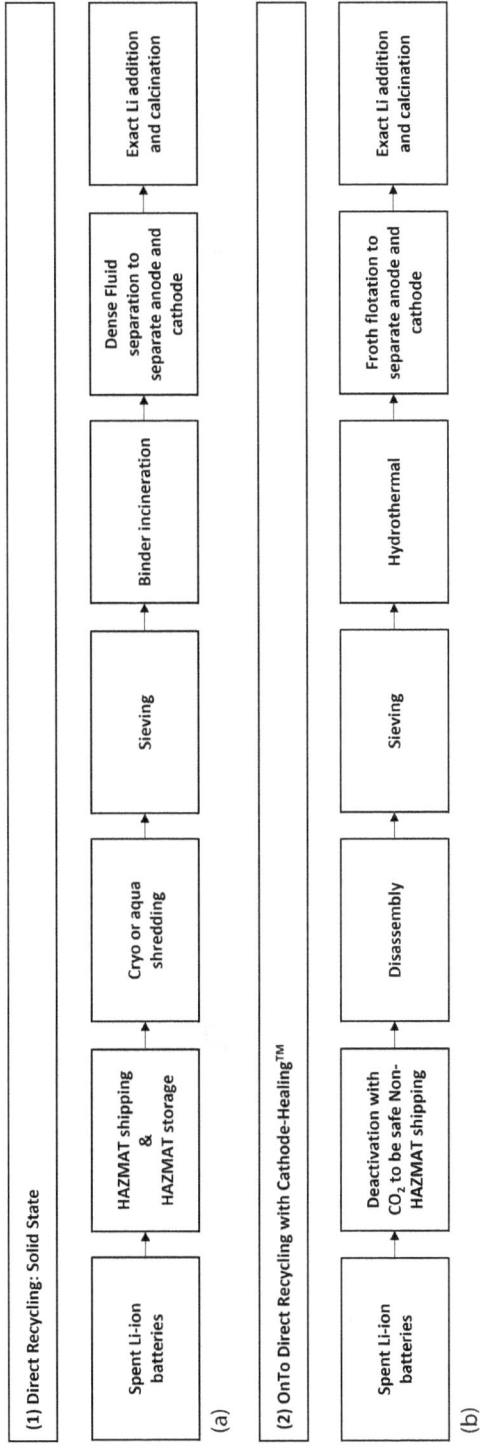

FIGURE 2.1 (a) Direct recycling and (b) OnTo direct recycling process [2].

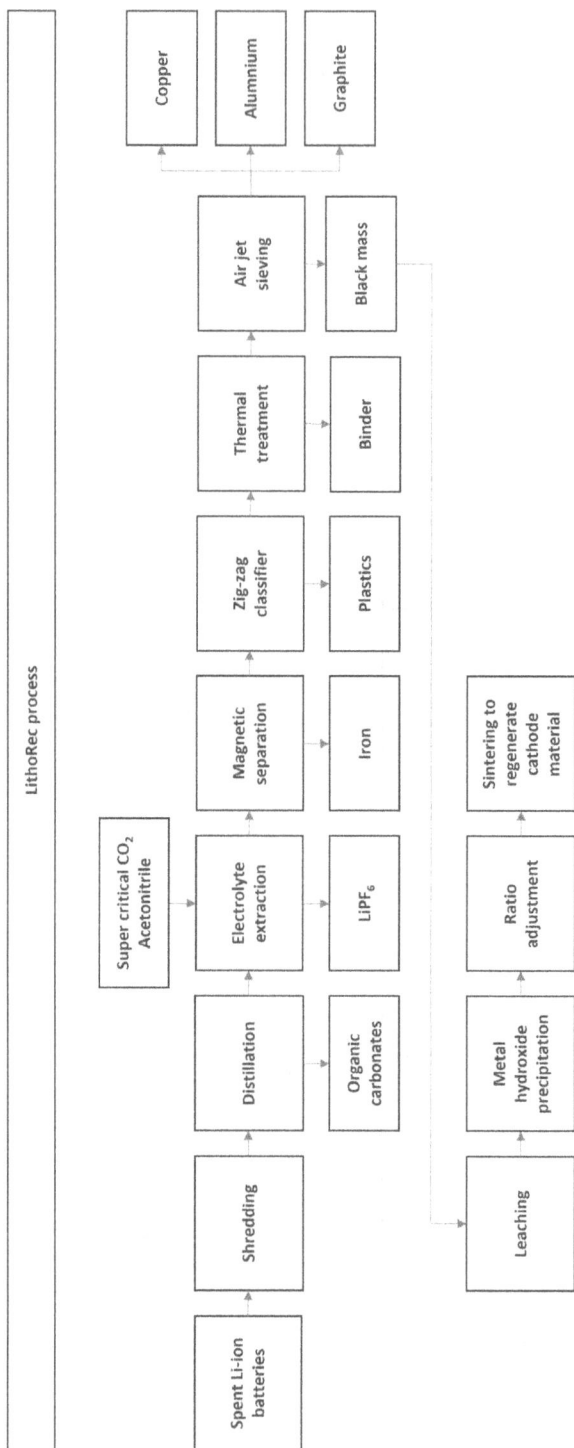

FIGURE 2.2 LithoRec process (LithoRec) [4, 5].

2.1.4 RecycLiCo™ Process

RecycLiCo™ Battery Materials Inc (formerly American Manganese Inc), a company headquartered in Vancouver, Canada, with its research partner Kemetco Research Inc., developed the RecycLiCo™ process. The RecycLiCo™ process is a patented, closed-loop, hydrometallurgical process utilizing SO_2 reductive leaching to extract valuable metals from cathode material scraps from lithium-ion battery (LIB) manufacturers and black mass concentrates produced by mechanical and thermal treatment end-of-life LIBs. The process achieves near 100% recovery and battery-grade purity of materials, such as lithium hydroxide and nickel-based ternary cathode precursors, including NCA and NMC.

2.1.5 Retriev Technologies

Retriev Technologies, a company in British Columbia, Canada, has developed a process involving disassembled and dismantled spent LIB packs, followed by feeding the separated spent LIB cells and smaller packs (i.e., laptop, power tool, and cell phone) by conveyor to an automated crusher. The crusher, which operates under a liquid solution to prevent fugitive emissions and to reduce the reactivity of processed batteries, produces three types of materials, which are metal solids, metal-enriched liquid, and plastic fluff [6, 7].

Depending on the chemistry of the spent LIBs, the metal solids may contain variable amounts of Cu, Al, and Co, which can all be used as raw materials in new products. The metal-enriched liquid is solidified using filtering technology and sent offsite for further metal purification.

2.1.6 Solvent Extraction Recycling Process

Hydrometallurgical recycling with solvent extraction recovers nickel and cobalt as nickel sulfate and cobalt sulfate. It is the most adopted commercial process currently.

Brunpt Recycling, a subsidiary of Contemporary Amperex Technology Co. Ltd (CATL), recycles spent cathode active material through a hydrometallurgical process with solvent extraction, and can achieve a recovery rate of nickel, cobalt, and manganese up to 99.3% [8].

Li-Cycle, a Canadian startup, uses a combination of mechanical size reduction and hydro-metallurgical recovery techniques to recycle LIBs. Their process combines solvent extraction with their patented shredding process to produce nickel sulfate and cobalt sulfate [9].

Lithion Recycling, a company headquartered in Quebec, Canada, recovers 95% of the nickel and cobalt through solvent extraction of the spent LIBs with battery chemistries from LCO, NMC, LMO, and NCA, and regenerates high-purity materials [10].

2.1.7 Sumitomo Recycling

Sumitomo Metal Mining Co. Ltd. (SMM), a company headquartered in Tokyo, has developed a process to recover and recycle Cu and Ni from the spent LIBs by smelting and refining the metals, as shown in Figure 2.3 [11].

The process that SMM developed selectively recovers nickel, cobalt, and copper as an alloy by using a pyrometallurgical refining process independent of the existing

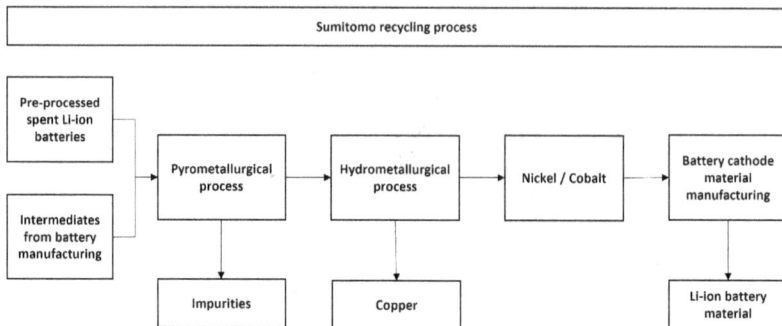

FIGURE 2.3 Sumitomo recycling process [11].

process to separate most of the impurities from the spent LIBs. Then the alloy is leached and refined using a hydrometallurgical process to recycle the nickel and cobalt for use as a battery material and the copper for electrolytic copper.

2.1.8 Umicore Recycling

Umicore, a company headquartered in Brussels, Belgium, has developed a unique pyrometallurgical treatment followed by hydrometallurgical processing, as shown in Figure 2.4 [12].

The pyrometallurgical phase converts spent LIBs into three fractions. The first is an alloy fraction that contains valuable metals like Co, Ni, and Cu designed for the downstream hydrometallurgical process. The second is a slag fraction that can be used in the construction industry or further processed for metal recovery. The third fraction is clean air released from the stack after it has been treated by the unique gas cleaning process. The alloy fraction is fed to the subsequent hydrometallurgical process, whereby the alloy is further refined to convert the metals into active cathode active materials for new LIBs [13].

The following Table 2.1 compares the available recycling technologies in the market.

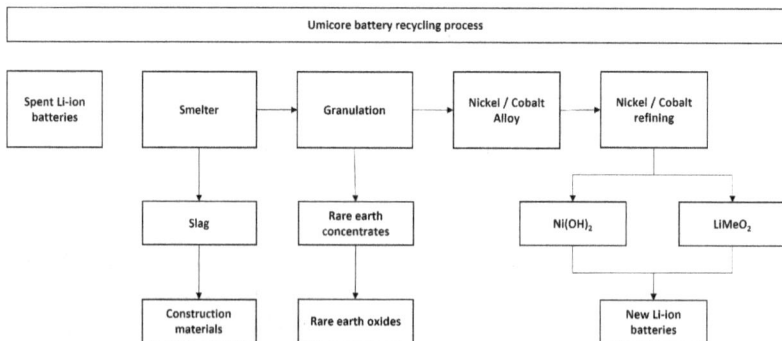

FIGURE 2.4 Umicore spent battery recycling process [12].

TABLE 2.1

Examples of Recycling Companies on the Market

No	Company	Pre-treatment	Process			Recycled Battery Element Form	Energy Required	Status @ 2022
			Mechanical	Pyro-metallurgy	Hydrometallurgy			
1	AkkuSer	Sorting	Size reduction, magnetic separation	N/A	N/A	Co Cu Fe	Low	Pilot plant
2	Ascent Elements	N/A	N/A	N/A	H_2SO_4, H_2O_2 reductive leach, solvent extraction	$NiSO_4$ $CoSO_4$ $MnCO_3$ Li_2CO_3	Medium	Pilot plant
3	Brunpt Recycling (CATL subsidiary)	Discharge, dismantling	N/A	N/A	H_2SO_4, H_2O_2 reductive leach, solvent extraction	$NiSO_4$ $CoSO_4$ $MnCO_3$ Li_2CO_3	Medium	In market
4	Duesenfeld	Discharge, disassembly	Crushing	Calcination	Unknown leaching reagent	Li_2CO_3 Metal Oxide	High	Pilot plant
5	GEM	Discharge, dismantling	N/A	N/A	H_2SO_4, H_2O_2 reductive leach, solvent extraction	$NiSO_4$ $CoSO_4$ $MnCO_3$ Li_2CO_3	Medium	In market
6	Li-Cycle	Discharge	Size reduction Magnetic separation Flotation	N/A	H_2SO_4, H_2O_2 reductive leach, solvent extraction	$NiSO_4$ $CoSO_4$ $MnCO_3$ Li_2CO_3	Medium	In market

(Continued)

TABLE 2.1 (*Continued*)
Examples of Recycling Companies on the Market

| No | Company | Pre-treatment | Process | | | Recycled Battery Element Form | Energy Required | Status @ 2022 |
			Mechanical	Pyro-metallurgy	Hydrometallurgy			
7	Lithion Recycling	N/A	N/A	N/A	Unknown leaching reagent	N/A	Medium	Pilot plant
8	OnTo Technology	Discharge, dismantling	Shredding, Supercritical CO_2 electrolyte removal	Brief Heating	Hydro-thermal	Cathode active material	Medium	Pilot plant
9	RecycLiCo™ Battery Materials	Sorting, active material/ current collector separation	Size reduction,		H_2SO_4, SO_2 reductive leach	LiOH, Li_2CO_3 Li_2SO_4 NMC811 NMC622 NMC532	Low	Demo plant
10	Redwood Materials	Discharge, dismantling	N/A	N/A	H_2SO_4, H_2O_2 reductive leach, solvent extraction	$NiSO_4$ $CoSO_4$ $MnCO_3$ Li_2CO_3	Medium	In market
11	Retriev Technologies	Dismantling	Wet crushing, screening, flotation	N/A	N/A	Al, Cu, Li, Ni, Co	Low	In market
12	Sumitomo	Sorting, dismantling	Heat Treatment	Smelting	Electro-winning H_2SO_4 leach	Ni, Co, Cu	High	In market
13	Umicore	Dismantling	N/A	Smelting	Leaching, solvent extraction Co and Ni refining	$Ni(OH)_2$ $LiMeO_2$	High	In market

2.2 CURRENT COMMERCIAL PRODUCTION OF BLACK MASS AS FEEDSTOCK FOR HYDROMETALLURGICAL RECYCLING PROCESS

Common feedstocks for hydrometallurgical recycling are cathode scrap from battery manufacturing and black mass from end-of-life batteries. Cathode scrap includes the cathode trimming and cathode material that failed to achieve the manufacturer's specifications. Figure 2.5 shows the cathode scrap.

Black mass is obtained from the shredded pre-electrolyte, defective LIBs, end-of-life LIBs, and/or cathode scrap. Once the LIBs are identified as defective or reach end-of-life, the batteries (containing aluminum, copper, anode, cathode, plastic, and separator) are collected, dismantled, and shredded. The shredded materials are then separated, where the anode active material and cathode active material are combined as black mass as shown in Figure 2.6. The black mass composition varies significantly, depending on its origin battery chemistry, and if the material was from primary or secondary LIBs. Black mass produced can contain 9–45 wt% graphite. Table 2.2 lists the typical black mass components for ternary-type LIBs. Black mass generated from cathode scrap and pre-electrolyte defective batteries does not require heat-treatment prior to shredding. For end-of-life LIBs, there are two commercial methods for producing black mass, depending on if the batteries were shredded with or without charge.

2.2.1 BLACK MASS PRODUCTION FROM CATHODE SCRAP AND PRE-ELECTROLYTE DEFECTIVE BATTERIES

Cathode scrap and pre-electrolyte defective batteries are fed into a rotary sheer shredder, which is suitable for shredding bulky materials to reduce and homogenize the cathode scrap and batteries to the desired particle size of around 10–15 mm. The shredded material is then passed through a magnetic separator to remove any magnetic materials, such as iron, before entering an air separator or a hydrocyclone.

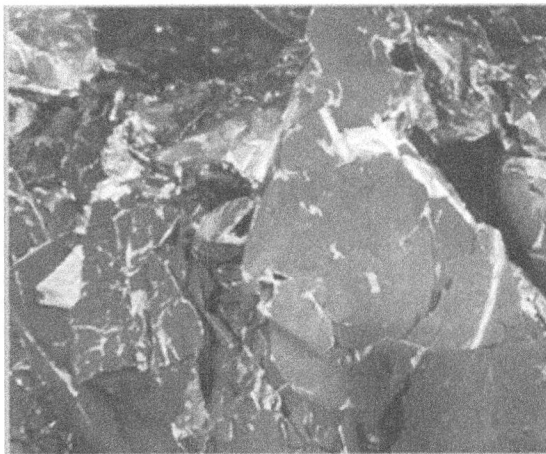

FIGURE 2.5 Cathode scrap from battery manufacturer.

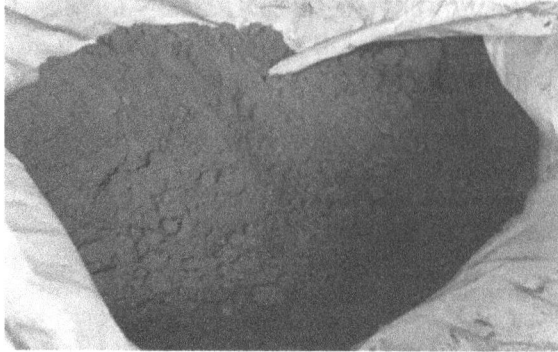

FIGURE 2.6 Black mass of LIB.

Lighter materials (i.e., plastic, separators, and some black mass) will be separated from the heavier materials (i.e., copper foil, aluminum foil, and most of the black mass). The lighter materials will then go through a vibration screen to further separate the black mass from the plastic and separator. The heavier materials collected from the air separator will be further grinded into smaller particle size and then go through another air separator to separate the black mass from the copper and aluminum. Figure 2.7 shows the general flowsheet.

TABLE 2.2
Example of Black Mass Composition

Black Mass					
Elements		mg/kg	Elements		mg/kg
Ag	Silver	<0.5	Mo	Molybdenum	<1.
Al	Aluminium	<1.	Na	Sodium	117
As	Arsenic	<2.	Ni	Nickel	494121
B	Boron	N/A	P	Phosphorus	<5.
Ba	Barium	1.8	Pb	Lead	<2.
Be	Beryllium	<0.2	S	Sulfur	572
Bi	Bismuth	6.6	Sb	Antimony	<2.
Ca	Calcium	126	Se	Selenium	26.6
Cd	Cadmium	<0.2	Si	Silicon	156
Co	Cobalt	62977	Sn	Tin	<2.
Cr	Chromium	<0.5	Sr	Strontium	0.6
Cu	Copper	<1.	Ti	Titanium	<1.
Fe	Iron	10.2	Tl	Thallium	<2.
K	Potassium	<5.	U	Uranium	<5.
Li	Lithium	<1.	V	Vanadium	1.9
Mg	Magnesium	101	Zn	Zinc	26.1
Mn	Manganese	59021	C	Carbon	10–45%

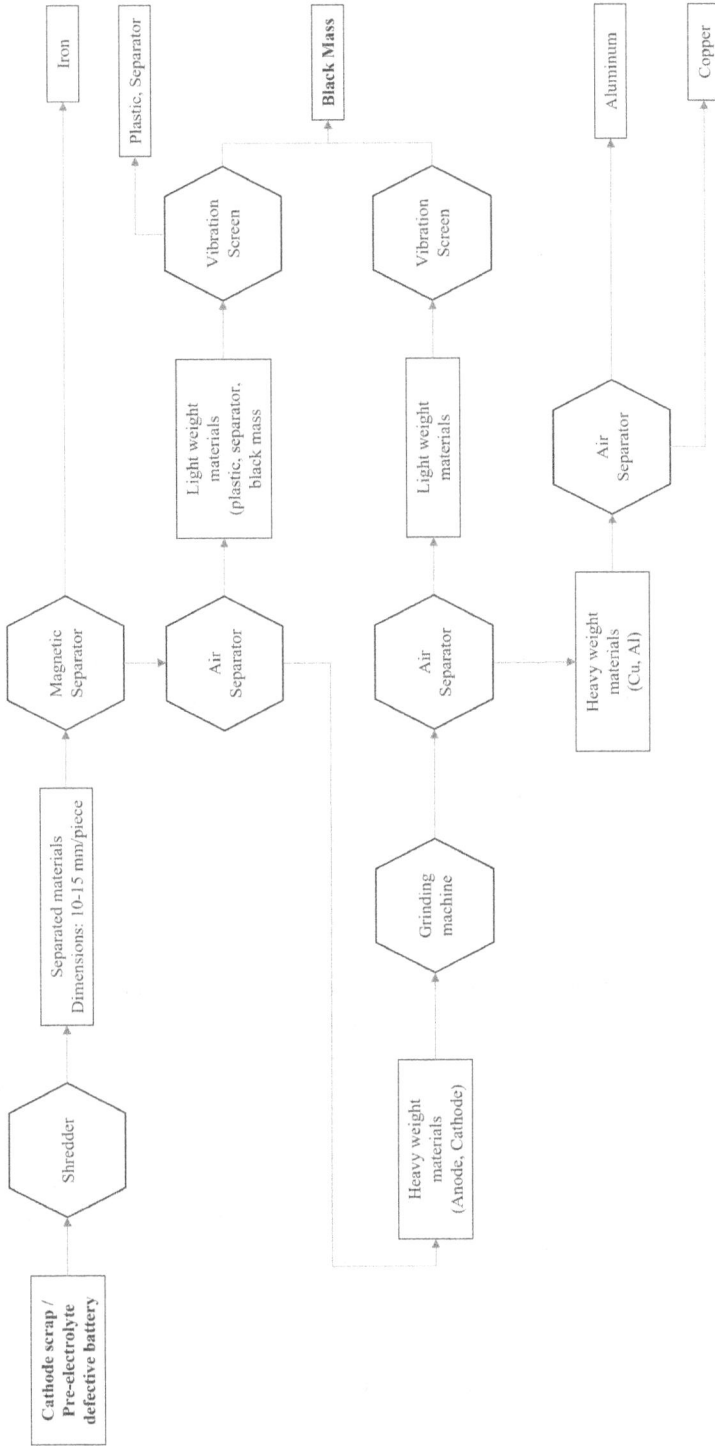

FIGURE 2.7 Flowsheet for black mass production generated from cathode scrap and/or pre-electrolyte defective batteries.

2.2.2   BLACK MASS PRODUCTION WITHOUT FIRST DISCHARGE THE BATTERY

To produce black mass without first discharging the battery is conducted by feeding the end-of-life battery, regardless of specification, without discharging into a demolition system in a vacuum environment. First, the vacuum removes any vaporized organic solvent and exhaust gases. Second, nitrogen is injected to deplete oxygen and create an oxygen-free environment for crushing. After crushing, the demolition system temperature is raised to approximately 140°C to evaporate off the solvent in the electrolyte, where the evaporated solvent enters the heating system as a combustible gas to assist burning and generates heat.

Once the electrolyte solvent is removed, the crushed material is magnetically separated, and the battery shell and other magnetic materials are removed. The anode, cathode, and separator are then entered into the high-speed rubbing-shredding system. The system is designed with a multi-row rubbing blade system. During the high-speed operation, strong airflow is generated, causing the anode and cathode to collide and rub with each other; this results in the anode active material falling off the copper foil and the cathode material falling off the aluminum foil. After shredding, the copper foil and aluminum foil naturally form round small particles. The shredded materials are then fed into a cyclone for further separation. The copper particles are heavier than aluminum particles whereas the shredded plastic has the smallest density. Based on density, the separation efficiency of copper powder, aluminum powder, anode and cathode material powder, and plastic powder can achieve more than 98%.

The obtained anode and cathode material powders then enter an anaerobic pyrolysis system, where the following are achieved: (1) the organic matter in the anode material powder and cathode material powder are vaporized and pyrolyzed; (2) the vaporized gases are combusted; (3) the active materials are discharged in high-temperature environment; and (4) the heat generated from combustion is recycled and reused.

Exhaust gases from the anaerobic pyrolysis system are recycled and treated. Carbonate from the battery electrolyte and hydrogen fluoride and phosphorus fluoride from the lithium hexafluorophosphate ($LiPF_6$) account for approximately 20% of the exhaust gases. These gases are discharged into a combustion furnace, through which the gases are joined with the vaporized organic solvent and oxygen for mutual incineration. Heat generated from combustion can be recycled to heat the anaerobic pyrolysis furnace.

Exhaust gas after combustion then passes through a scrubber with acute cooling, multi-channel spraying, aerosol separation, and a honeycomb adsorption system to absorb the harmful gases in the exhaust gas. The treated exhaust gas then proceeds through aerosol separation to remove moisture and then is discharged through the inducer and chimney. Figure 2.8 presents the flowsheet for black mass production from undischarged batteries.

2.2.3   BLACK MASS PRODUCTION FROM DISCHARGED BATTERIES

To produce black mass from discharged batteries involves three main processes: disassembly, stripping, and concentration. These three main processes are described as follows and Table 2.3 lists the equipment used in the black mass production.

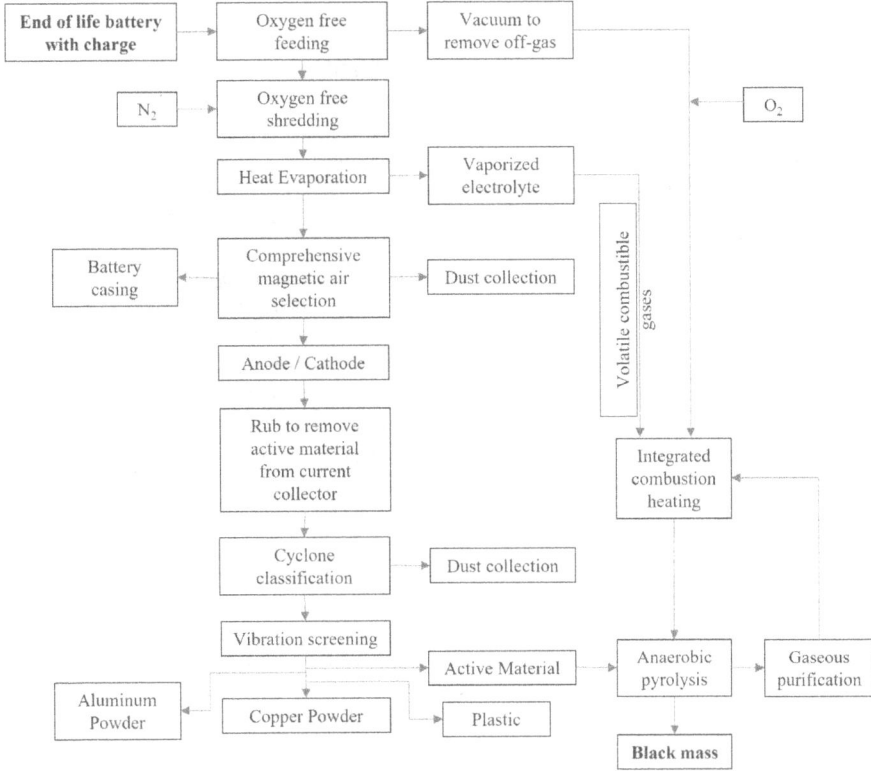

FIGURE 2.8 Flowsheet of black mass production without first discharging the battery.

2.2.3.1 Disassemble and Cutting

The end-of-life battery pack is transported to the dismantling and cutting section. The battery pack goes through a capacity analyzer to determine the leftover capacity in the battery pack, and then a discharge device is used to discharge the battery pack. After discharge, the wiring harness, shell, and other accessories are removed. Once the battery pack is disassembled, the battery module and battery cells are then further deep discharged to 1 V before entering the shredder. Shredding is conduced in a nitrogen-filled cutting room by a shredder with the battery module and battery cells fed by a conveyer belt, to avoid an electrolyte fire. As hydrogen fluoride can be generated when the battery electrolyte, lithium hexafluorophosphate, exposed in oxygen, the entire cutting room is continuously filled with nitrogen to create an anaerobic atmosphere to eliminate the oxygen exposure. The exhaust contains nitrogen and evaporated organic solvent in the electrolyte which needs to go through a scrubber to remove the organic solvent.

The outer battery steel shell or aluminum alloy shell of the batteries are shredded into 10–15 mm and collected into a collection box. The shredded anode, cathode, and separator enter the stripping process. For production anode scrape and cathode

TABLE 2.3

Equipment List of a Black Mass Production Plant

No	System	Equipment
1	Feeding system to automatic feed single LIB cell and scrap	a. Insulated vibration feeding, b. Sealed conveying belt, metering, c. Vacuum feeding and other equipment.
2	Oxygen free environment crushing and shredding system	a. Crusher, b. Shredder with explosion-proof system, c. Protective gas automatic input system, d. Oxygen content automatic display system, e. Automatic vacuum system, f. Exhaust gas filtration system, g. Exhaust gas outflow system.
3	Oxygen free electrolyte vaporization system	a. Pressure gauge, b. Oxygen meter, c. Automatic nitrogen replenishment system, d. Automatic extraction vacuum system to ensure that the oxygen content is less than 1%, e. Automatic temperature control system, f. Explosion-proof combustion system to prevent equipment chain explosion, g. Nitrogen automatic heating system, h. 140 degrees Celsius steam generator, i. Distilled water system,
4	Material separation system	a. Magnetic separator, b. Gravity separator, c. Negative pressure air separation, d. Temperature meter, e. Automatic nitrogen supplement system, f. negative pressure air regulation system, g. Air circulation system.
5	Separator rubbing and cleaning system	a. Separator dry rubbing equipment, b. Gas spray cleaning equipment.
6	High-speed rubbing separation system	a. Water automatic cooling system, b. nitrogen protection system, c. multi-group water-cooled cyclone aggregate system.
7	Enclosed screen system	a. Vibratory screenings, b. Enclosure housing.
8	Specific gravity separation system	a. Specific gravity separator.

scrap, there is no need to disassemble and can be directly fed to the stripping process after shredding.

2.2.3.2 Stripping

The stripping process is conducted in a nitrogen atmosphere. The shredded pieces are mixed and impregnated in an alkaline solution to form a slurry. The slurry is then fed into a rotating stripper to achieve separation between anode active material and copper foil, cathode active material and aluminum foil, and separator. The copper foil, aluminum foil, and separator are collected through a sieve while the anode active material and cathode active material are filtered, washed, and dried to form a black mass. The copper foil and aluminum foil are cleaned and can be sold as a product.

2.2.3.3 Evaporation Concentration

A waste-treatment system to treat organic waste is required as the stripping solution and wash water contains organic solvent and electrolyte. When the stripping solution contains 30% of organic solvent and electrolyte, the stripping efficiency will decrease. The stripping liquid needs to be evaporated and concentrated for disposal. The concentrated stripping solution contains approximately 43wt% of the electrolyte solvent (carbonate), 4% electrolyte solute ($LiPF_6$), 3wt% alkaline salt, and 50wt% water.

The engineering process flow diagram is detailed in Figure 2.9(a). Figure 2.9(b) shows a 3D rendering of a black mass production plant developed by Shenzhen Tsingtech Equipment Technology Co. and Research Institute of Tsinghua University in Shenzhen. The production process includes the following steps: (1) discharging the battery cell in a solid salt bath; (2) deactivation the battery with liquid nitrogen at low temperature; (3) hierarchical crushing; (4) organic matter removal; (5) magnetic separation and vortex current sorting; (6) separation of anode and cathode; (7) separation of active material from aluminum foil or copper foil with vibration screen; and (8) electrolyte solvent recovery. This production process differs from using salt water to discharge the battery and eliminates the issues of polluted saltwater disposal, long production cycle, and changes to smoke, fire, or explosion. It is reported that the process can achieve 98.5% separation on anode active material and cathode active material as well as 96.3% on electrolyte recovery.

2.3 HYDROMETALLURGICAL RECYCLING OF LITHIUM IRON PHOSPHATE BLACK MASS

Lithium iron phosphate black mass often contain lithium iron phosphate, copper, and aluminum. Hydrometallurgical recycling of lithium iron phosphate black mass focus on recycling lithium and iron phosphate. As shown in Figure 2.10, the lithium iron phosphate black mass is first oxidation leached with sulfuric acid and an oxidant such as oxygen and hydroxide peroxide at 60°C heated by steam. During leach, lithium is leached into the solution together with some iron and aluminum while the majority of the iron phosphate remained as solid. After the iron phosphate solid is filtered, the leachate is purified by adding sodium hydroxide to raise the pH and precipitate aluminum and iron as aluminum hydroxide and iron hydroxide. The purified lithium sulfate solution is then added with sodium carbonate solution to precipitate the majority of the lithium as lithium carbonate at 80 to 85°C. Crude lithium carbonate contains a small amount of sodium sulfate and floating Li^+, CO_3^{2-} and SO_4^{2-}, which need to be washed. The washed wet lithium carbonate is dried by hot air to produce a lithium carbonate product with a moisture content less than 0.5%.

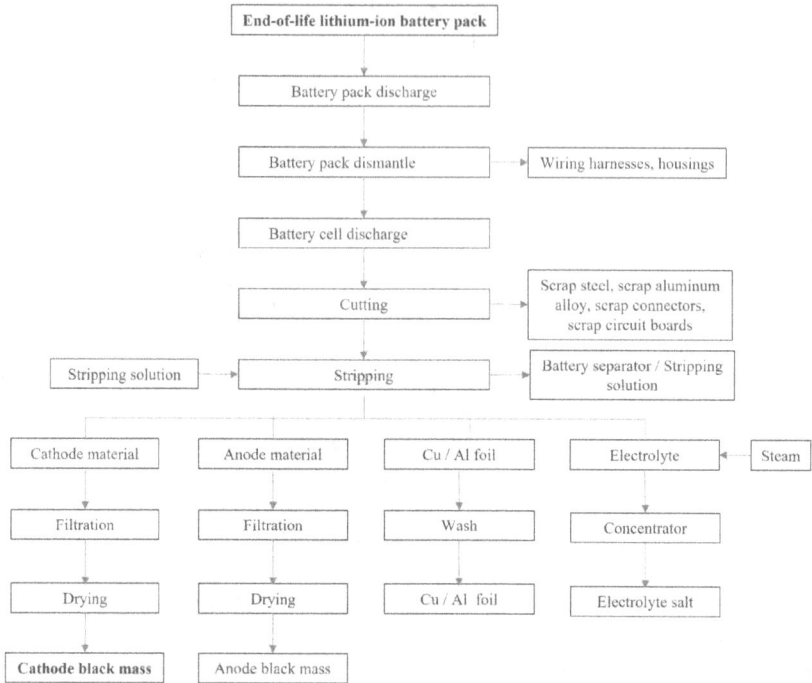

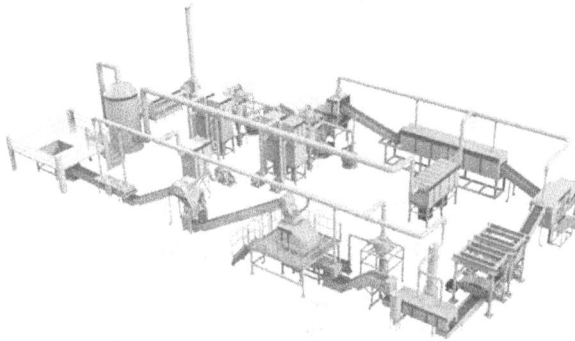

FIGURE 2.9 Black mass production with first discharge the battery pack.

The filtrate from the lithium carbonate precipitation process and the crude lithium carbonate wash solution both contain low concentration of lithium sulfate and sodium sulfate. These two solutions are then combined. To further recover lithium sulfate, the pH of the combined solution is adjusted to 7 by sulfuric acid and then evaporate by mechanical vapor recompression (MVR) evaporation system and cool to generate sodium sulfate crystal. The solid-liquid separation of sodium sulfate is carried out by the centrifuge to obtain anhydrous sodium sulfate wet products and mother liquor containing concentrated lithium sulfate and sodium sulfate. The mother liquor is then recycled back to the lithium carbonate circuit to further recover the lithium as lithium carbonate.

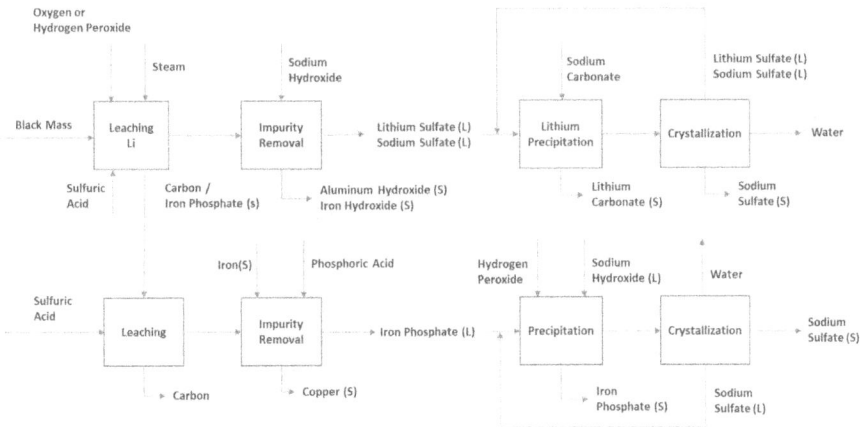

FIGURE 2.10   Hydrometallurgical recycling of lithium iron phosphate flowsheet.

In recent generation of lithium iron phosphate, small particles of lithium iron phosphate are coated on the carbon/graphite to improve the material conductivity. The filtered iron phosphate solid as shown in Figure 2.10 is further leached with excess sulfuric acid to dissolve iron phosphate. The leach solution is then filtered by a filter press to obtain iron phosphate solution and carbon/graphite wet powder. The carbon/graphite powder is dried in an oven and can be sold as a graphite carburizer product. The filtered iron phosphate solution is then purified by adding excess iron powder to reduce divalent copper ions in the solution as copper. Once the copper is removed, the ratio of phosphorus and iron in the solution is determined and then phosphoric acid is added to adjust the molar ratio of phosphate and iron to 1:1 to produce iron phosphate stock solution for precursor generation.

Iron phosphate precursor is produced by adding the iron phosphate stock solution, hydrogen peroxide, and sodium hydroxide at fixed flow rates at 40–50°C with solution pH controlled at 2–3. After the reaction is completed, the iron phosphate slurry is transferred into the aging tank for 2–3 hours before being filtered by a filter press.

The filtered iron phosphate precursor is repulp-wash with 2–5% phosphoric acid, liquid-to-solid ratio 1:2 for an hour twice. After the two acid washes, repulp-wash using pure water with liquid-to-solid ratio 1:2, retention time of 1 hour is conducted before filtered by filter press. The wet iron phosphate precursor is first dried in a dryer to remove water at 120°C for 2 hours, then calcined in a calcination kiln at 550°C for 1 hour to obtain battery-grade iron phosphate precursor product.

2.4   HYDROMETALLURGICAL RECYCLING OF LITHIUM-ION BATTERY CONTAINING NICKEL, COBALT, MANGANESE, AND ALUMINUM BY SOLVENT EXTRACTION

Since 1990, LIBs with lithium cobalt oxide as cathode material were widely used in cell phones and laptops. The early battery recycling program was aimed to recover Co from these batteries. In mining, cobalt is often a co-product from copper mining and is often retrieved by solvent extraction. As such, the early LIB hydrometallurgical recycling process to recover cobalt is based on using solvent extraction to extract cobalt as cobalt sulfate.

Due to lithium cobalt oxide cathode material having low power density despite having high energy storage capability, it is easily overheated during high power discharge thus not suitable for electric vehicle (EV) applications. In recent years, LIBs to power EV use lithium nickel cobalt aluminum oxide cathode (NCA), lithium nickel manganese cobalt oxide cathode (NMC), or lithium iron phosphate cathode (LFP).

The hydrometallurgical recycling of NMC and NCA batteries generally involve leaching, impurity removal, metal extraction, and lithium recovery.

Ammonia and several acids such as hydrochloric acid, nitric acid, or sulfuric acid are commonly used as leaching reagent. Impurity removal is often conducted using sodium hydroxide (NaOH) [8, 15, 17–21] to precipitate metal ions such as aluminum to aluminum hydroxide or using ion exchange resin such as MTS9500 and MTX7010 from Purolite to remove impurities such as iron, zinc, and calcium. MTX7010 resin is a D2EHPA-impregnated product that can separate Ca from Co and Ni. The pH for the MTX7010 to absorb Ca and Mn are very close and separation would require careful manipulation of the pH. The efficiency will also depend on the relative concentrations.

Nickel and cobalt are commercially recovered by using solvent extraction to recover cobalt and nickel as cobalt sulfate and nickel sulfate.

Lithium recovery is often conducted by adding sodium carbonate (Na_2CO_3) into the lithium enriched solution to precipitate lithium as lithium carbonate (Li_2CO_3).

2.4.1 COMMON EXTRACTANT USED IN SOLVENT EXTRACTION

To extract nickel and cobalt from the leach solution with solvent extraction, the commonly used extractants are shown in Table 2.4. There are extractants used in an acidic environment such as CYANEX 272, CYANEX 301, DEPHA, IONQUEST 290, Versatic 10, and ACORGA K2000 as well as extractants used in an alkaline environment such as ALAMIN 308.

The acidic extractants DEHPA, IONQUEST801, or CYANEX272 are based on phosphorus-based, where the loading of the metal ions is based on a cation exchange mechanism as shown in the following question:

$$2RH + M^{2+} \leftrightarrow R_2M + 2H^+ \tag{2.1}$$

The following Table 2.4 shows the chemical composition and the extraction medium of phosphoric acid (DEHPA), phosphonic acid (IONQUEST801), and phosphinic acid (CYANEX272). For the selectivity of cobalt over nickel, the selectivity increases in the following order:

Phosphoric < Phosphonic < Phosphinic

Current Commercial Hydrometallurgical Recycling Process 57

TABLE 2.4

Common Used Solvent Extractant for Nickel and Cobalt

Extractant Class	Extractant Type	Extractant Structure		Commercial Examples	Extraction Medium	Extracted Complexes	Reference
Acidic	Phosphinic Acid		(LH)	CYANEX™ 272 Ionquest™ 291	Sulfate	[Co(LHL)$_2$]	[7]
Acidic	Dithiophosphinic Acid		(LH)	CYANEX™ 301	Sulfate	[NiL$_2$(H$_2$O)$_2$] [CoL$_2$]	[7]
Acidic	Carboxylic Acid		(LH)	Versatic™ 10	Sulfate	[CoL$_2$]·4LH [CoL$_2$·2LH]$_2$ [NiL$_2$]·4LH [NiL$_2$·2LH]$_2$	[8]
Acidic	Phenolic oxime		(LH)	ACORGA™ K2000 Lix™84-1	Sulfate	[NiL$_2$]	[9]
Basic	Tertiary amine		(LH)	Alamine™ 308	Sulfate	(LH)$_2$[CoCl$_4$]	[10]

Figure 2.11(a), (b), and (c) illustrate the metal extraction rate of DEHPA, CYANEX272, and IONQUEST801 at different pH. Among these three extractants, CYANEX272 is specifically designed to separate nickel and cobalt thus exhibits the highest selectivity for cobalt over magnesium, calcium, and nickel.

Another commonly used phosphinic acid extractant as shown in Figure 2.12(a) is CYANEX301, which was commonly used for co-extraction of nickel and cobalt in hydrochloric acid medium. CYANEX301 is highly selective for nickel/cobalt over calcium/magnesium but easy to be poisoned by copper, cadmium, chromium, lead, silver, mercury, and molybdenum as shown in Figure 2.12(b).

The acidic extractant ACORGA K2000 is based on Phenolic Oxime, which can extract nickel as nickel tetramine or nickel hexamine from an ammonia environment according to Eqs. 2.2 and 2.3:

$$Ni(NH_3)_4^{2+} + 2RH \leftrightarrow R_2Ni + 2NH_3 + 2NH_4^+ \qquad (2.2)$$

$$Ni(NH_3)_6^{2+} + 2RH \leftrightarrow R_2Ni + 4NH_3 + 2NH_4^+ \qquad (2.3)$$

With cobalt and nickel in the solution, Co^{2+} will be co-extracted with nickel by Phenolic Oxime and oxidized to a Co^{3+} chelate. Two stages of stripping are required with first to strip the nickel and then followed by a reductive strip to recover the cobalt and regenerate the organic. Figure 2.13 shows the extraction efficiency over ammonia concentration.

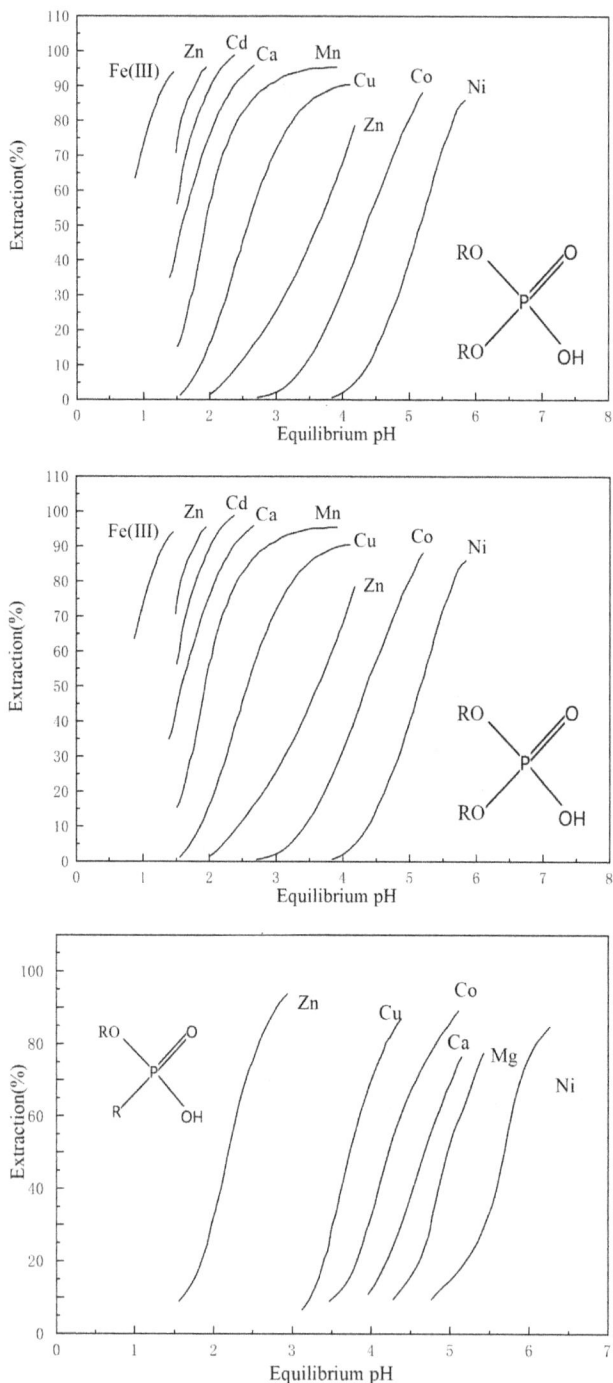

FIGURE 2.11   The metal extraction rate of DEHPA, CYANEX272, and IONQUEST801 at different pH.

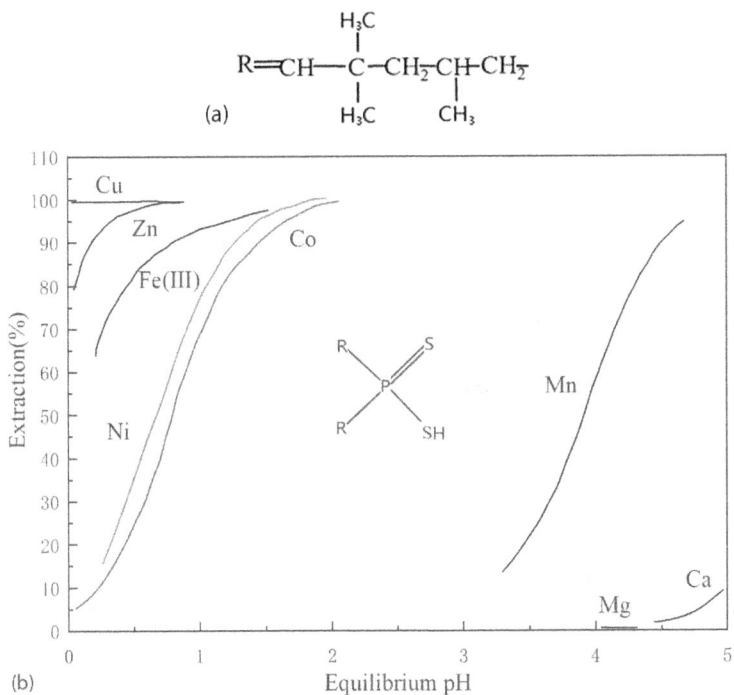

FIGURE 2.12 (a) Composition of CYANEX301 and (b) metal extraction rate at various pH.

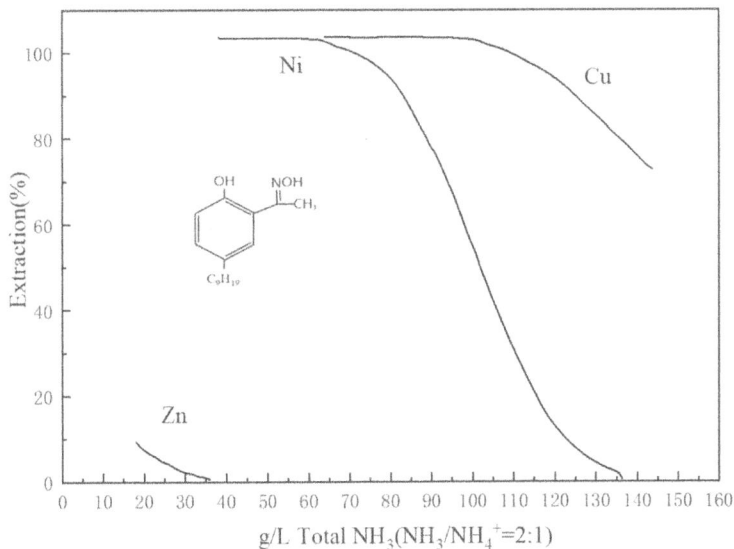

FIGURE 2.13 Extraction efficiency of phenolyic oximes on metal at various ammonia concentration.

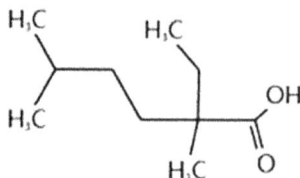

FIGURE 2.14   C9-C10 carboxylic acid.

The acidic extractant Versatic 10 is based on carboxylic acid, which has C9-C10 chain links as shown in Figure 2.14.

Carboxylic Acid can bulk extract cobalt and nickel with high selectivity against solution containing more than 10 g/L of magnesium. Figure 2.15 shows the order of the selectivity of extraction of base metal cations at various pH conditions by carboxylic acid. By raising the pH from acidic to alkaline, the metal selectivity is $Fe^{3+} > UO_2^{2+} > Sn^{2+} > Al^{3+} > Cu^{2+}, Zn^{2+} > Pb^{2+}, Cd^{2+}, RE^{3+} > Mn^{2+} > Ni^{2+} > Co^{2+} > Fe^{2+} > Ca^{2+} > Mg^{2+}$.

Comparing to CYANEX 272, the carboxylic acid will extract nickel before it extracts cobalt but with a much lower selectivity for nickel relative to cobalt. The pH to extract nickel and cobalt is around 6 to 7. Carboxylic acid also has a much higher selectivity for nickel and cobalt relative to magnesium.

The alkaline extractant ALAMIN 308 is based on Tertiary Amines, which is a trialkylamines R_3N. The normal combination is the octyl/decyl chain links with

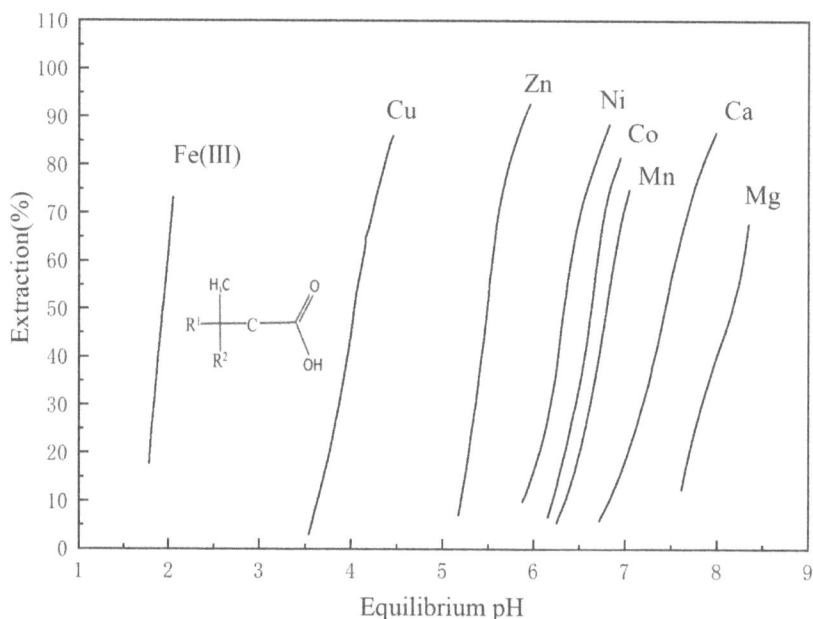

FIGURE 2.15   Metal extract by carboxylic acid at various pH.

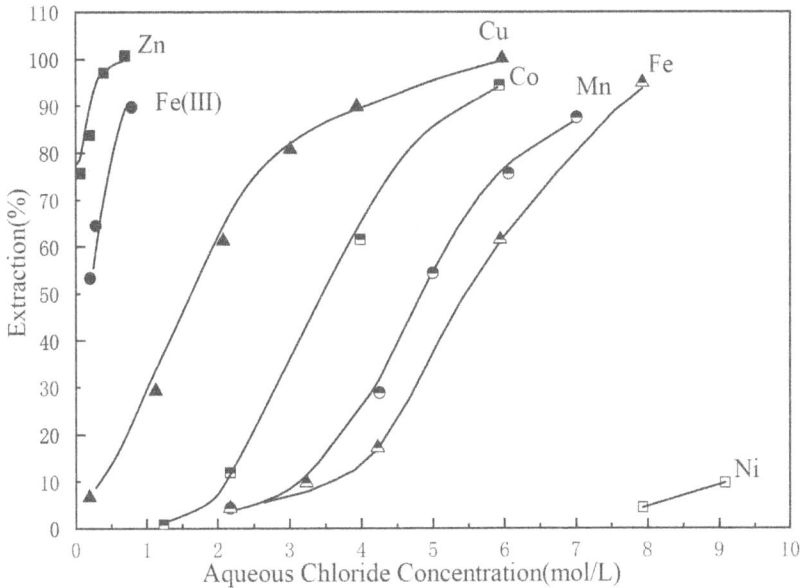

FIGURE 2.16 Metal selectivity of the tertiary amines at various chloride concentration.

some are triisoctyl and trioctyl chain links. To use Tertiary Amines as extract-ant, the amines are first protonated in acid solution as shown in Eq. 2.4. Once the amines are protonated, it can be used to extract cobalt from cobalt chloride accord-ing to Eq. 2.5.

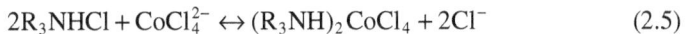

$$R_3N + HCl \leftrightarrow R_3NHCl \tag{2.4}$$

$$2R_3NHCl + CoCl_4^{2-} \leftrightarrow (R_3NH)_2 CoCl_4 + 2Cl^- \tag{2.5}$$

Figure 2.16 shows the metal selectivity of the Tertiary Amines at various chloride concentration.

The disadvantage of using hydrochloric acid and Tertiary Amine solvent to extract cobalt is the presence of chloride often causes corrosion and special construction of material for the reactor is needed.

2.4.2 SOLVENT EXTRACTION PROCESS TO RECOVER NICKEL AND COBALT AS NICKEL SULFATE AND COBALT SULFATE

The current commercial hydrometallurgical recycling process focus on recover lith-ium, cobalt, and nickel from battery scrap and spent battery. It uses sulfuric acid and hydrogen peroxide to leach the cathode scarp and black mass to extract cathode metals. During leaching, metal ions such as Ni^{2+}, Co^{2+}, Mn^{2+}, Al^{3+}, Fe^{2+}, and Li^+ in the spent cathode material dissolve into the leach solution. The leachate is then feed to solvent extraction process to recover nickel, cobalt, and manganese. The common extraction medium used to recovery nickel and cobalt is listed in Table 2.5.

TABLE 2.5

Chemical Composition and Extraction Medium of DEHPA, CYANEX272, and IONQUEST801

Extractant	DEHPA	CYANEX272	IONQUEST801
Chemical Composition			
Extraction medium	HCl	H_2SO_4	H_2SO_4

2.4.2.1 Example Extractant – CYANEX272

With the leaching solution in sulfate medium and the need to separate nickel and cobalt, CYANEX 272 is one promising extractant as the solvent extraction extractant.

The pH and metal selectivity order of CYANEX272 where 50% of the target metal is extracted is shown in the following:

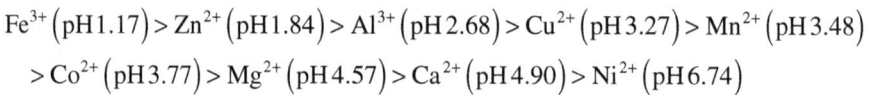

$$Fe^{3+}(pH\,1.17) > Zn^{2+}(pH\,1.84) > Al^{3+}(pH\,2.68) > Cu^{2+}(pH\,3.27) > Mn^{2+}(pH\,3.48)$$
$$> Co^{2+}(pH\,3.77) > Mg^{2+}(pH\,4.57) > Ca^{2+}(pH\,4.90) > Ni^{2+}(pH\,6.74)$$

The chemical and physical properties of CYANEX272 are shown in Table 2.5, where the solubility is a function of pH, temperature, and ionic strength.

As shown in Table 2.6 (row 8 and 9) and Table 2.7 (row 5, 6, and 7), the solubility of CYANEX272 in aqueous solution increases with pH, especially when the pH is higher than 5. On the other hand, the solubility of CYANEX272 is also affected by the total nickel and cobalt salt concentration in the solution. As shown in Table 2.7 (row 1, 2, 3, and 4), when the total salt concentration in the solution is too high, it will limit the solubility of CYANEX272 despite the pH.

As previous described in Eq. 2.1 and shown in Eq. 2.6, the extraction of cobalt with CYANEX272 is via a cation exchange mechanism where 2 moles of CYANEX272 are required to extract 1 mole of divalent metal.

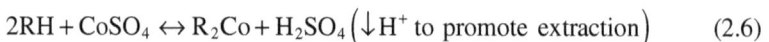

$$2RH + CoSO_4 \leftrightarrow R_2Co + H_2SO_4\left(\downarrow H^+ \text{ to promote extraction}\right) \qquad (2.6)$$

where the extraction kinetics approaches to 90% equilibrium after 60 seconds.

As the extraction of cobalt will generate acid, lowering the acidity with a base solution can promote the extraction. The common base solution used is sodium hydroxide or ammonia hydroxide.

As shown in Figure 2.17, extraction of Co by CYAEX272 is best carried out in the pH region between 4 to 5, controlling the pH in the region without overdosing the

TABLE 2.6

Chemical and Physical Properties of CYANEX272 by Andrew Bekle MAusIMM, BSc Chem, MChE (Sales and Business Development Manager Asia Pacific, Solvay Mining Solutions) https://youtu.be/E3ocpzilajo

No	Typical Properties	
1	Chemical	Bis(2,4,4-Trimethylpentyl) Phosphinic Acid
2	Concentration	85%
3	Appearance	Colorless to light amber
4	Molecular weight	290 g/mol
5	Specific gravity a5 24°C	0.92 g/cm^3
6	Viscosity (Brookfield) @ 25°C	142 cp
7	Viscosity (Brookfield) @ 50°C	37 cp
8	Solubility in H$_2$O at pH 2.6	16 µg/ml
9	Solubility in H$_2$O at pH 3.7	38 µg/ml
10	Boiling point	> 300°C
11	Pour point	−32°C
12	Flash point (closed cup)	> 108°C

base is critical for the separation of nickel and cobalt. Controlling the pH can be done by interstage pH control and/or pre-neutralization. When using sodium hydroxide as the base, interstage pH control is conducted by adding the base directly into the mixing tank as illustrated in Figure 2.18(a) and the following Eqs. 2.7, 2.8, and 2.9, where 80–100 g/L sodium hydroxide is added into the mix of aqueous solution and organic solution. This method allows a pH profile and targets to be established at various extraction stages.

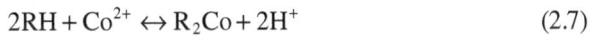

$$2RH + Co^{2+} \leftrightarrow R_2Co + 2H^+ \tag{2.7}$$

$$2H^+ + 2NaOH \leftrightarrow 2Na^+ + 2H_2O \tag{2.8}$$

TABLE 2.7

Concentration and pH Effect on the Solubility of CYANEX272 in Aqueous Solution by Andrew Bekle MAusIMM, BSc Chem, MChE (Sales and Business Development Manager Asia Pacific, Solvay Mining Solutions) https://youtu.be/E3ocpzilajo

			Solution Composition (g/L)		
No	Ni	Co	Total Salt Concentration	Equilibrium pH	CYANEX272 Solubility (µg/ml)
1	100	2	300	3.5	0.5–1.5
2	25	25	133	4.6	2
3	25	25	133	5.3	2
4	25	25	133	6.2	2
5	5	5	27	4.6	3
6	5	5	27	5.5	8
7	5	5	27	6.5	25

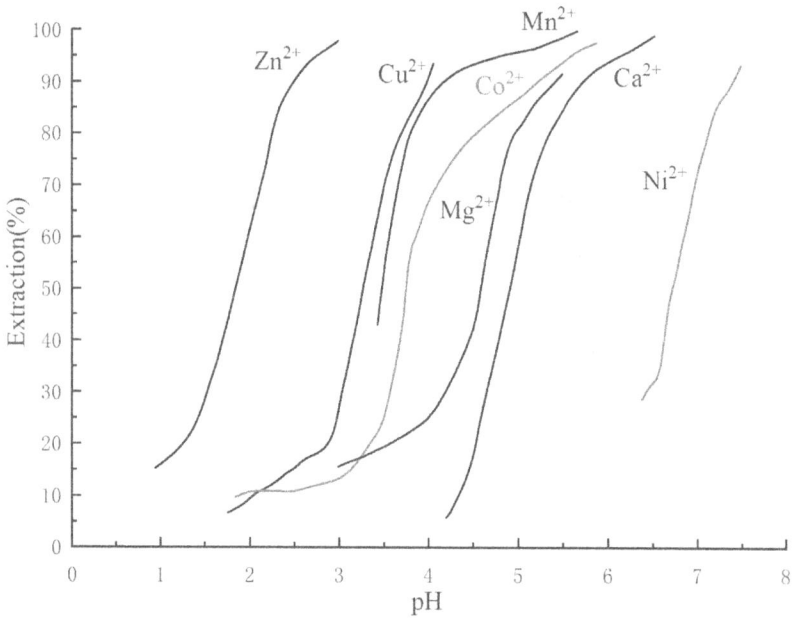

FIGURE 2.17   pH region of nickel and cobalt extraction by CYANEX272.

Overall:

$$2RH + Co^{2+} + 2NaOH \leftrightarrow R_2Co + 2Na^+ + 2H_2O \tag{2.9}$$

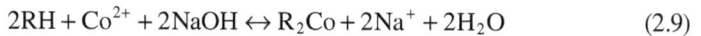

Pre-neutralization is conducted by salt conversion, where the organic solution is first mixed with 300–400 g/L sodium hydroxide solution to load the sodium onto the organic. The sodium loaded organic is then mixed with the aqueous solutions, where the sodium is exchanged for cobalt as shown in the following Eq. 2.10, Eq. 2.11, and Figure 2.18(b).

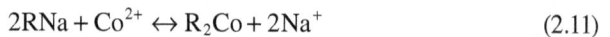

$$RH + NaOH \leftrightarrow RNa + H_2O \tag{2.10}$$

$$2RNa + Co^{2+} \leftrightarrow R_2Co + 2Na^+ \tag{2.11}$$

The choice of base, sodium hydroxide or ammonia hydroxide, depends on the economic and process treatment. Sodium hydroxide is inexpensive but generates

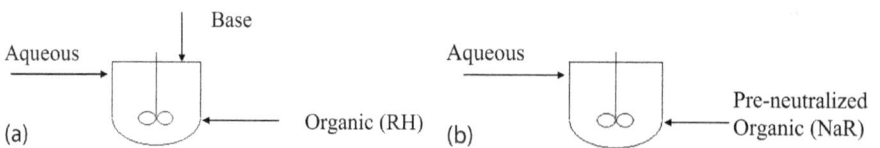

FIGURE 2.18   (a) Interstage pH control and (b) pre-neutralization.

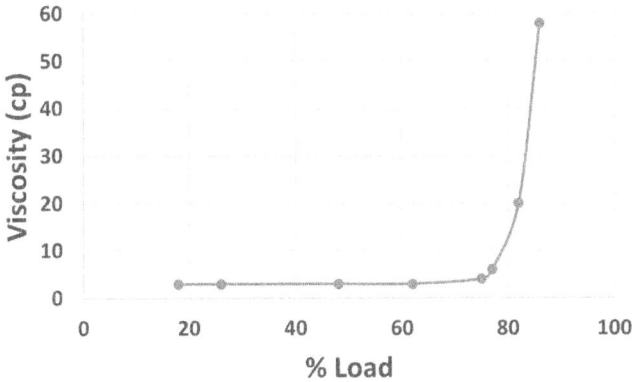

FIGURE 2.19   CYANEX272 viscosity vs % load at 25°C.

sodium sulfate by-product with very little market value. Ammonia hydroxide is more costly but generates ammonia sulfate by-product which can be sold to fertilizer industry.

To achieve good phase separation of the organic phase and the aqueous phase, it is important to maintain a low organic phase viscosity. There are two major factors influence the organic viscosity, which are extractant concentration and metal loading. The higher the extractant concentration in the organic phase, the more viscous of the organic phase will be. The typical extractant concentration to use CYANEX272 is 10–20vol% while some others having the range between 5–40%.

For CYANEX272, the maximum loading of cobalt onto the ligands in the organic phase is recommended to be 60–65% as the viscosity will increase exponentially and form gels or emulsions when over 70% of the ligands are loaded with cobalt as shown in Figure 2.19. When the loading is over 70%, the stoichiometry of the organic/metal complex changes and induces a polymerization reaction resulting organic loss due to it becoming very viscous.

Once the Co is extracted by the CYANEX272 from the aqueous phase into the organic phase, the Co can be stripped from the organic using sulfuric acid as shown in Eq. 2.12.

$$R_2Co + H_2SO_4 \leftrightarrow CoSO_4 + 2RH\left(\uparrow H^+ \text{ to promote stripping}\right) \qquad (2.12)$$

2.4.2.2   Example of Solvent Extraction Circuit to Separate Nickel and Cobalt

The following Figure 2.20 illustrates the separation of nickel and cobalt, where Figure 2.20(a) illustrates the extraction of the cobalt from a cobalt/nickel enrich aqueous solution, and Figure 2.20(b) illustrates the stripping of cobalt from organic using sulfuric acid to produce cobalt sulfate aqueous solution.

An example of solvent extraction circuit to separate cobalt from nickel and remove impurities as shown in Figure 2.21 is composited of three main unit operations, extraction, scrub, and strip. Extraction is where the cobalt and impurities load

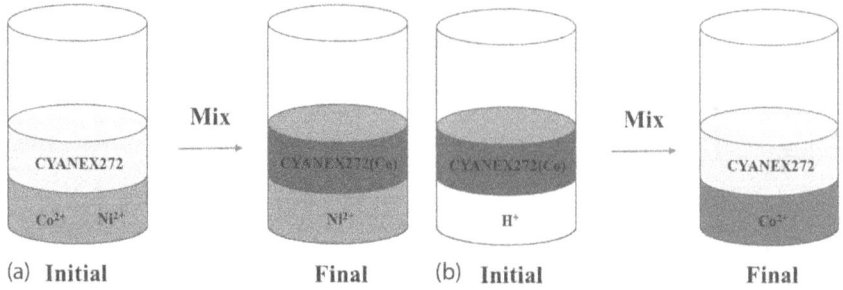

FIGURE 2.20   (a) Solvent extraction to separate Co and Ni and (b) stripping of Co from solvent.

onto the organic phase while nickel remains in the aqueous phase. Scrub is where impurities are removed from organic phase. Strip is where acid strip the cobalt in the organic phase back into aqueous phase.

2.4.2.2.1 Extraction
The nickel-cobalt enriched solution is fed into the extraction unit operation and mixed with the organic. Extraction usually contains 3 to 6 stages as illustrates in Figure 2.22. The example in Figure 2.21 shows the extraction of cobalt from a

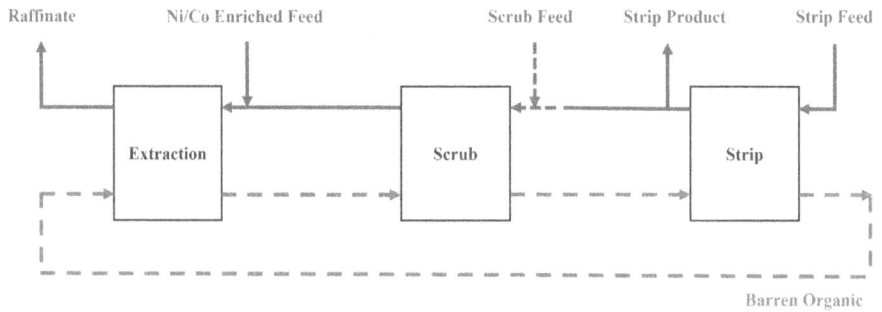

FIGURE 2.21   Example of a general solvent extraction circuit.

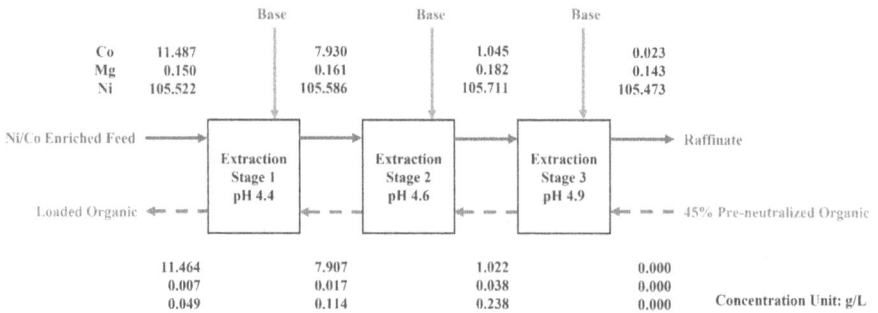

FIGURE 2.22   Example of 3 stages extraction circuit in cobalt solvent extraction.

feed aqueous solution containing 105.522 g/L of nickel, 0.150 g/L of magnesium, and 11.487 g/L of cobalt. After 3 stages of extraction, the cobalt in the raffinate aqueous solution contains 105.473 g/L of nickel, 0.143 g/L of magnesium, and 0.023 g/L of cobalt with higher than 99% of cobalt extracted into the organic phase. During the 3 stages of extraction, the pH of each stage gets higher with the third stage where the raffinate exits the highest. The flow of the organic phase is countercurrent to the flow of aqueous feed solution where the loaded organic solution exits at the stage with the lowest pH.

2.4.2.2.2 Scrubbing

The loaded organic solution contains impurities such as magnesium and nickel, which can be removed through scrubbing unit operation. Multiple scrubbing stages, as shown in Figure 2.23, will increase the purity of the cobalt sulfate. 3–7 scrubbing stages are usually required pending on the purity to be reached. In multiple scrubbing stages, the stage where loaded organic entering has the highest pH and the stage where the scrub feed entering has the lowest pH. Scrubbing can be done by two mechanisms, which are metal-metal exchange and selective stripping. Metal-metal exchange uses cobalt sulfate solution as the scrub solution. Selective stripping use acid solution as the scrub solution.

The metal-metal exchange mechanism, as shown in Eq. 2.13, also called the crowding effect, is conducted by scrubbing the organic solution with a cobalt sulfate solution recycled from the stripping stage. The extractant in the organic phase will preferentially want to load cobalt over magnesium and nickel. When the organic solution encounters with a solution with high concentration of cobalt, the extractant will exchange the magnesium and nickel with the cobalt. With high cobalt sulfate purity requirement, the cobalt that being recycle can be up to 30%. The cobalt being recycled can be lower with more scrub stages, which is a tradeoff between having higher capital with more scrubbing stages versus lower operating costs down the line with a reduced recycle of cobalt.

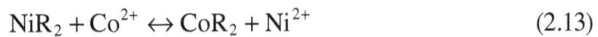

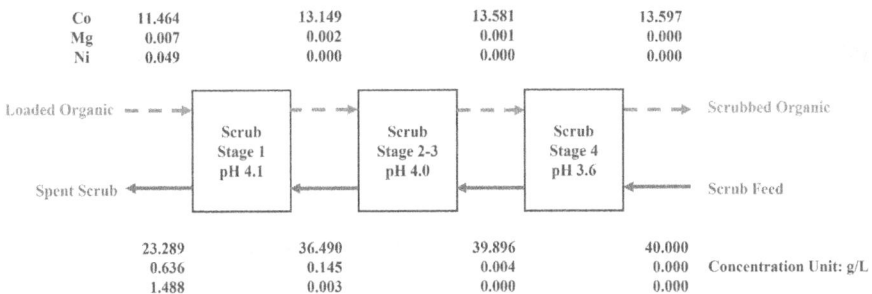

$$NiR_2 + Co^{2+} \leftrightarrow CoR_2 + Ni^{2+} \qquad (2.13)$$

FIGURE 2.23 Example of 3 stages metal-metal exchange scrub circuit in cobalt solvent extraction.

Co	11.464	11.349	11.010	10.890
Mg	0.007	0.003	0.002	0.000
Ni	0.049	0.000	0.000	0.000

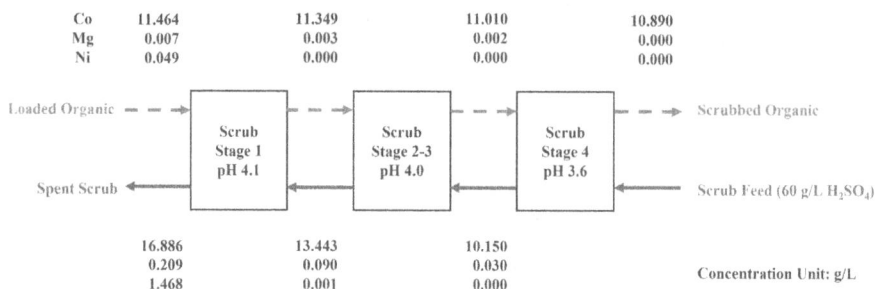

FIGURE 2.24 Example of 3 stages of selective stripping scrub circuit in cobalt solvent extraction.

The spent scrub after exiting the circuit will rejoin the extraction feed so no cobalt is loss. As shown in Figure 2.23, the loaded organic contains 11.464 g/L of cobalt, 0.007 g/L of magnesium, and 0.049 g/L of nickel. After 3 stages of scrubbing with cobalt sulfate solution containing 40 g/L of cobalt, the scrubbed organic contains 13.597 g/L of cobalt and no magnesium and nickel.

Selective stripping mechanism, as shown in Eq. 2.14, is another mechanism where a weak acid is used to gently strip off the metals. The weak acid will preferentially strip the magnesium and nickel first.

$$NiR_2 + H_2SO_4 \leftrightarrow 2RH + NiSO_4 \tag{2.14}$$

Figure 2.24 shows an example of selective stripping using 60 g/L of sulfuric acid as the scrub solution. With the same loaded organic containing 11.464 g/L of cobalt, 0.007 g/L of magnesium, and 0.049 g/L of nickel, the scrubbed organic after 3 stages of scrubbing containing 10.150 g/L of cobalt and without any magnesium and nickel.

The benefit of selective stripping is it reduces the cobalt recycle in the system and can achieve similar selectivity and purities.

2.4.2.2.3 Stripping

Stripping typically contains 2–4 stages, the cobalt is stripped quantitatively from the CYANEX272 with strong sulfuric acid (i.e., 120 g/L) without recirculating. Stripping maybe conducted in one stage provided that the pH being sufficiently low (pH<2) or having excess acid. Multiple stages are used to reduce free acid levels in the final cobalt sulfate solution product. The final concentration of cobalt in the cobalt sulfate solution is targeted to be 110 g/L.

2.4.2.2.4 Full Circuit

Figure 2.25 shows the full circuit that combing the 3 stages of extraction unit operation, 3 stages of scrubbing unit operation, and 2 stages of stripping unit operation.

As shown in Figure 2.25, the spent scrub recombined with the nickel cobalt enriched feed resulting a feed containing 11.662 g/L of cobalt, 0.152 g/L of magnesium, and 102.221 g/L of nickel. To control the pH in the range of 4–5 where the cobalt is extracted, pre-neutralization the barren organic with sodium carbonate is

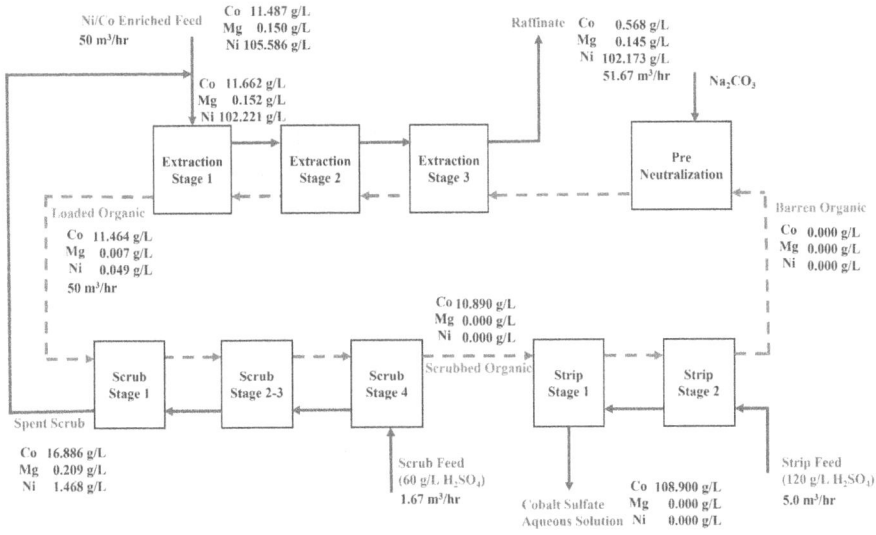

FIGURE 2.25 Example of full circuit of cobalt solvent extraction.

conducted. More than 99% of the cobalt is extracted into the organic phase and after scrubbing and stripping, a cobalt sulfate solution containing 108.900 g/L cobalt can be obtained. Cobalt sulfate crystal can be generated by evaporating with an mechanical vapor recompression (MVR) evaporation system.

Nguyen et al [14] developed a flowsheet of using direct precipitation and solvent extraction to recover manganese, cobalt, nickel, and lithium from the sulfate leach liquor as shown in Figure 2.26.

The aluminum, cobalt, iron, lithium, manganese, and nickel containing leachate is first treated with potassium permanganate to precipitate manganese as manganese oxide. The precipitation of manganese oxide is highly selective at room temperature within the equilibrium pH range of 2.0–3.0 and is relatively pure. After manganese is removed, aluminum and iron can then be selectively separated from nickel, cobalt and lithium using 10% (v/v) D2EHPA at an equilibrium pH of 2.5.

The solution containing cobalt, nickel and lithium can then be treated for cobalt separation with 60% Na-0.56 mol·dm−3 PC-88A at an equilibrium pH of 4.5. A small amount of nickel and lithium co-extracted into the organic phase can be removed by scrubbing with a 2.0 g/L cobalt sulfate solution. The selective scrubbing of these impurities is associated with the "crowding effect" of cobalt replacing nickel and lithium in the organic phase. High purity (99.9%) cobalt sulfate is then recovered using 0.2 M H_2SO_4 to strip the organic solution. The raffinate containing nickel and lithium after cobalt extraction can be treated in two different ways. Nickel can be recovered using solvent extraction with 5.0% (v/v) PC-88A or can be precipitated as nickel hydroxide by adding sodium hydroxide. Lithium in the final lithium enriched solution can be recovered by adding sodium carbonate to precipitate lithium in the form of lithium carbonate at high temperature (398 K).

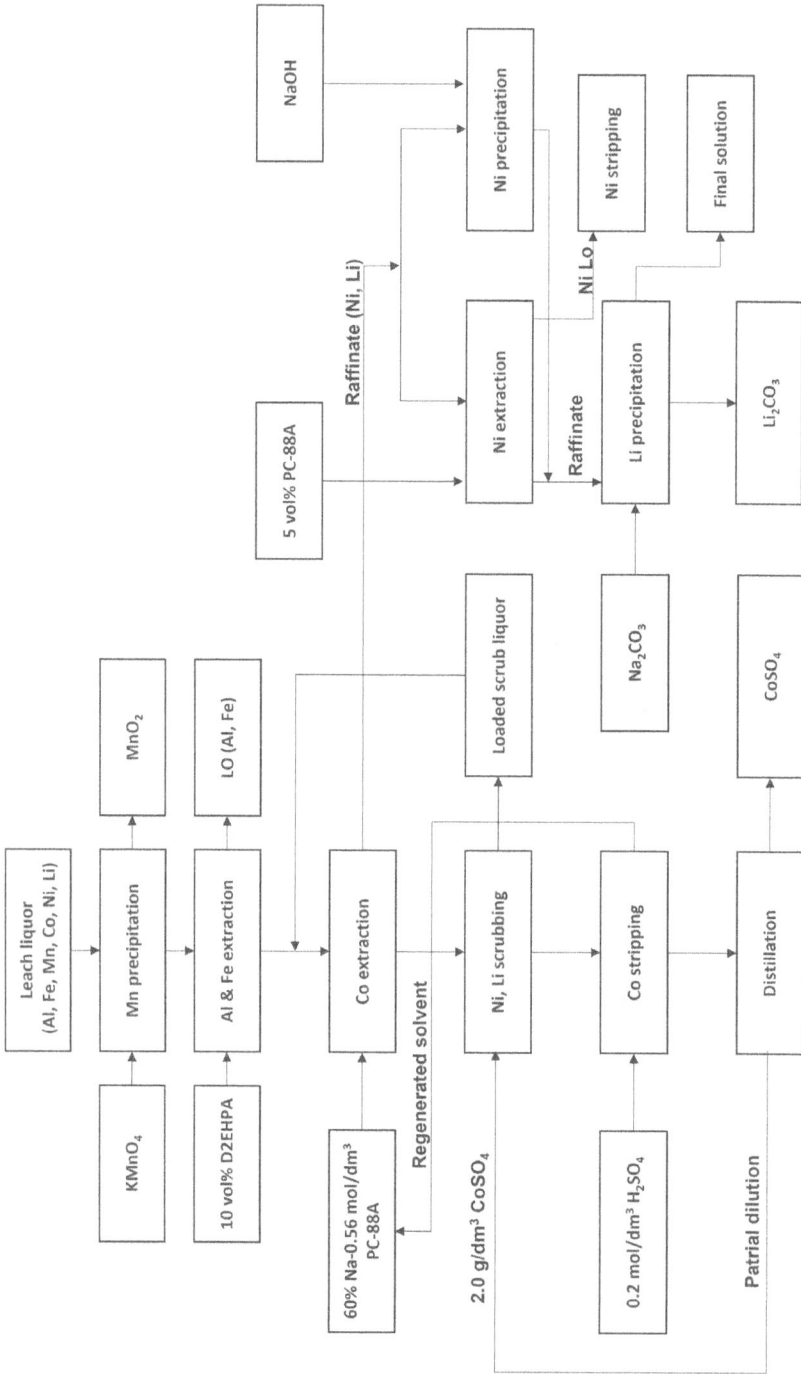

FIGURE 2.26 Flowsheet to recover cobalt, nickel, and lithium from sulfate leach solution [14].

2.5 CHALLENGES IN RECYCLING LITHIUM-ION BATTERY CATHODE ACTIVE MATERIALS

2.5.1 ECONOMY CHALLENGES

Using $NaOH$ for impurity removal and metal extraction together with adding Na_2CO_3 to recover lithium can result in a solution enriched with Na^+ ions. The sodium containing solution needs to be treated before discharge or reuse; this further complicates the hydrometallurgical recycling process and generates wastewater.

Currently, the main options for the treatment of spent LIBs are landfilling, reuse, and recycling with pyrometallurgical process and hydrometallurgical processes or direct recycling. Landfilling may cause pollution as metal like Co, Li, Fe, Mn, and Cu may slowly leaching into soil, groundwater, or surface water. Reusing the retired electric vehicle batteries is of great interest as these batteries usually holding 70–80% of residual energy capacities and can support proper application for long-term [15]. However, to reuse, the retired batteries need to be tested, recategorized, reconfigured, and fitted with a suitable battery-management system (BMS). The cost for these exercises could be higher than new batteries. Recycling with pyrometallurgical process recovers the valuable metals like Ni, Co, Mn, and Cu and let go of Li, electrolyte, and anode material. Due to the high energy consumption, the economy of pyrometallurgical process depends on the metal prices especially Co. As current trend of research is to reduce the quantity of Co in LIBs, the economic challenges is greater down the road [16].

Recycling with hydrometallurgical processes is largely price driven as technology is not a critical differentiating factor. Despite the choice of leaching agents or metal recovery methods, the hydrometallurgical processes offer the same level of final products. Therefore, the key differentiating factor becomes price. Although the cost to process spent cathode active material is lower than the cost to process virgin raw material when fabricating new cathode active material, recycling with hydrometallurgical process is still not economically feasible after including the cost of transportation, dismantle, and disassemble [17]. Furthermore, current cost to product lithium from brine solution is around USD $1800 per ton and from hard rock is USD $5000 per ton [18]. The low-cost of lithium production creates a barrier for recycling lithium from spent LIBs. LIBs contain only a small fraction of lithium as a percent of weight. The average lithium cost associated with LIB production is less than 3% of the production cost. As such, although lithium is 100% recyclable through recycling spent cathode active material, electrolyte and LTO anode, recycling lithium will not be the focus until the volume of the spent LIBs grows exponentially.

Another challenge hindering the industry is the long-term nature of financial investments required by market participants to develop specialized waste disposal services. As the market is still unexplored, specialized processes and dedicated small scale recycling plants closer to vehicle manufacturers or LIB recycling facilities are likely to be the trend soon. These specialized and customized processes increase the required financial investments and render the overall profitability of investments unknown, thereby create ambiguity and uncertainty about making such commitments.

2.5.2 FEEDSTOCK SUPPLY AND LOGISTICS

The various LIB chemistries, (i.e., LFP, NCA, NMC, LCO, LMO etc.), scatter size, format, and complicated distribution channels challenge recycling plant to collect a stable feedstock. A more flexible and universal recycling process is required for the recovery of different types of spent LIBs, as well as a process including the recycle of electrolyte and anode materials [19].

Furthermore, the spent LIBs, especially damaged/defective batteries can possibly undergo thermal runaway, typically resulting from internal shorting, leading to fire or explosion. Governments around the world are modifying the transportation regulations year to year as they recognize the potential hazardous of the spent LIBs.

Currently in North America, spent LIBs are classified as Class 9, miscellaneous hazardous materials, under United States regulations (40 CFR 173.21) [20]. The transportation of spent LIBs needs to be in specification packaging (UN Packaging), which increases the cost, the effort is needed to achieve regulatory compliance, and the complexity of transportation spent LIBs to the recycling plant.

2.5.3 SAFETY AND STORAGE

LIBs when over-charged, over-discharged, or heated might cause electrolyte leakages or explosions [21]. To ensure the safety of spent LIB recycling process, the spent batteries first need to be deep discharged to avoid violent reaction from the charged electrodes releasing the stored energy when exposed to air. Secondly, battery electrolyte, lithium hexafluorophosphate ($LiPF_6$), needs to be extracted without contacting water to avoid the release of poisonous hydrofluoric acid (HF) by the following reaction 2–15:

$$LiPF_6 + H_2O \rightarrow HF + LiOH + PF_5 \qquad (2.15)$$

Furthermore, if electrolyte is not extracted, high heat may be generated during battery disassemble due to partial internal short circuits and water moisture which turns to steam and reacts with $LiPF_6$ to generate highly corrosive HF and phosphorus oxyfluoride (POF_3) by the following reaction 2–16:

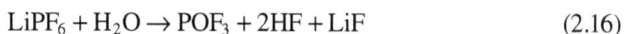

$$LiPF_6 + H_2O \rightarrow POF_3 + 2HF + LiF \qquad (2.16)$$

In addition, during decomposing if internal short circuit and thermal runaway happen, it will cause explosion. To avoid thermal runaway, the safest approach is to have the lowest amount of spent LIBs on site as possible. However, this is contradictory to the requirement to secure stable feedstock for processing purposes. The development of safe storage is further complicated by various forms of factors not known if the batteries are damaged or not. As a result, strict protocols must be implemented regarding the pallet/container spacing, total storage density and application of appropriate fire suppression systems within any LIB storage space in order to mitigate the risk associated with thermal runaway and fire [22].

These concerns and requirements increase the upfront capital investment, operation cost, and risk for spent LIB recycling. As such, it is important to develop a simple, environmentally acceptable, and economic recycling process to recycle the valuable metals from spent LIBs.

REFERENCES

1. AkkuSer Oy, https://www.akkuser.fi/en/home/
2. Direct Recycling and OnTo Technology LLC, https://www.bestmag.co.uk/content/play-it-again-steve%E2%80%94-how-direct-recycling-can-repair-battery-cathodes
3. Sloop S., Crandon L., Allen M., Koetje K., Reed L., Gaines G., Sirisaksoontorn W., Lerner M., A direct recycling case study from a lithium-ion battery recall, Sustainable Materials and Technology, 25, e00152, (2020).
4. Nowak S., Winter M., Recycling of lithium ion batteries, Wiley Analytical Science, https://analyticalscience.wiley.com/do/10.1002/gitlab.15680/ March (2018).
5. Nowak S., Winter M., The role of sub- and supercritical CO_2 as "processing solvent" for the recycling and sample preparation of lithium-ion battery electrolytes, Molecules, 22, 403, March (2017).
6. Retriev Technologies Inc., https://www.retrievtech.com/lithiumion
7. Smith W., Swoffer S., Process for recovering and regenerating lithium cathode active material from Lithium-ion Batteries, US Patent # 8,882,007 B1, Nov (2014).
8. Brunpt Recycling, https://www.catl.com/en/solution/recycling/
9. Li-Cycle Corp., https://li-cycle.com/technology/
10. Lithion Recycling Inc., https://www.lithionrecycling.com/lithium-ion-battery-recycling-process/
11. Sumitomo Metal Mining Co. spent battery recycling process for lithium-ion batteries https://www.recyclingtoday.com/article/japanese-company-develops-lithium-ion-battery-recycling-process/, March (2019).
12. Elwert T., Goldmann D., Romer F., Buchert M., Merz C., Schueler D., Sutter J., Current developments and challenges in the recycling of key components of (hybrid) electric vehicles, Recycling, 1, 25–60, (2016).
13. Umicore battery recycling process, https://csm.umicore.com/en/battery-recycling/our-recycling-process/
14. Nguyen V.T., Lee J.C., Jeong J., Kim B.S., Pandey B.D., Selective recovery of cobalt, nickel, and lithium from sulfate leachate of cathode scrap of li-ion batteries using liquid-liquid extraction, Metals and Materials International, 20, 357–365, (2014).
15. Lih W.C., Yen J.H., Shieh F.H., Liao Y.M., Second use of retired lithium-ion battery packs from electric vehicles: technological challenges, cost analysis and optimal business model, 2012 International Symposium on Computer, Consumer and Control, (2012).
16. Gaines L., Profitable recycling of low-cobalt lithium-ion batteries will depend on new process developments, One Earth, 1(4), 413–415, (2019).
17. Steward D., Mayyas A., Mann M., Economics and challenges of Li-ion battery recycling from end-of-life vehicles, Procedia Manufacturing, 33, 272–279, (2019).
18. Low Cost, High Margins: Lithium Brine Extraction | INN (investingnews.com), https://investingnews.com/innspired/lithium-brine-extraction-electric-vehicle-market/ Nov (2018).
19. Yao Y., Zhu M., Zhao Z., Tong B., Fan Y., Hua Z., Hydrometallurgical processes for recycling spent lithium-ion batteries: A critical review, ACS Sustainable Chemistry & Engineering, 6, 13611–13627, (2018).
20. Gaines L., The future of automotive lithium-ion battery recycling: Charting a sustainable course, Sustainable Materials and Technologies, 1–2, 2–7, (2014).

21. Park Y.J., Kim M.K., Kim H.K., Lee B.M., Risk assessment of lithium-ion battery explosion: Chemical leakages, Journal of Toxicology and Environmental Health, Part B, 21(6), 370–381, (2019).
22. Hogg S., Batteries need to be 'renewable' too: why recycling matters now, Energy Storage News, Sept (2019).

3 RecycLiCo™ Recycling Process for Lithium-Ion Battery Cathode Active Materials

Joey Jung

CONTENTS

DOI: 10.1201/9781003269205-3

3.1 INTRODUCTION

The commercialization of lithium-ion batteries (LIBs) has resulted in vigorous development of various proprietary designs and formulations for their cathode active materials. A LIB cathode is made up of an active material such as lithium-cobalt oxide ($LiCoO_2$ also referred to as LCO) commonly used in laptop batteries, cell phones, and electric vehicles; lithium-ion phosphate ($LiFePO_4$ also referred to as LFP) used in energy storage and electric vehicles; lithium nickel manganese cobalt oxide ($LiNi_{1/3}Co_{1/3}Mn_{1/3}O_2$ also referred to as NMC) used in electric vehicles; lithium nickel cobalt aluminum oxide ($LiNi_{0.8}Co_{0.15}Al_{0.05}O_2$ also referred to as NCA) used in electric vehicles; or lithium-manganese dioxide ($LiMnO_2$ or LMO) used in electric power tools and plug-in hybrid electric vehicles. In general, LIBs having relatively high energy/power densities and long life cycle are suitable for portable electronics, electric vehicles, and stationary power stations. However, their charge capacities will slowly degrade with regular use and will eventually need to be replaced. The limited nature abundance of the electrode materials such as lithium, cobalt, and nickel can limit their sustainable usage, particularly in large-scale applications where a large quantity of electrode material is necessary. To mitigate this limitation, recycling the end-of-life electrode materials for recovering valuable raw materials are definitely needed.

Regarding the recycling of the spent LIB cathode active materials such as Co-containing ones, the incorporated other metal components such as aluminum, calcium, copper, magnesium, and zinc, can add some complexity to the recycling of the spent batteries. The range of materials involved require a multi-stage treatment approach to separate metals when hydrometallurgical methods are applied to the mixed cathode active materials. Normally, the hydrometallurgical treatments can break down the battery components by leaching and producing tailored lithium-ion materials. The most common reagent reported for leaching cathode active material is sulfuric acid (H_2SO_4). In addition to acid, the leach generally requires a reducing agent to control the oxidation state of the cathode metals. The cobalt, nickel and manganese in cathode active materials are typically present at higher oxidation states [1]. To accelerate the rate of dissolution of these metals from battery material, various reducing agents have been applied. Among all reducing agents used in leaching LIB cathode active materials with acid, H_2O_2 presents the best option [2–5]. Recently, ferrous and ferrous-cuprous catalyzed leaching systems [6, 7] and sodium metabisulfite [8] were reported as the effective alternatives to the conventional peroxide method. Leaching with sodium metabisulfite could not only dissolve the cathode metals in acid but also minimize the dissolution of impurities like aluminum (Al) and copper (Cu) in leach solution [8]. Impurities like Al and Cu need to be monitored because their dissolution could increase reagent requirements and add complexity to the subsequent metal separation and purification stages.

To recover dissolved metals from acid leach solutions, solvent extraction methods have been applied and demonstrated by several researchers [9, 10]. However, great care needs to be taken to avoid cross-contamination when using multiple reagents, and the cost and complexity of solvent extraction processes are hard to justify for relatively small-scale LIB cathode re-processing operations [1].

RecycLiCo™ Battery Materials Inc. (formerly American Manganese Inc), with its research partner Kemetco Research Inc., develop a patented, RecycLiCo™ process that separates the LIB cathode materials from aluminum current collectors, uses sulfur dioxide (SO_2) in combination with H_2SO_4 to leach cobalt, lithium, manganese, and nickel from the cathode materials, and co-precipitates the metals into metal hydroxides which can be used to remake LIB cathode precursors.

SO_2 is a widely used reducing agent used in the mining industry in combination with H_2SO_4 to leach cobalt and manganese from higher valent resource materials such as cobalt (III) oxide and manganese (IV) dioxide. The main advantages of using SO_2 as a reducing agent over hydrogen peroxide (H_2O_2) are:

- H_2O_2 catalytically decomposes in the presence of metal oxides potentially leading to high reagent consumption.
- H_2O_2 hazards have been extensively documented with sensitivity to contamination that can lead to rapid decomposition producing water, oxygen, and heat. The pressures in large scale storage and handling systems can result in explosions capable of destroying structures and injuring personnel.
- The reaction of SO_2 with the active cathode materials are near stoichiometric in practice.
- Although SO_2 is hazardous, it is not explosive. As such, engineering controls can provide for safe and reliable operation.
- SO_2 is produced by reacting elemental sulfur with oxygen. The elemental sulfur is a by-product of oil production, thus provides a reliable, low-cost reagent that can be safely transported in large quantities.

Compared to solvent extraction, precipitation methods are simpler and can be efficient if the conditions are selected carefully. Precipitation methods recover metals as mixed metal products with different compositions (hydroxides, carbonates, sulfides). With this method, a complete separation of Ni and Co is difficult as these metals (of the same oxidation state) are situated next to each other in the stability diagram [11]. Separation of the mixed metal products can be made possible through the re-solubilization of mixed Co-Ni hydroxides under oxidative conditions in a pH range of 2.3 to 5.5, resulting in an almost complete nickel dissolution while leaving cobalt in the precipitate [12]. The RecycLiCo™ process co-precipitates the Ni, Mn, and Co to metal hydroxide without individually separating the metals.

In this chapter, the basic concept and development of the RecycLiCo™ process are described through recycle and recover three different cathode compositions, NMC, NCA, and LMO, in acid with the addition of sulfur dioxide, followed by a series of precipitation stages to recover the solubilized metals either simultaneously or sequentially. A number of process options are evaluated and optimized in bench-type, locked-cycled semi-continuous experiments using LCO cathode material to result the RecycLiCo™ process flowsheet. The patented RecycLiCo™ process achieves a 99% leach extraction efficiency of lithium, cobalt, nickel, and manganese from the battery waste and upcycles it into high-value lithium chemicals and cathode precursors for direct integration into the re-manufacturing of new LIBs. Using minimal processing steps, RecycLiCo™ process reduces the conventional LIB supply chain environmental footprint into a closed-loop process that enables a sustainable and circular solution for LIB materials.

3.2 SEPARATION OF LITHIUM-ION CATHODE MATERIAL FROM ALUMINUM FOIL CURRENT COLLECTOR

Waste and spent lithium cathode obtained from a recycler has waste and spent cathode active material bonded on the aluminum foil [13]. Cathode active material can be separated from aluminum foil using organic solvents such as dimethylacetamide, N-methyl-pyrrolidinone, or N-Methyl-2-pyrrolidone (NMP) [14], choline chloride-glycerol [15], and trifluoroacetic acid [16] to dissolve the binder. However, using organic solvent causes contamination and complicates the recycling process [17]. RecycLiCo™ process adapted an alternative method to separate waste and spent lithium cathode active material from aluminum foil by using weak H_2SO_4. Research from understanding the stability of the organic binder, experimenting various acid concentrations to remove binder, and separating waste and spent cathode active material from aluminum foil in both beaker and pilot scales to examine the efficiency of separation is described below:

3.2.1 BINDER STABILITY

The stability of the PVDF/NMP binder can loss its adhesion to the aluminum foil under acid condition. As shown in Figure 3.1(a)–(d), a layer of PVDF/NMP binder with the dimension of 4 cm x 4 cm was coated on a piece of 5 cm x 5 cm aluminum foil and immersed in 1 M H_2SO_4. The PVDF/NMP lost its adhesion after 4 hours and was floating in the H_2SO_4 solution.

A series of 8 experiments were conducted to examine the effect of H_2SO_4 concentration on PVDF/NMP binder adhesion. As shown in Figure 3.2, the time required for PVDF/NMP layer to detach from aluminum foil is decreased with increasing concentration of sulfuric acid.

As shown in Table 3.1, for Samples 7 and 8 immersed in 240 g/L of H_2SO_4, the PVDF/NMP layer can be detached within 50 minutes and 60 minutes, respectively; For Samples 5 and 6 immersed in 180 g/L of H_2SO_4, the PVDF/NMP layer can be completely separated after 220 minutes; Samples 3 and 4 immersed in 120 g/L

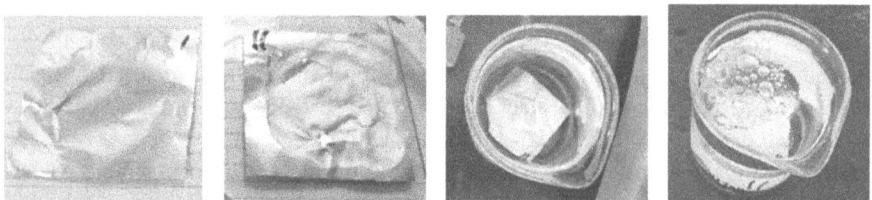

FIGURE 3.1 PVDF/NMP binder detached from Al foil in H_2SO_4.

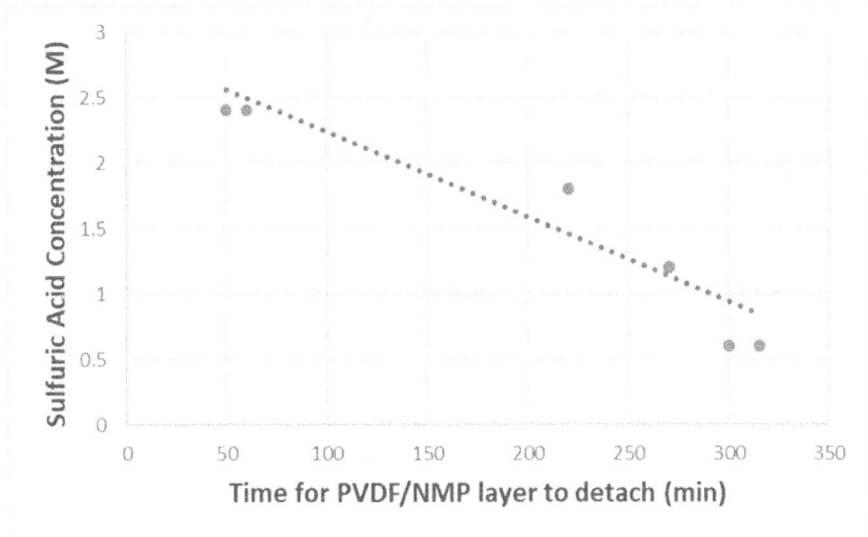

FIGURE 3.2 Impact of H_2SO_4 concentration on time required for PVDF/NMP layer to detach from Al foil.

of H_2SO_4 are separated after 270 minutes; finally, Samples 1 and 2 immersed in 60 g/L of H_2SO_4 were separated within 300 minutes and 315 minutes, respectively. Table 3.1 also shows that the aluminum foils are not leached despite the presence of H_2SO_4, evidenced by that the concentrations of aluminum ions in H_2SO_4 samples collected after the experiments were below ICP-OPES detection limit.

TABLE 3.1
Time for PVDF/NMP Coating to Detach from Aluminum Foil

Sample ID	Al Foil (g)	PVDF/NMP Weight After Curing (g)	H_2SO_4 Concentration (M)	Time for PVDF/ NMP Layer to Detach from Al Foil (min)	Al Concentration in H_2SO_4 After PVDF/ NMP Layer Detached (mg/L)
1	0.225	0.025	0.60	300	<10
2	0.236	0.022	0.60	315	<10
3	0.233	0.035	1.20	270	<10
4	0.227	0.042	1.20	270	<10
5	0.220	0.043	1.80	220	<10
6	0.202	0.030	1.80	220	<10
7	0.199	0.044	2.40	50	<10
8	0.195	0.025	2.40	60	<10

3.2.2 Separation of Cathode Active Material from Cathode Current

3.2.2.1 Acid Concentration Effect on Leaching Aluminum Foil Collector

The effect of H_2SO_4 concentration on leaching aluminum from cathode waste was investigated by immersing 1 cm x 4 cm pieces of cathode waste into various concentration of H_2SO_4. Pulp density of 25 wt% cathode waste was immersed into 0.10 M, 0.25 M, 0.30 M, 0.35 M, 0.41 M, 0.60 M, 2.17 M, 2.78 M, 3.43 M, and 4.10 M H_2SO_4, respectively. All experiments were conducted at 60°C. Hourly check was conducted to monitor whether or not the cathode active material was detached from the aluminum current collector. Once the separation of cathode active material from aluminum current collector was observed, a solution sample was collected and filtered. The filtered solution was analyzed by inductively coupled plasma optical emission spectroscopy (ICP-OES) to determine the metal ion contents in the solution. Two types of cathode wastes, $LiNi_{0.5}Mn_{0.3}Co_{0.2}O_2$ (NMC) and $LiNi_{0.80}Co_{0.15}Al_{0.05}O_2$ (NCA) were examined. The cathode wastes were obtained from Interco Trading, Inc., a material recycling company in Illinois, USA.

The effect of H_2SO_4 concentration on the separation of aluminum foil from cathode active material was studied at 15 wt% pulp density and 20 wt% pulp density. Pulp density of 15 wt% of cathode waste was immersed in H_2SO_4 concentrations of 0.8 M, 1.0 M, 1.2 M, 1.4 M, and 1.6 M, respectively. Pulp density of 20 wt% of cathode waste was immersed in H_2SO_4 concentrations of 1.4 M, 1.6 M, 1.8 M, 2.0 M, and 2.2 M, respectively. Each test was mixed with an overhead mixer for 2 hours to see if the cathode active material was separated from aluminum foil.

Figure 3.3 shows the effect of H_2SO_4 concentration on leaching aluminum foil during the separation of NMC and NCA cathode active material from aluminum foil. It can be seen that at H_2SO_4 concentrations lower than 0.6 M, minimum amounts of

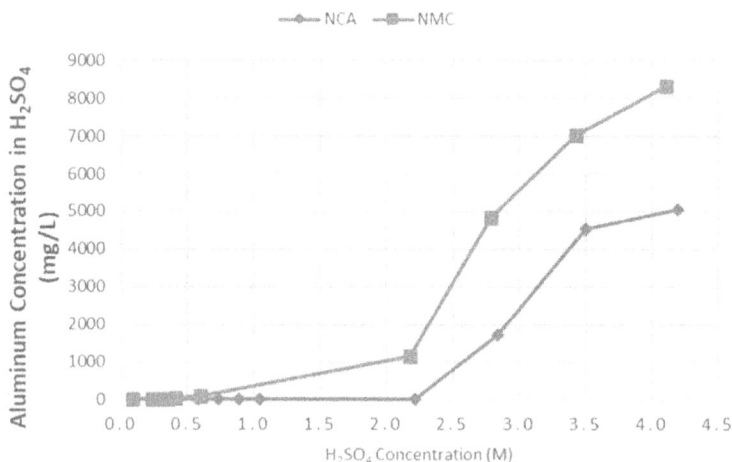

FIGURE 3.3 Effect of H_2SO_4 concentration on leaching Al during separation of NMC and NCA cathode active material from Al foil.

Al^{3+} could be detected after the NMC cathode active materials detached from aluminum foil. However, using higher H_2SO_4 concentration could result in aluminum leached into the solution. On the other hand, with NCA cathode waste material, Al^{3+} was detected with H_2SO_4 concentration higher than 2.2 M.

Results from Figure 3.3 show aluminum foil being leached out at 0.5 M H_2SO_4 with NMC cathode active material and at 2.2 M H_2SO_4 with NCA cathode active material, respectively. The concentration of H_2SO_4 to leach out aluminum is lower than the finding from the binder stability study, which shows that aluminum foil is not leached with 2.4 M H_2SO_4. The experiment temperature was believed to be the factor contributing the effect. When immersing NMC and NCA cathode waste materials into H_2SO_4 solution, the temperature of the solution was controlled at around 60°C, which promoted the leaching of aluminum foil.

3.2.2.2 Acid Required to Separate Cathode Active Material from Aluminum Foil

Figure 3.4 shows the separation experiments of 15 wt% NCA and 15 wt% NMC cathode waste materials at various H_2SO_4 concentrations. As shown in Figure 3.4(b), NCA

FIGURE 3.4 H_2SO_4 Concentration on separation of cathode active material and aluminum foil. (a) cathode immersed in various H_2SO_4 concentration, (b) separation of 15 wt% NCA cathode active material and aluminum foil at various H_2SO_4 concentration, and (c) separation of 15 wt% NMC cathode active material and aluminum foil at various H_2SO_4 concentration.

TABLE 3.2

Weight Ratios of the Required 98% H_2SO_4 to Separate Cathode Active Material from Aluminum Foil

Pulp Density	Weight Ratio of 98 wt% H_2SO_4 to Cathode Waste to Achieve Separation	
	NCA	NMC
15 wt%	0.96	0.80
20 wt%	0.94	0.78

cathode active material is not able to separate from aluminum foil with 0.8 M H_2SO_4. A low degree of separation is achieved with 1.0 M H_2SO_4 and a high degree separation iss achieved with H_2SO_4 concentration at 1.4 M and higher. Experiments with NMC cathode waste material also demonstrated that H_2SO_4 concentrations at 1.4 M and higher could achieve separation of NMC cathode active material from aluminum foil.

Experiments were also conducted to determine the amount of H_2SO_4 required to separate 20 wt% of NCA and NMC cathode waste materials. Both NCA and NMC cathode active materials were separated from aluminum foil with H_2SO_4 concentration at 2.04 M and higher.

The analyzed experimental data showed that NCA cathode was composed of approximately 20 wt% aluminum foil and 80 wt% NCA cathode active material. NMC cathode was composed of approximately 9 wt% of aluminum foil and 91 wt% of NMC cathode active material.

As shown in Table 3.2, despite the pulp density, in order to achieve separation between cathode active material and Al foil, the weight ratio of the required 98 wt% H_2SO_4 to NCA cathode is around 0.95; and the weight ratio of the required 98 wt% H_2SO_4 to separate NMC cathode is around 0.79.

On the other hand, ICP-OES analysis identified the H_2SO_4 solutions contained low level of aluminum ions, as shown in Table 3.3, which indicates that aluminum foil can be leached to achieve the separation of cathode active material from aluminum foil.

TABLE 3.3

Aluminum Concentration in Leachate

	H_2SO_4 Concentration (M)	Al Concentration in Leachate (ppm)
15 wt% NCA	1.4	681
20 wt% NCA	2.0	631
15 wt% NMC	1.4	1191
20 wt% NMC	2.0	1366

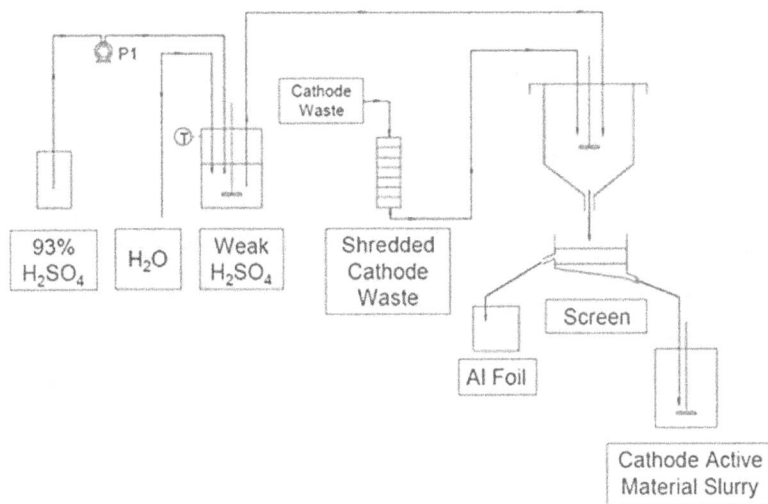

FIGURE 3.5 Experimental setup for pilot scale separation of cathode active material and aluminum foil.

3.2.2.3 Separation of Cathode Active Material from Aluminum Foil in 100 L Scale

Pilot scale separation of cathode active material from aluminum foil was conducted by placing the required amount of weak acid into a 100 L reactor followed by feeding the pre-weighted, shredded cathode waste into the reactor under agitation. The reactor temperature was maintained at 50°C. Figure 3.5 shows the experiment setup.

The separation experiment was monitored for 3 hours to ensure the cathode active material separated from aluminum foil. Once the separation was achieved, the slurry containing cathode active material and aluminum foil contained slurry was put into a vibrate screen to retain all the aluminum foil while letting the cathode active material contained slurry passing through. A solution sample was collected, filtered, analyzed by ICP-OES to determine the metal ion contents in the solution.

Pilot scale separation of cathode active material from aluminum foil was conducted by mixing 20 kg of NCA or NMC cathode waste with 80 Kg of 2.0 M H_2SO_4 in a 100 L reactor under agitation at 50°C for 3 hours. As shown in Figure 3.6 where aluminum foil can be successfully separated from NCA cathode waste and NMC cathode waste.

3.3 LEACHING WITH SO_2 IN COMBINATION WITH H_2SO_4

3.3.1 EVALUATING LEACHING CONDITIONS FOR VARIOUS CATHODE ACTIVE MATERIALS

Three different cathode active materials, NMC, NCA, and LMO, were leached using sulfur acid with the objective of extracting valuable metals. The cathode active materials (>99.99% pure), for validating the methods were obtained as

(a)

(b)

FIGURE 3.6   (a) Aluminum foil from NCA cathode waste and (b) aluminum foil from NMC cathode waste.

battery-grade reagents from MTI Corp. All chemicals [H_2SO_4, NaOH, Na_2CO_3, NaHS, Na_2S, LiOH, $(NH_4)_2CO_3$] were analytical grade, and all solutions were prepared with deionized water. Leaching tests were conducted in a 1000 ml baffled reactor equipped with a condenser and overhead agitator. The reactor was placed in a heating mantle for temperature control. The cathode active material was mixed with sulfuric acid solution and SO_2 was sparged into the resulting slurry, targeting a pre-determined Oxidation-Reduction-Potential (ORP) set point. Parameters including acid concentration, leaching time, pH and ORP were monitored and controlled as necessary. At the end of the leach stage, the products (leachate and reaction residue) were filtered, and the leach residue was thoroughly washed with DI water. Subsequent precipitation tests were done using the same type of reactor. At the beginning of a precipitation test, the solution was sparged with nitrogen gas to remove the dissolved oxygen. Typically, 200 ml of the leachate was used for batch tests studying different reagents and conditions. At the end of each test, the products were filtered, and the precipitate was washed thoroughly and dried at 70°C. A sketch of the reactor is given in Figure 3.7.

The quantitative metal contents for Al, Mn, Ni, Co, and Li were analyzed by ICP-OES. The concentration of sulfate was determined using the method of turbidimetric (EPA 9038), where sulfate ion was converted to a barium sulfate suspension under the controlled conditions. The resulting turbidity was measured using a HACH TU5200 Laser turbidimeter. The concentration of dithionate was directly measured by HPLC using sodium perchlorate eluent along with an AS17 ion exchange column.

Typical leach conditions were 10% solids, 1.2M H_2SO_4 and SO_2 sparging targeting an ORP of 350 mV. To determine the optimal reaction time, kinetic samples were taken at time intervals. Figure 3.8(a), (b), and (c) presents examples of the leaching kinetics of NMC, NCA, and LMO material.

Figure 3.8(a) shows that 80% of metals (Li, Co, Ni, Mn), from NMC are extracted within 30 minutes, and that almost a complete dissolution of cathode active material can be seen within 2 hours of the reaction time. A repetitive test shows 97.8% reproductivity. Further leaching tests with NMC indicated that the acid concentration

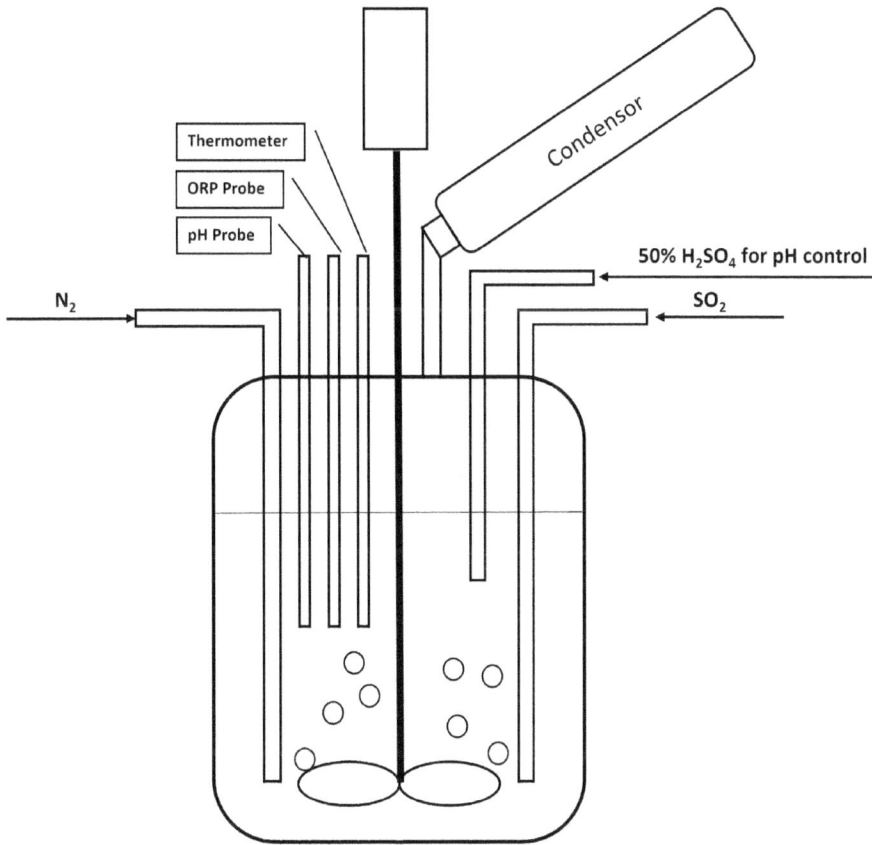

FIGURE 3.7 Reactor setup used for leach and precipitation testing.

could be reduced to 1.0 M H_2SO_4 without sacrificing metal extraction. The same is true for leaching tests with NCA shown in Figure 3.8(b) and LMO shown in Figure 3.8(c). The total metal extractions were over 99% in all tests with NMC and NCA materials. The extractions were similar for LMO material, and it was found that less acid (~0.8M) was required to achieve the same overall metal extraction. The selected test results are given in Table 3.4.

Along with metal extraction values, Table 3.4 presents the sulfate and dithionate concentrations in leach solutions. The dithionate concentration varies between 21–30 g/L and sulfate concentrations ranges from 150 to 198 g/L. These values are important in monitoring and avoiding exceeding the saturation levels and generating undesirable precipitates, particularly in the closed systems with the circulating solution streams. Examination of the complete test results revealed that higher acid concentrations and less SO_2 addition could generate less dithionate. Although the SO_2 addition was not crucial for lithium dissolution, this reagent was necessary to dissolve other metals. The SO_2 addition could impact the oxidation state of the transition metals, maintaining them in their more soluble forms.

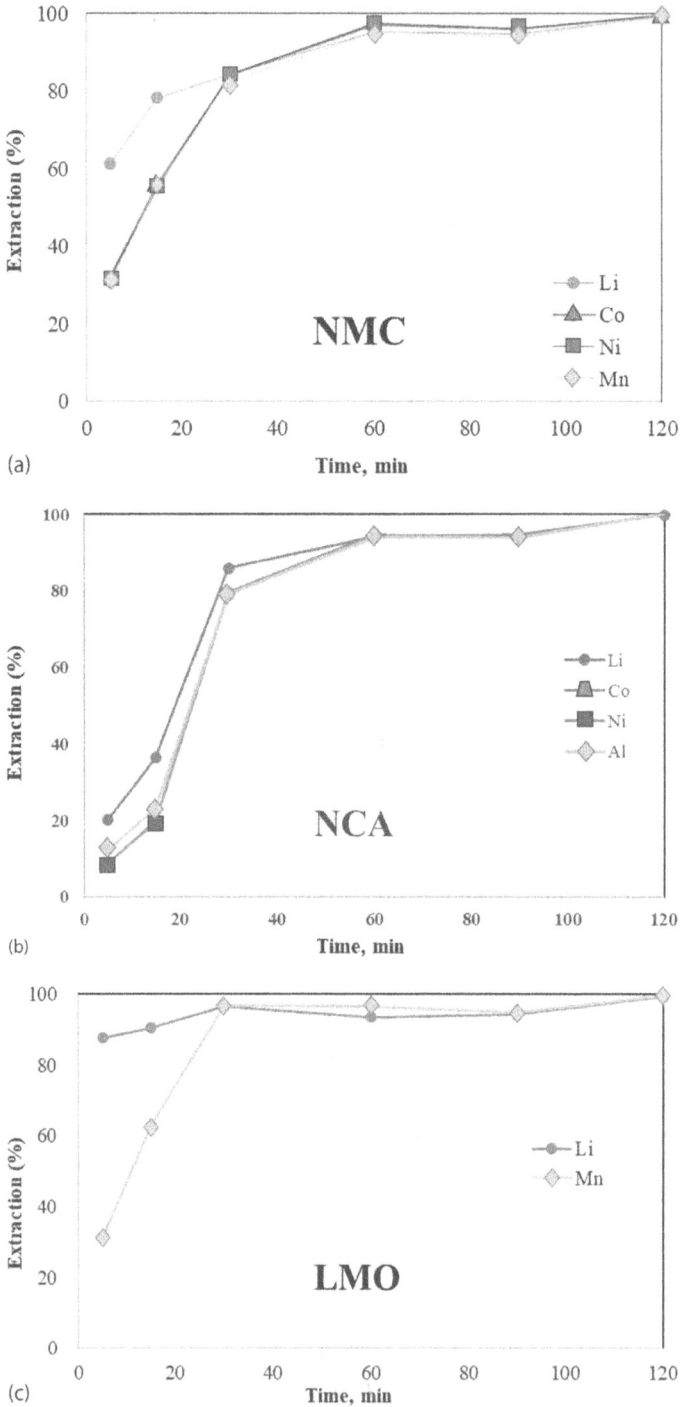

FIGURE 3.8 Leaching kinetics: (a) NMC, (b) NCA, and (c) LMO.

TABLE 3.4

Leach Tests Results with NMC, NCA, and LMO Cathode Active Materials

	Operation Conditions				Extraction				Concentration		
									Total		
Test ID	Solids	H_2SO_4	SO_2	pH	ORP	Li	Co	Ni	Mn	Acid	$S_2O_6^{2-}$	SO_4^{2-}
	%	M	g		(mV)	%	%	%	%	g	g/L	g/L
NMC-2-LT7	10%	1.0	119	1.0	486	99.8	100	99.9	100	87.8	23.9	184
NMC-2-LT8	8%	1.0	59.6	1.0	487	99.7	99.9	99.9	99.9	34.4	25.1	150
NCA-LT4	10%	1.0	59.6	1.5	412	100	100	100	100	33.1	29.9	166
NCA-LT5	10%	1.2	57.3	1.0	562	99.4	99.8	99.8	99.8	32.3	25.2	198
LMO-LT7	10%	0.8	69.1	0.6	637	97.9			98.3	23.3	21.2	188
LMO-LT8	10%	0.6	69.8	1.2	630	99.1			99.8	16.9	22.5	177

3.3.2 SULFUR DIOXIDE/SULFURIC ACID LEACH OF CATHODE ACTIVE MATERIAL COLLECTED FROM CATHODE SCRAP

3.3.2.1 Batch Leach to Extract >99% Ni, Co, Mn, and Li

Results from Chapter 3.3.1 shows that leaching commercial grade NMC, NCA, LMO, and LCO lithium cathode active materials with H_2SO_4/SO_2 can reach >99% extractions of Li, Co, Ni, and Mn with controlling leaching solution pH around 1.5 and ORP around 350–630 mV.

Batch leaching 20 wt% spent NCA and NMC cathode active materials were conducted in a 50 L reactor by controlling the pH of the slurry at 1.5 with 98 wt% H_2SO_4 in a reductive environment for 7 hours. The reductive environment was created by metering SO_2 gas at 1.5 L/min to maintain solution ORP at 550. The reaction temperature was maintained at 80°C. Solution samples were collected hourly and analyzed by ICP-OES to determine the metal contents in the solution. SO_2 was purchased from Praxair Inc. Leaching efficiency was calculated with the following equation 3.1:

$$\frac{\text{Weight of residue } (Kg) \times [M] \text{ in residue } (mg/Kg)}{\text{Weight of spent cathode active material } (Kg) \times [M] \text{ in spent cathode active material } (mg/Kg)}$$

$$(3.1)$$

where [M] is the concentration of Ni, Co, Mn, or Li.

Figure 3.9 shows the concentration of Ni, Co, Mn, Al, and Li remained in the residues during leaching. It can be found that a leaching retention time of 6 hours for spent NCA material shown in Figure 3.9(a), and 7 hours for spent NMC material shown in Figure 3.9(b) can achieve >99% extractions of Ni, Co, Mn, Al, and Li.

3.3.2.2 Leaching Using H_2SO_4/SO_2 Compared with Leaching Using H_2SO_4/H_2O_2

In order to compare leaching using H_2SO_4/SO_2 with leaching using H_2SO_4/H_2O_2, a leaching test with 20 wt% spent NMC cathode active material was leached using 98 wt% H_2SO_4 to control leaching solution pH at 1.5 and using 30 wt% H_2O_2 to control leaching solution ORP at 550 mV for 7 hours.

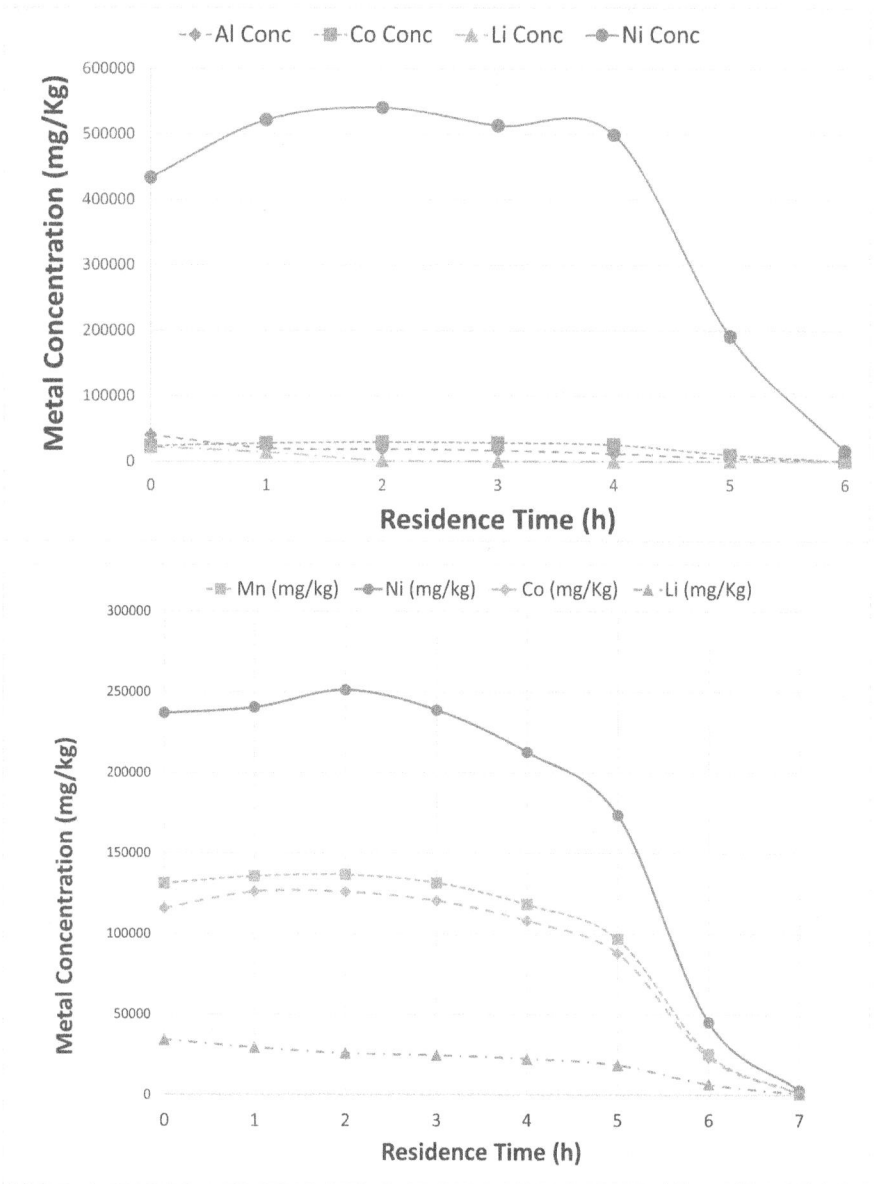

FIGURE 3.9 Retention time required for H_2SO_4/SO_2 leaching to achieve >99% leaching efficiency.

Table 3.5 shows the mass balance comparison between H_2SO_4/H_2O_2 leach and H_2SO_4/SO_2 leach of NMC cathode active material. Leaching with H_2SO_4/SO_2 results 15% less total leachate volume, reducing the degree of water management needed in the downstream impurity removal, metal extraction, and lithium recovery, which provides a better opportunity to achieve a closed-loop process.

TABLE 3.5

Mass Balance Comparison Between H_2SO_4/H_2O_2 Leach and H_2SO_4/SO_2 Leach of NMC Cathode Active Material

Leach chemicals	NMC (Kg)	Pulp Density (%)	H_2SO_4 Consumption (Kg)	H_2O_2 Consumption (Kg)	SO_2 Consumption (Kg)	Total Leachate (Kg)	(L)
H_2SO_4/H_2O_2	6.35	20	6.08	4.51		42.39	35.87
H_2SO_4/SO_2	6.35	20	2.88		1.80	36.48	31.21

3.3.2.3 Effect of SO_2 Dosage on Dithionate Generation

As reported in Chapter 3, leaching with H_2SO_4/SO_2 can generate dithionate, $S_2O_6^{2-}$ as a byproduct. The H_2SO_4/SO_2 leaching chemical reactions can be expressed in equation 3.2–equation 3.7 as follows:

For $LiNi_{0.5}Mn_{0.3}Co_{0.2}O_2$:

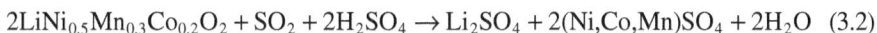

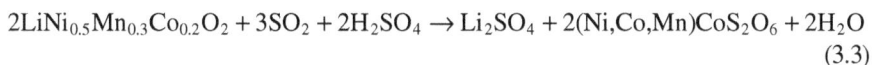

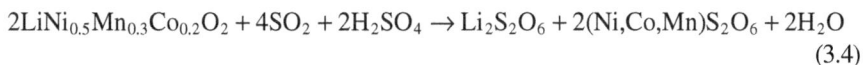

$$2LiNi_{0.5}Mn_{0.3}Co_{0.2}O_2 + SO_2 + 2H_2SO_4 \rightarrow Li_2SO_4 + 2(Ni,Co,Mn)SO_4 + 2H_2O \quad (3.2)$$

$$2LiNi_{0.5}Mn_{0.3}Co_{0.2}O_2 + 3SO_2 + 2H_2SO_4 \rightarrow Li_2SO_4 + 2(Ni,Co,Mn)CoS_2O_6 + 2H_2O \quad (3.3)$$

$$2LiNi_{0.5}Mn_{0.3}Co_{0.2}O_2 + 4SO_2 + 2H_2SO_4 \rightarrow Li_2S_2O_6 + 2(Ni,Co,Mn)S_2O_6 + 2H_2O \quad (3.4)$$

$(Ni,Co,Mn)SO_4$ and $(Ni,Co,Mn)S_2O_6$ represent a mixed metal sulfate and a mixed metal dithionate respectively.

For $LiNi_{0.8}Co_{0.15}Al_{0.05}O_2$:

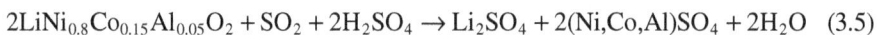

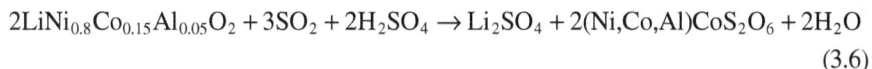

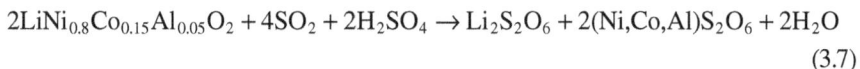

$$2LiNi_{0.8}Co_{0.15}Al_{0.05}O_2 + SO_2 + 2H_2SO_4 \rightarrow Li_2SO_4 + 2(Ni,Co,Al)SO_4 + 2H_2O \quad (3.5)$$

$$2LiNi_{0.8}Co_{0.15}Al_{0.05}O_2 + 3SO_2 + 2H_2SO_4 \rightarrow Li_2SO_4 + 2(Ni,Co,Al)CoS_2O_6 + 2H_2O \quad (3.6)$$

$$2LiNi_{0.8}Co_{0.15}Al_{0.05}O_2 + 4SO_2 + 2H_2SO_4 \rightarrow Li_2S_2O_6 + 2(Ni,Co,Al)S_2O_6 + 2H_2O \quad (3.7)$$

$(Ni,Co,Al)SO_4$ and $(Ni,Co,Al)S_2O_6$ represent a mixed metal sulfate and a mixed metal dithionate respectively.

Figure 3.10 shows the effect of SO_2 dosage on $S_2O_6^{2-}$ generation in H_2SO_4/SO_2 leach of NMC cathode active material. When dosing with 1.5 L/min SO_2, approximately 6.5 g/L of $S_2O_6^{2-}$ and 138 g/L of SO_4^{2-} are detected in the leachate, which is 40.6% more $S_2O_6^{2-}$ and 14.9% less SO_4^{2-} comparing to dosing with 1.0 L/min SO_2.

When dosing with 1.5 L/min SO_2, the leaching solution ORP is around 550 mV. The retention time required to achieve >99 leaching efficiency is 7 hours as shown

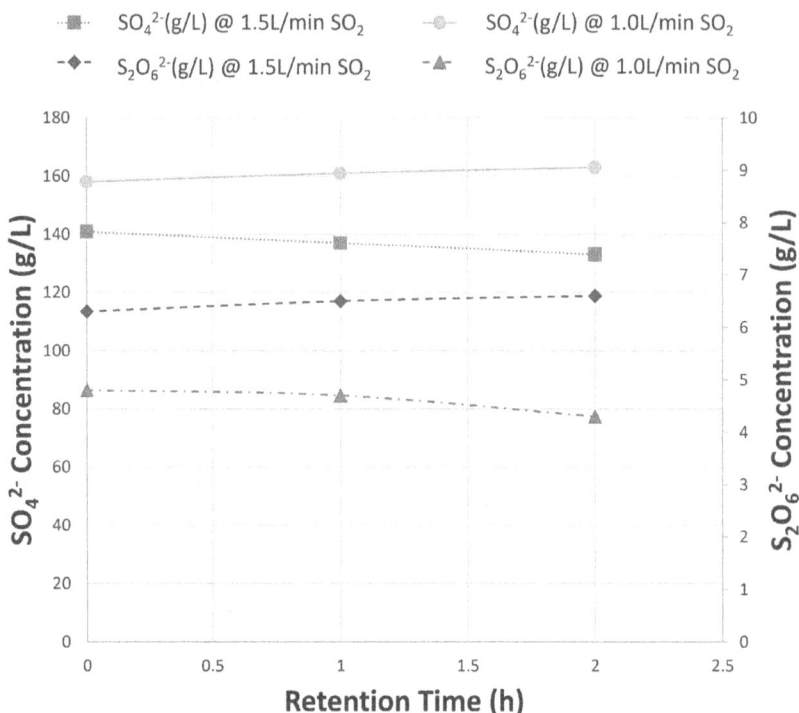

FIGURE 3.10   Effect of SO_2 dosage on dithionate generation.

in Figure 3.9(b). When dosing with 1.0 L/min SO_2, the leaching solution ORP is around 350 mV. A longer retention time, 9 hours, is required to achieve >99 leaching efficiency.

Although a longer retenion time is required when dosing with 1.0 L/min SO_2 to achieve >99% extraction of Ni, Co, Mn, and Li, the leachate can result in a lower concentration of $S_2O_6^{2-}$. The presence of dithionate can affect the leach solution oxidation-reductive potential which impacts the ORP control, leaching efficiency, and SO_2 consumption. In a closed-loop system, the concentration of $S_2O_6^{2-}$ will potentially build up and impact the downstream metal extraction and lithium recovery.

3.3.2.4   Continuous Leaching Operation Cost vs. Capital Cost

Continuous leaching was conducted by continuing feeding the spent cathode active material slurry into the reductive leach reactor at 5 L/h, 10 L/h, or 15 L/h while maintaining the slurry pH at 1.5 by 98 wt% H_2SO_4 and controlling ORP at 550 by metering SO_2 for 4 hours. The reaction temperature was maintained at 70°C during the leaching.

The leachate could overflow through three different overflow ports at 30 L, 40 L, and 50 L, respectively, which allowed different retention times for the leaching reaction. The overflow solution was collected in a 150L agitated reactor and purged

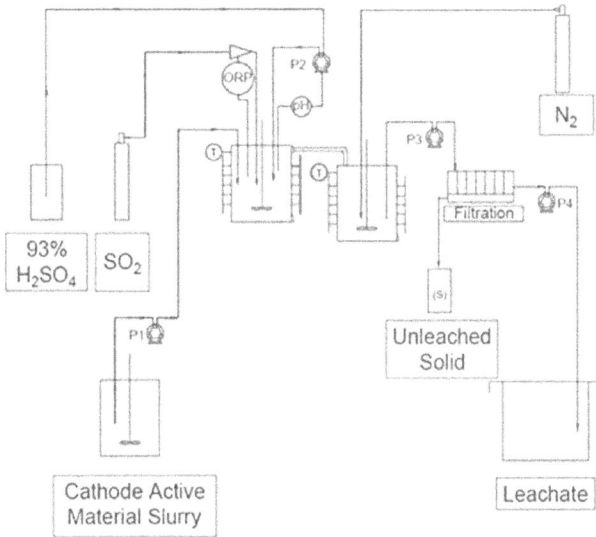

FIGURE 3.11 Experiment setup for continuous leach with sulfur dioxide and sulfuric acid.

with nitrogen gas to dissipate any remaining SO_2 in the solution. The overflow solution sample was collected hourly and filtered. The filtered solution was analyzed by ICP-OES to determine the metal contents in the solution. The filtered residue was washed, dried, and analyzed by ICP-OES to determine the metal contents. The final leach solution was filtered by a filter press to separate the unleached solid and leachate. Figure 3.11 shows the experiment setup.

Experiments shown in Table 3.6 investigates the impact of the retention time. When shortening the retention time from 7 hours to 6 hours and 5 hours, the size of the reactor can reduce by 14% and 29%, respectively, which save the initial capital expenditure (CapEX) cost. On the other hand, the consumption for SO_2 is decreased by 15% and 17%, respectively, and the consumption of H_2SO_4 is increased by 37% and 47%, respectively, which results in an overall increase of operation expenses (OPEX). In the meantime, the leaching efficiencies for Co, Ni, and Mn are decreased slightly while Li remains complete leached.

TABLE 3.6
Leaching Parameters at Different Retention Times

No	Retention Time h	Total SO_2 L/min	Total SO_2 Kg	H_2SO_4 Kg/hr	Total H_2SO_4 Kg	Ratio of Reactor Size	Temp °C	Co (%)	Ni (%)	Mn (%)	Li (%)
1	7	1.5	1.80	0.30	2.1	1.00	70	99.9	99.9	99.9	99.9
2	6	1.5	1.54	0.48	2.9	0.86	70	98.8	98.7	98.6	99.9
3	5	1.5	1.29	0.62	3.1	0.71	70	97.5	97.4	97.3	99.9

The reduction of retention time could decrease the initial CapEX but increase the overall OPEX, which in long run has a negative impact on the economic. A detail economic study on the overall spent LIB cathode active material recycling process should be conducted to further analyze the impact.

3.4 EVALUATING OPTIONS FOR RECOVERING SOLUBILIZED METALS

Three different precipitating agents were used to recover the dissolved metals from leach solutions. Sodium hydroxide (NaOH) was used to recover metals as hydroxides. Sodium carbonate could produce carbonate compounds and was mainly used for lithium recovery. Sodium sulfide could form insoluble metal sulfides and was tested as an option to minimize the Mn in Ni/Co mixed metal precipitates. Some tests were used for recovering all metals simultaneously and some for a sequential precipitation approach, and various conditions were tested for each approach. Metal recovery results from NMC leach solutions are given in Table 3.2.

The results shown in Table 3.7 indicate that increasing the temperature (50°C in NMC-2-PT3 versus ~20°C in NMC-2-PT2) at pH ~8 can only increase Co and Ni recoveries slightly whereas Mn recovery to the Ni-Co product is increased by 15%.

TABLE 3.7
Precipitation Test Results Using NMC Leachate

	Conditions					Distribution				
Test	NaOH g/kgᵃ	Na₂S g/kgᵃ	Na₂CO₃ g/kgᵃ	pH	Temp °C	Dry Wt. g	Li %	Co %	Ni %	Mn %
NMC-2-PT 2										
Stage 1 (Ni, Co)	538			8.0	20	19.0	2.9	95.5	97.8	21.9
Stage 2 (Mn)			359	10.0	20	4.7	1.4	2.8	1.4	49.5
Stage 3 (Li)			523	10.6	95	2.6	27.7	0.01	0.01	0.05
NMC-2-PT 3										
Stage 1 (Ni, Co)	550			7.8	50	20.9	3.4	98.8	99.5	36.9
Stage 2 (Mn)			933	10.0	20	3.2	1.6	0.64	0.27	32.9
Stage 3 (Li)			523	10.6	95	0.2	1.8	0.01	0.01	0.08
NMC-2-PT 5										
Stage 1 (Ni, Co)	176	339		5.0	20	25.0	3.1	99.6	99.9	6.6
Stage 2 (Mn)			348	10.0	20	5.9	2.0	0.11	0.03	68.6
Stage 3 (Li)			523	10.5	95	2.1	25.2	0.002	0.001	0.04
NMC-2-PT 7										
Stage 1 (Ni, Co, Mn)	655			10.0	20	28.7	12.6	100.0	100.0	100.0
Stage 2 (Li)			563	11.0	95	2.8	29.2	0.002	0.001	0.004

ᵃ g reagent/kg cathode material treated.

Complete removal of all three metals (Co, Ni, and Mn) was achieved when the pH was adjusted to 10 in Test NMC-2-PT7.

In this test the consumption of caustic was increased but the overall reagent requirement was reduced. A small amount of caustic was still needed in Test NMC-2-PT5 using Sodium sulfide precipitation at pH 5. Co and Ni recoveries for this test were over 99.5%. A small amount of Mn was co-precipitated with the Co and Ni sulfides, but overall, this test showed the most effective separation of Mn from the Co-Ni product. The remaining Mn and Li could then be precipitated as two separate products using the controlled sodium carbonate addition and temperature.

The NCA leach solution required an initial treatment stage to remove Al separately from other metals. The Al was removed through a neutralization with caustic at pH 4.5 to 5.5. Within this range, higher pH gave better Al removal. After Al removal, the pH was further increased to above 10 with caustic to collect the Co and Ni as the mixed metal hydroxides. Following that, sodium carbonate was added to recover the lithium. The test results are given in Table 3.8.

Table 3.8 shows that close to 99% of Al can be precipitated from solution at pH 5.5 (NCA-PT4), while just over half of Al (56%) is removed at pH 4.5. Just over 90% of Co and Ni are recovered at both pH 10.5 and 11.5 with a balance lost to entrainment in the voluminous aluminum oxyhydroxide [AlO(OH)] precipitate. In Test NCA-PT 6,

TABLE 3.8
Precipitation Test Results Using NCA Leachate

| | Conditions | | | | | Distribution | | | | |
| | NaOH | NaHS | Na$_2$CO$_3$ | pH | Temp | Dry Wt. | Li | Co | Ni | Al |
Test	g/kg[a]	g/kg[a]	g/kg[a]		°C	g	%	%	%	%
NCA-PT 2										
Stage 1 (Al)	117			4.5	20	0.58	0.09	0.38	0.48	56.1
Stage 2 (Co, Ni)	643			11.5	20	22.3	3.97	90.4	90.0	42.5
Stage 3 (Li)			527	11.5	95	2.33	26.8	0.00	0.00	0.07
NCA-PT 4										
Stage 1 (Al)	131			5.5	20	1.33	0.03	1.13	1.40	98.7
Stage 2 (Co, Ni)	484			10.5	20	22.7	1.9	90.8	90.9	0.3
Stage 3 (Li)			527	10.7	95	2.88	31.5	0.02	0.02	0.08
NCA-PT 6										
Stage 1 (Al, Co, Ni)	505			10.5	20	19.4	0.99	100.0	100.0	99.2
Stage 2 (Li)			527	10.6	95	2.09	31.2	0.03	0.03	0.11
NCA-PT 8										
Stage 1 (Al)	130			5.03	20	1.13	0.1	0.70	0.84	96.3
Stage 2 (Co, Ni)	101	528		6.12	20	35.2	4.2	92.0	91.8	2.7
Stage 3 (Li)			527	10.8	95	2.85	32.5	0.00	0.00	0.03

[a] g reagent/kg cathode material treated.

a simultaneous co-precipitation of Al, Co, and Ni as hydroxides can be successfully demonstrated at pH 10.5. Sulfide addition in Test NCA-PT 8 shows similar levels of Co and Ni recovery to test NCA-PT 4, but the precipitate is significantly larger.

A similar approach was taken in treating LMO leach solution, although the composition was much simpler. Both caustic and carbonate precipitation methods were used to recover Li and Mn from solution. Test results are shown in Table 3.9.

From Table 3.9, it is clear that Mn recovery is completed at pH 10 using either caustic or carbonate. The lower concentrations can make the recovery of lithium particularly challenging with LMO, and only about 20% of the initial Li feed is recovered as carbonate whether the reagent was added as sodium carbonate powder or as CO_2 gas sparges in alkaline solution.

Figure 3.12 summarizes the metal recovery from the leach solutions produced from all three cathode active material types (NMC, NCA, and LMO) as a function of pH. Good recovery values for Co, Ni, and Mn can be achieved at pH 10 using the hydroxide system. Similar recovery of Co and Ni can be seen at a lower pH (5.0) with aqueous sulfide addition (not included in Figure 3.3). The lithium recovery is low, but the expectation is that the recovery would be increased by recycling lithium-rich process solutions back to the leaching stage, therefore no alternative lithium recovery methods were included in this part of the investigation.

TABLE 3.9
Precipitation Test Results with LMO Leachate

	Conditions					Distribution		
Test	NaOH g/kg[a]	Na₂CO₃ g/kg[a]	CO₂ Sparging Time min	pH	Temp °C	Dry Wt. g	Li %	Mn %
LMO-PT 2								
Stage 1 (Mn)	740			10.0	20	19.9	0.83	100.0
Stage 2 (Li)		297		10.6	95	0.91	18.3	0.01
LMO-PT 4								
Stage 1 (Mn)	760			10.1	50	18.9	0.69	100.0
Stage 2 (Li)		297		11.3	95	0.91	18.0	0.00
LMO-PT 6								
Stage 1 (Mn)		1246		10.0	20	31.0	23.3	100.0
Stage 2 (Li)		297		11.0	95	0.00	0.00	0.00
LMO-PT 9								
Stage 1 (Mn)	805		34	11.7–8.59	20	29.5	16.8	100.0
Stage 2 (Li)		297		11.6	95	0.24	4.87	0.04

[a] g reagent/kg cathode material treated.

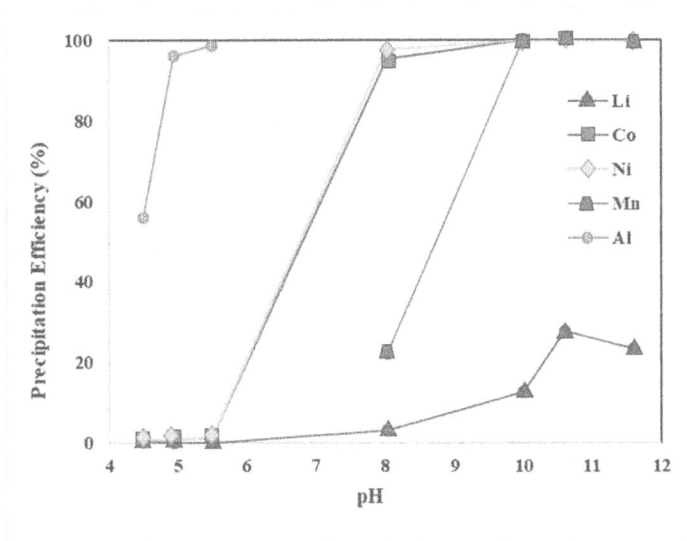

FIGURE 3.12   Precipitation efficiency versus pH.

3.5   RECYCLICO™ FLOWSHEET DEVELOPMENT THROUGH LOCKED-CYCLE TESTING OF LCO CATHODE ACTIVE MATERIAL

Following the positive results from bench scale tests, a locked-cycle campaign was initiated to evaluate the effects of the recycled process solutions on overall efficiency. Two different campaigns were run, with each campaign consisting of 10 cycles. Both campaigns were based on the same conditions for leaching an LCO active cathode active material using sulfuric acid and sulfur dioxide under the same target conditions. The main difference between these campaigns was in the subsequent metal recovery flowsheet. The first locked-cycle campaign recovered the leached metals sequentially using sodium hydroxide to remove Co followed by sodium carbonate for Li recovery. The second locked-cycle campaign used a combination of sodium hydroxide and sodium carbonate to recover both valuable metals simultaneously. After metal recovery, the remaining metal-deficient solutions were treated by chilling for sodium sulfate removal before being recycled back to the next cycle's leaching stage. The chiller was added to lower the equilibrium sulfate concentration and prevent uncontrolled precipitation of sodium sulfate in the system [19]. Evaporation of the recycled solutions was incorporated as needed to maintain the overall water balance. The locked-cycle campaigns used 100 grams of cathode active material per cycle at a pulp density of 8% with 1.2M H_2SO_4 to maintain a pH close to 1.5 and an ORP target of 350 mV. The leach residence time was 2 hours per cycle. The temperature was maintained at 50°C after the initial exothermic phase to help ensure that the leach reaction went to completion.

The first campaign used sodium hydroxide addition under nitrogen sparging to precipitate cobalt as cobalt hydroxide. The residence time was 2 hours. The filtrate from the cobalt hydroxide precipitation stage was heated to 95°C and mixed with 1.2 times the stoichiometric requirement of sodium carbonate for 0.5 hours to precipitate lithium carbonate (Li_2CO_3). The precipitated lithium carbonate was filtered and washed with the saturated Li_2CO_3 at 95°C. Except for Cycle 1, all the saturated Li_2CO_3 wash solution was prepared using Li_2CO_3 solids generated from previous cycles. The filtrate from the lithium precipitation stage was cooled to 5°C to initiate Na_2SO_4 crystal formation. The crystallization stage was run for 2 hours before the resulting crystals were filtered. The crystallization filtrate and leach residue wash solution were then combined with evaporated wash solution from the cobalt precipitation stage and the resulting solution was acidified for use as the leach solution for the next cycle.

In the second locked-cycle campaign, the Co and Li were precipitated together by adding 1.2 times of stoichiometric quantity of sodium carbonate at 95°C to the leach solution after first adjusting the pH to 11 with sodium hydroxide to form cobalt hydroxide. The combined precipitate was filtered and washed with a saturated Li_2CO_3 solution (using reagent Li_2CO_3). As with the first campaign, the filtrate was then chilled to 5°C for 2 hours before filtering the resulting sodium sulfate crystals. For this campaign, the filtered crystals were also subjected to a re-pulp wash in the saturated Na_2SO_4 solution, also at 5°C. The crystallization filtrate was then combined with the crystal wash solution, Co-Li precipitate wash solution and the leach residue wash solution. This combined solution was evaporated to the target leach volume and acidified for use as the leach solution for the next cycle.

Figure 3.13 shows the schematic of the first locked-cycle campaign. The optional of nanofilter showed in Figure 3.13 was not executed during the locked-cycle

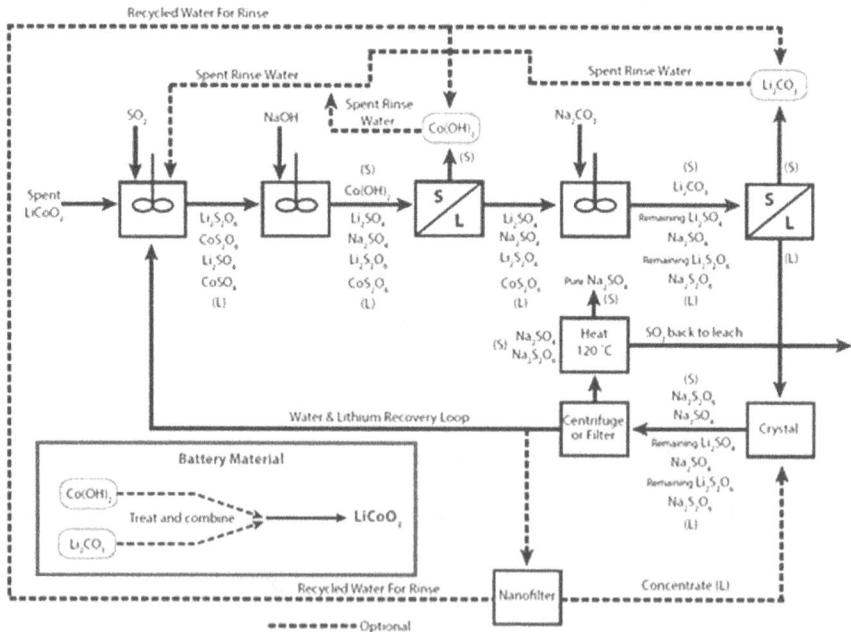

FIGURE 3.13 First lock cycle flowsheet with sequential recovery.

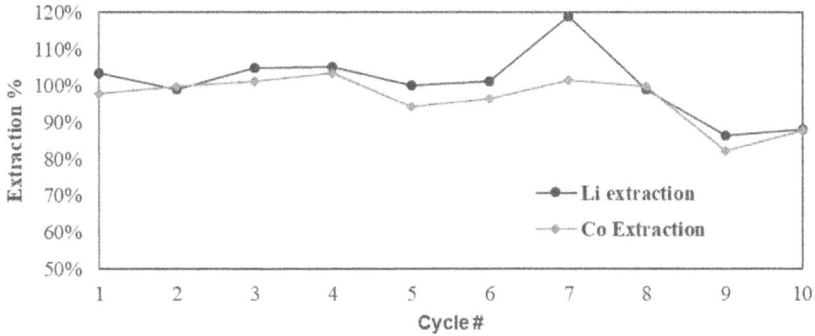

FIGURE 3.14 LCO leaching results – first locked-cycle campaign.

campaign. The extraction values for the first locked-cycle campaign are shown in Figure 3.14. Near 100% extractions can be achieved for both metals for the initial 8 cycles.

The recovery values are declined in Cycles 9 and 10 to some extent (82–88% for both Li and Co). This can be attributed to the changing composition of the recycle streams and will be discussed further. The recoveries of leached metals from solution are shown in Figure 3.15. Cobalt is precipitated with sodium hydroxide and lithium is recovered with sodium carbonate. Results indicate a consistent cobalt recovery through all cycles of 90% or greater, and a steady rise in lithium recovery with each cycle as is predicted.

To crystalize the Na_2SO_4 and remove it from the recycle stream, the filtrate from lithium precipitation was cooled to 5°C. Generally, the sodium removal was approximately 30% of the total sodium in the system in each cycle.

Overall, the first locked-cycle test showed high extraction of metals from LCO cathode active material and 93–98% cobalt recovery from acid leach solution. Lithium recovery from solution was around 55% initially and exceeded 85% once the locked-cycle circulating load was established in later cycles. The material fed and products recovered in each cycle and their compositions are given in Table 3.10.

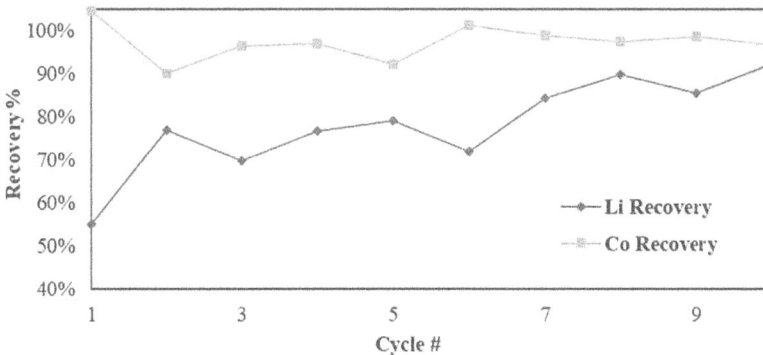

FIGURE 3.15 Metal recoveries in the first locked-cycle campaign.

TABLE 3.10

First Locked-Cycle Campaign – (a) Feed and (b) Products

(a)

	Leaching Stage				Hydroxide Product			
	Feed Li	Feed Co	Recycle Li	Residue	Mass	Li	Co	CO Rec.
Cycle ID	g	g	g	g	g	%	%	%
Cycle 1	6.84	58.4		0.92	96	0.02	63.4	104
Cycle 2	6.82	58.4	3.64	0.70	96	0.35	54.3	90
Cycle 3	6.86	58.4	2.89	1.00	98	0.20	57.1	96
Cycle 4	6.86	58.4	3.19	0.88	92	0.12	61.5	97
Cycle 5	6.85	58.4	3.52	6.27	87	0.07	61.7	92
Cycle 6	6.87	58.4	2.65	0.89	97	0.17	60.7	101
Cycle 7	6.85	58.4	2.50	0.75	94	0.12	61.0	99
Cycle 8	6.85	58.4	4.13	0.85	95	0.15	59.6	97
Cycle 9	6.79	58.4	3.84	0.94	93	0.08	61.7	99
Cycle 10	6.94	58.4	4.30	0.69	93	0.12	60.4	97

(b)

	Carbonate Product				Sulfate Crystals		
	Mass	Co	Li	Li Rec.	Mass	Li	Co
Cycle ID	g	mg/Kg	%	%	g	g/Kg	mg/Kg
Cycle 1	20	629	19.3	55	123		
Cycle 2	29	1220	18.2	77	256	1.28	5.0
Cycle 3	28	170	17.1	70	251	1.71	5.0
Cycle 4	29	398	18.4	77	199	1.03	5.0
Cycle 5	30	552	18.3	79	286	2.24	5.0
Cycle 6	27	238	18.4	72	200	1.05	5.0
Cycle 7	33	116	17.6	84	231	0.89	5.0
Cycle 8	48	61	12.8	90	266	0.76	5.0
Cycle 9	34	51	17.0	85	270	2.88	6.7
Cycle 10	34	45	18.6	92	273	0.54	5.0

The sulfate and dithionate concentrations at different stages in the process over the 10 cycles are shown in Figures 3.16 and 3.17, respectively. It is clear that the sulfate level can quickly stabilize around 300 g/L in the leach solution, which is reduced to approximately 50 g/L after removing sodium sulfate crystals in the final crystallization stage. In Cycle 6, the crystallization treatment leaves higher sulfate (100 g/L), but this does not significantly affect the sulfate level in the subsequent leach, which remains near 300 g/L like most other cycles. In contrast, the dithionate level is

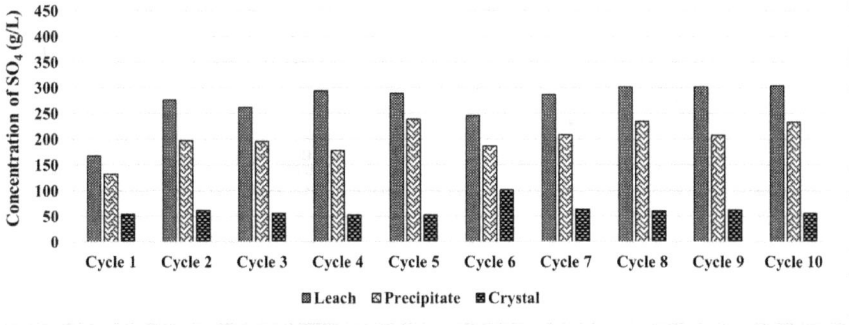

FIGURE 3.16 Sulfate concentration profile in the first locked-cycled campaign.

consistently higher after sulfate crystallization (dithionate does not reach saturation to crystalize), and reaches a level of 14–16 g/L in Cycles 8 to 10, which coincides with the reduced leaching efficiency in those cycles as noted above. The presence of dithionate affects the leach solution oxidation-reductive potential and can contribute to some variations in cobalt extraction. The reactions responsible for dithionate formation and the important reactions for metal dissolution and recovery are as follows [18]:

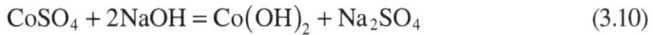

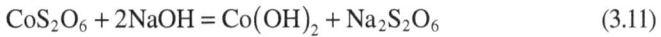

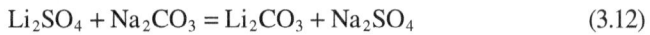

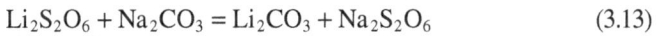

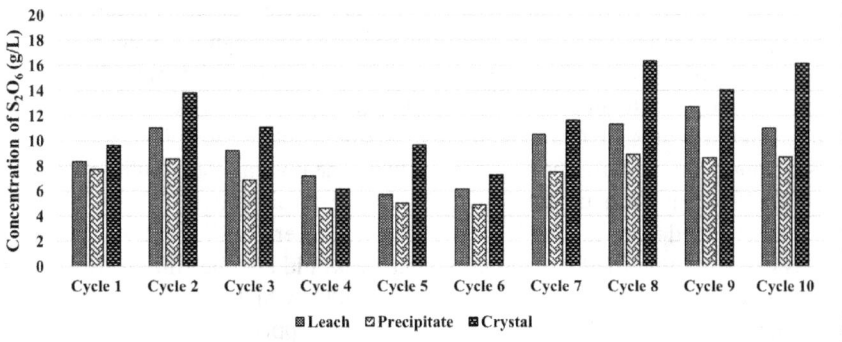

$$2LiCoO_2 + SO_2 + 2H_2SO_4 = Li_2SO_4 + 2CoSO_4 + 2H_2O \tag{3.8}$$

$$2LiCoO_2 + 4SO_2 + 2H_2SO_4 = Li_2S_2O_6 + 2CoS_2O_6 + 2H_2O \tag{3.9}$$

$$CoSO_4 + 2NaOH = Co(OH)_2 + Na_2SO_4 \tag{3.10}$$

$$CoS_2O_6 + 2NaOH = Co(OH)_2 + Na_2S_2O_6 \tag{3.11}$$

$$Li_2SO_4 + Na_2CO_3 = Li_2CO_3 + Na_2SO_4 \tag{3.12}$$

$$Li_2S_2O_6 + Na_2CO_3 = Li_2CO_3 + Na_2S_2O_6 \tag{3.13}$$

FIGURE 3.17 Dithionate concentration profile in the first locked-cycled campaign.

TABLE 3.11

Sodium Sulfate Concentrations (Calculated) by Cycle in the First Locked-Cycle Campaign

	Na_2SO_4 (g/L) estimated based on Na in solution			
Cycle#	Recycled into Leach	Leach Filtrate	Precipitation Filtrate	Crystallization Filtrate
1			242	126
2		113	346	169
3	102	106	364	165
4	119	122	342	172
5	141	141	402	153
6	81	78	325	138
7	81	78	385	183
8	151	156	425	211
9	153	131	352	151
10	151	143	398	181

Table 3.11 shows the concentrations of sodium sulfate through the process for each cycle, calculated based on the sulfate and sodium concentrations measured in each solution. By comparing the sodium sulfate concentrations in the recycled solution with those in the leach filtrate, it can be seen that they are very close, suggesting that there is little or no formation of sulfate crystals in the leach during this locked-cycle campaign.

The second locked-cycled campaign repeated the leach conditions of the first (also using LCO cathode active material as feed), followed by combined precipitation of Co and Li as hydroxides and carbonates. Figure 3.18 shows the schematic of the second locked-cycle campaign.

As with the first campaign, the final filtrate was treated by chilling for sodium sulfate removal prior to being recycled back to the next cycle's leach. For the second campaign, a limited re-pulp wash was added for the recovered sodium sulfate crystals and this wash solution was added to the recycled solution, along with other wash streams, which were also recycled in the first campaign. As there were higher volume of wash water in the second campaign, to control the water balance, the combined recycle stream was evaporated to the target volume before the next leach cycle, whereas in the first campaign only the dilute cobalt precipitate wash solution was evaporated to control the solution volume. The optional of nanofilter showed in Figure 3.18 was not executed during the locked-cycle campaign.

Since the feed material and leaching conditions were similar in both campaigns, it was expected that the leaching performance would also be similar. Initially, Co extraction was near 100% and Li was above 90%, which was similar to the first campaign, but after Cycle 2 the cobalt extraction dropped significantly, as shown in Figure 3.19. The lost cobalt could be recovered from the leach residue by washing with large amounts of water, which is shown in Figure 3.19 as the "Leach Wash."

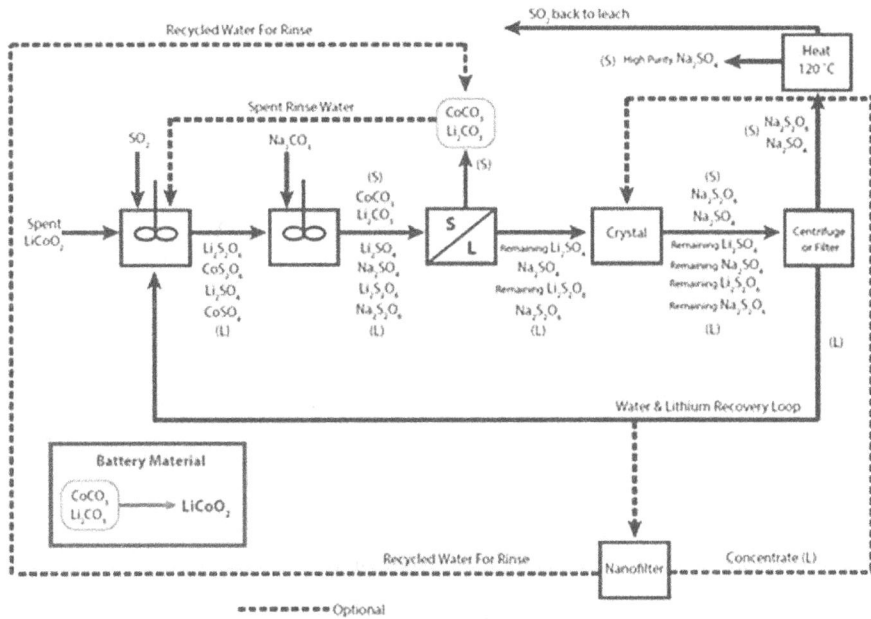

FIGURE 3.18 Second locked-cycle flowsheet with simultaneous recovery of metals.

Following leaching, the solubilized metals were recovered with sodium hydroxide and carbonate as shown in Figure 3.20. Overall, the recovery of lithium was risen rapidly and was good through all cycles except the first one. Lithium recoveries were likely slightly overestimated in these tests due to the requirement for addition

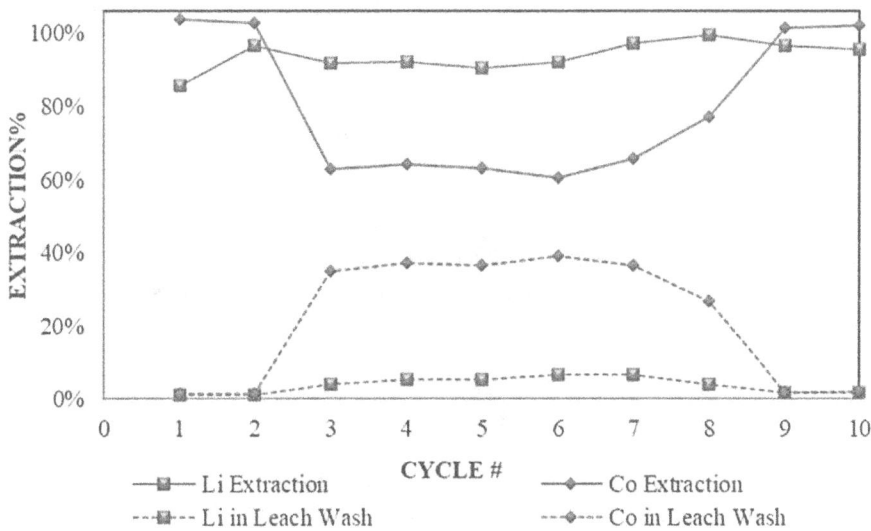

FIGURE 3.19 Second campaign – leach results.

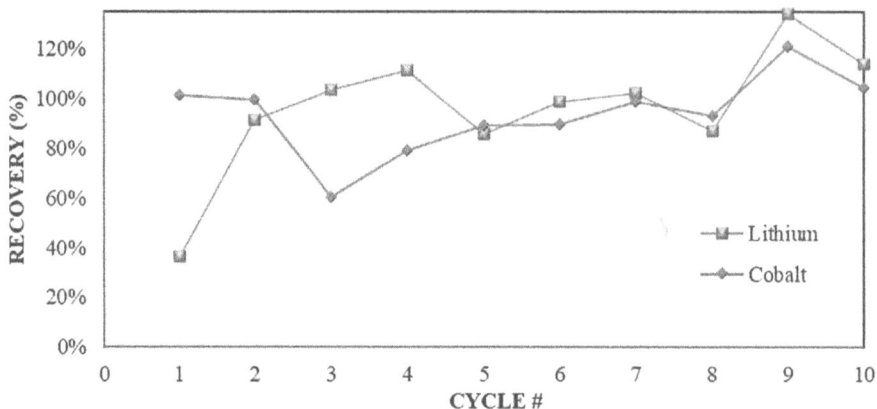

FIGURE 3.20   Metal recovery in the second campaign, utilizing single stage precipitation.

of a fresh saturated Li_2CO_3 solution to the second wash of the co-precipitated filter cake from each cycle. A theoretical saturation of lithium was used in accounting for this addition, but temperature variations likely resulted in higher concentrations. Despite this uncertainty, the overall lithium recovery appeared to be comparable to that achieved in the first campaign. Cobalt recovery was affected in early cycles by losses to crystal formation in the leach but increased in later cycles as this material was recycled back into the process. The excess recoveries for both metals in Cycle 9 reflected recovery of losses from previous cycles after decreasing the pulp density in the later cycles.

As in the first locked-cycle campaign, the second campaign showed consistent extraction of the metals from LCO material. The main issue was the entrapment of leached metal into sulfate crystals formed during leaching. Another indication of crystal formation was the higher mass of leached residues for Cycle 3 through Cycle 7 shown in Table 3.12.

Figures 3.21 and 3.22 show sulfate and dithionate levels in the leach, precipitation filtrates after sulfate crystal removal. The sulfate concentrations in leach filtrates are consistently higher (> 300 g/L SO_4), when compared to the first campaign (Figure 3.16). In addition, it is noteworthy that high levels of dithionate in Cycles 3 and 4 coincide with the lowest extractions of Co in the leach stage.

Table 3.13 presents the calculated sodium sulfate levels in the principal process streams for each cycle. The sodium sulfate in the recycled solution is over 220 g/L while the leach filtrate is significantly lower, with concentrations between 145 and 164 g/L. This confirms the formation of sodium sulfate crystals during leaching. The high sulfate concentrations in the recycle stream are due to the evaporation of the combined process solutions to control the water balance.

As simple as sodium sulfate chemistry might seem, its crystallization behavior can be a complex phenomenon. As many as three phases may form under certain condition: stable thenardite (Na_2SO_4) and mirabilite ($Na_2SO_4 \cdot 10H_2O$) as well as a metastable heptahydrate, $(Na_2SO_4 \cdot 7H_2O)$. In a system involving more than a single solid phase, metastable $Na_2SO_4 \cdot 7H_2O$ may be expected to crystallise and it is not

TABLE 3.12
Second Locked-Cycle Campaign – Feed and Products

	Leaching Stage				Hydroxide / Carbonate Product					Sulfate Crystals		
	Feed Li	Feed Co	Recycle Li	Residue	Mass	Li	Li Rec.	Co	Co Rec.	Mass	Li	Co
Cycle ID	g	g	g	g	g	%	%	%	%	g	g/Kg	mg/Kg
Cycle 1	6.80	58.4		0.51	108	2.88	36	54.5	101	109	0.23	5.0
Cycle 2	6.80	58.4	4.25	0.73	129	5.46	91	44.9	100	201	0.18	5.0
Cycle 3	6.80	58.4	3.94	10.09	98	7.84	103	35.9	60	271	0.24	5.0
Cycle 4	6.80	58.4	4.87	2.00	121	6.90	112	38.1	79	262	0.24	5.0
Cycle 5	6.80	58.4	4.13	2.18	122	5.45	86	42.4	89	282	0.24	5.0
Cycle 6	6.80	58.4	4.25	3.85	122	5.93	99	42.5	90	211	0.67	6.0
Cycle 7	6.80	58.4	4.43	3.42	143	5.17	102	39.9	99	221	0.35	5.0
Cycle 8	6.80	58.4	3.67	0.69	161	4.01	87	33.6	93	255	0.34	8.2
Cycle 9	6.80	58.4	3.98	1.27	168	5.99	134	41.8	121	393	1.59	5.0
Cycle10	6.80	58.4	5.15	1.48	147	5.79	114	41.3	105	314	0.34	5.0

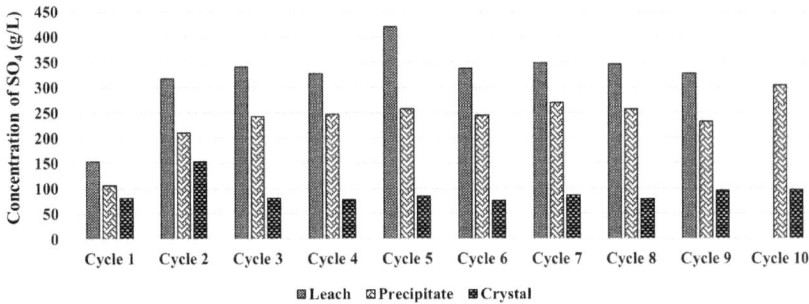

FIGURE 3.21 Sulfate concentration profile in the second locked-cycled campaign.

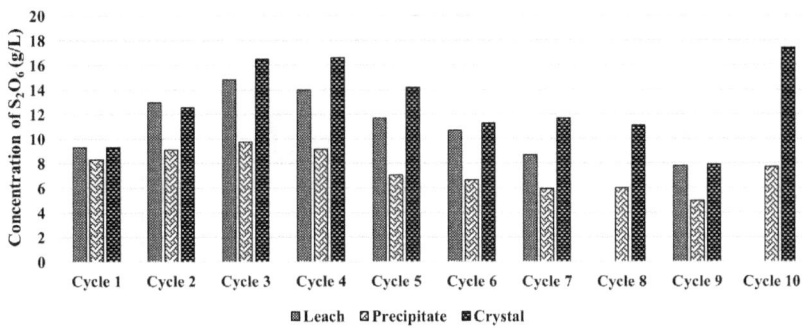

FIGURE 3.22 Dithionate concentration profile in the second locked-cycled campaign.

TABLE 3.13

Sodium Sulfate Concentration (Calculated) Per Cycle in the Second Locked-Cycle Campaign

	Na_2SO_4 (g/L) estimated based on Na in solution			
Cycle#	Recycled into Leach	Leach Filtrate	Precipitation Filtrate	Crystallization Filtrate
1		0.1	188	102
2	161	163	342	203
3	228		362	95
4	227	164	276	94
5	226	145	375	89
6	229	149	354	78
7	224	161	385	91
8	214	179	357	71
9	183		331	101
10	203	195	422	86

always reproducible [19]. The residence time is also an important factor in the formation of precipitates. Considerable care must therefore be taken to ensure solution concentrations are well below the saturation levels. High levels of hydration with sodium sulfate crystals can also increase crystallization by removing water from the system. Figure 3.23 shows the solubility curves for all three phases of sodium sulfate and it shows that sodium sulfate levels over 200 g/L, as observed in the second locked-cycle campaign, could have generated crystals in the leach solution.

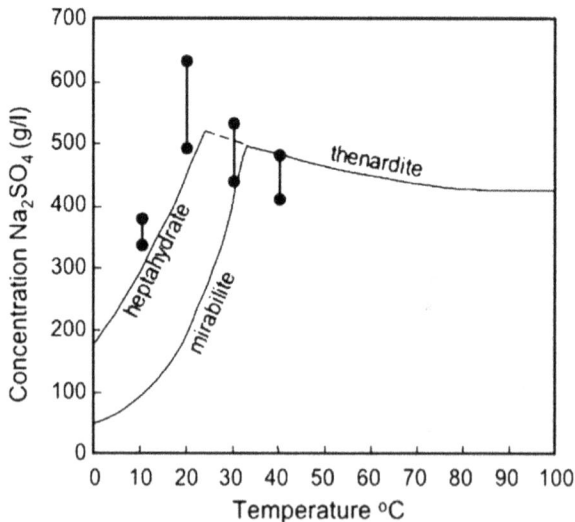

FIGURE 3.23 Solubility curves for thenardite, mirabilite, and heptahydrate phases [19].

The experimental evidence collected in both campaigns showed that both simultaneous and sequential precipitation methods were successful, recovering over 90% of cobalt and lithium from LCO active cathode active material, although the grades of cobalt and lithium in the products were higher (60% for Co and 18% for Li, as listed in Table 3.5), using sequential precipitation with hydroxide for Co and carbonate for Li compared with the single combined product (about 40% Co and 6% Li average for 10 cycles, as listed in Table 3.7). The control of sulfate levels in the recycle stream with crystallization should be an important addition to the earlier bench optimization study and it could prove to be a necessary step to prevent losses associated with sulfate crystal formation due to circulating sulfate loads in recycle streams. In addition to sulfate, the dithionate generated in the leach could increase with recycle but did not reach saturation levels. When dithionate concentration reached saturation, it could also be controlled by crystallization along with sulfate [18].

The principal difference between the first locked-cycle campaign and the second locked-cycle campaign as shown in Figures 3.4 and 3.9 lies in the metal precipitation stage. The flowsheet in Figure 3.4 recovers cobalt separately from lithium, while the flowsheet in Figure 3.9 uses a single stage to produce a combined Co/Li product. Both flowsheets include the crystallization stage and recycled streams (remaining lithium and sodium) to minimize water consumption and maximize lithium recovery in the system. On a larger scale, dilute wash recycle streams can be treated by nanofiltration to produce clean water for washing products and to control the overall water balance rather than using evaporation. The concentration from nanofiltration may be recycled back to the crystallizer to maximize sodium sulfate removal.

In summary, the RecycLiCo™ process with sulfuric acid/sulfur dioxide leach system to recover metals from NMC, NCA, LMO, and LCO lithium cathode active materials is developed and validated. Excellent (>99%) extractions for all metals (Li, Co, Ni, and Mn) are achieved within 2 hours at 10% solids, 0.8 -1 M H_2SO_4, pH 1 to 1.5 and with sulfur dioxide targeting ORP in the range of 350–630 mV. Along with the extracted metals, the method generates sulfate and dithionate in leach solutions that are needed to be controlled in circulating loads, which are required for high lithium recovery.

Cobalt and nickel can be effectively precipitated from leach solutions by modifying the pH to 10 with NaOH, or with sodium sulfide after neutralizing to pH 5 with NaOH. At pH 10, a complete recovery of leached manganese is also achieved, with the most effective separation of manganese from cobalt and nickel obtained with sulfide precipitation. The optimal pH for selectively precipitating aluminum from leach solutions is 5.5. The lithium recovery is incomplete in batch experiments, but with process solution recycling in the locked-cycle campaigns, it is shown that lithium recovery can be greatly improved, from 30% or less in batch tests to levels exceeding 90% after establishing the circulating load. The cobalt recovery from LCO cathode active material in two locked-cycle semi-continuous campaigns is consistent and essentially completed in each cycle. However, when sodium sulfate concentrations in the recycle stream exceed 200 g/L, a portion of the leached cobalt is entrained by sodium sulfate crystallization during the leach. Sodium sulfate crystallization is the result of sulfate ions from the leach acid combining with the sodium ions from neutralizing the acid. Although different cathode active materials might end up consuming different amounts of acid, sodium sulfate crystallization will be problematic for all cathode active materials as sulfate ion and sodium ion concentrations are built up during the locked-cycle close-loop process. The sodium sulfate crystals have to be solubilized with additional water in order to recover the entrapped cobalt. Therefore, one of the significant conclusions from these campaigns is the importance of optimizing the sodium sulfate crystallization stage prior to recycling the process solutions back to the leach stage when sulfuric acid and sodium-based reagents are being utilized in the recycling of cathode active materials. On the other hand, the two locked-cycle campaigns demonstrate an environmental friendlier closed-loop hydrometallurgical process that reduces discharges and minimizes environmental impact for recycling LIB cathode active materials.

In summary, immersing cathode active material waste into weak H_2SO_4 can separate aluminum foil and cathode active material. For NCA cathode, the weight ratio of required 98 wt% H_2SO_4 to NCA cathode waste is around 0.95; for NMC cathode, the weight ratio of required 98 wt% H_2SO_4 to NMC cathode waste is around 0.79.

To leach Ni, Co, Al, Mn, and Li from NCA and NMC cathode waste, >99% leaching efficiencies of Ni, Co, Mn, and Li were demonstrated by using H_2SO_4/SO_2. Based on the experiments conducted, the best conditions for H_2SO_4/SO_2 leach is to control pH at 1.5, ORP at 550 mV, and retention time of 7 hours with 20 wt% pulp density. Leach with H_2SO_4/SO_2 generates approximately 6.5 g/L of $S_2O_6^{2-}$, which affects the leach solution oxidation-reductive potential.

The concentration of $S_2O_6^{2-}$ can potentially build up, which might impact the ORP control, leaching efficiency, and SO_2 consumption, the downstream metal extraction, and Li recovery process.

Compared to leaching with using H_2SO_4/H_2O_2, leaching with H_2SO_4/SO_2 results 15% less total leachate volume, which reduces the degree of water management needed in the downstream impurity removal, metal extraction, and lithium recovery.

3.6   SUMMARY OF RECYCLICO™ PROCESS

The cathode materials for LIBs are layered crystals of lithium intercalated metal oxides. The metals in modern high energy cathodes are nickel based with a mix of cobalt, manganese, and aluminum. Example chemical formulations include $LiNi_{0.8}Mn_{0.1}Co_{0.1}O_2$ (NMC-811) and $LiNi_{0.8}Co_{0.15}Al_{0.05}O_2$ (NCA). The metal mixture is in the 3+ valent state in a fully discharged battery and in the 4+ valent state in a fully charged battery. Higher valent metal oxides (> 2+ valent), typically have low solubility in sulfuric acid solutions and thus require a reducing agent for dissolution.

RecycLiCo™ Battery Materials Inc. in partnership with Kemetco Research Inc. (Vancouver, Canada), developed US Patent 10246343 employing SO_2 as a reducing agent to extract the valuable metals from spent cathode materials followed by a closed loop process to produce high purity metal oxides and lithium carbonate. These compounds could be further processed to produce cathode precursors and lithium hydroxide if desired. US Patent 10308523 was further developed to incorporate separation of aluminum foil from cathode active materials for treatment of cathode scraps, separation of carbon and graphite, and separation of fluorides from cathode active materials for treatment of black mass.

REFERENCES

1. Harris B. The future of Ni-Co-Cu processing in the age of lithium-ion batteries, ALTA Conference Proceedings (2019).
2. Zheng X., Zhu Z., Lin X., Zhang Y., He Y., Cao H., Sun Z., A mini-review on metal recycling from spent lithium ion batteries, Engineering, 4(3), 361–370 (2018).
3. Lee C.K., Rhee K., Reductive leaching of cathodic active materials from lithium ion battery wastes, Hydrometallurgy, 1–3(68), 5–10 (2003).
4. Xu J., Thomas H.R., Francis R., Lum W., Wang R., Liang J., A review of processes and technologies for the recycling of lithium-ion secondary batteries, Journal of Power Sources, 177, 512–527 (2008).
5. Larouche F., Tedjar F., Amouzegar K., Houlachi G., Bouchard P., Demopoulos G.P., Zaghib K. Progress and status of hydrometallurgical and direct recycling of Li-ion batteries and beyond, Materials, 13, 801, doi:10.3390/ma13030801 (2020).
6. Ghassa S., Farzanegan A., Gharabaghi M., Abdollahi H., The reductive leaching of waste lithium ion batteries in presence of iron ions: Process optimization and kinetics modelling, Journal of Cleaner Production, 262, 121312 (2020).
7. Porvali A., Shukla S., Lundstrom M., Low-acid leaching of lithium-ion battery active materials in Fe-catalyzed Cu-H2SO4, Hydrometallurgy, 195, 105408 (2020).
8. Vieceli N., Nogueira C.A., Guimaraes C., Pereira M.F.C., Durao F.O., Margarino F., Hydrometallurgical recycling of lithium-ion batteries by reductive leaching with sodium metabisulphite, Waste Management, 71, 350–361 (2018).
9. Chagnes A., Pospiech B., A brief review on hydrometallurgical technologies for recycling spent lithium-ion batteries, Journal of Chemical Technology & Biotechnology, 88(7), 1191–1199 (2013).
10. Dhiman S., Gupta B., Partition studies on cobalt and recycling metals from waste Li-ion batteries via solvent extraction and chemical precipitation, Journal of Cleaner Production, 225, 820–832 (2019).
11. Monhemius A., Precipitation diagrams for metal hydroxides sulfides arsenates and phosphates, Transactions of the Institution of Mining and Metallurgy, 86, C202 (1977).
12. Vaughan J., Hawker W., White D. Chemical aspects of mixed Nickel-Cobalt hydroxide precipitation and refining, Conference Proceeding (2011).
13. Zubi G., Dufo-Lopez R., Carvalho M., Pasaoglu G., The lithium-ion battery: State of the art and future perspectives, Renewable and Sustainable Energy Reviews, 89, 292–308 (2018).
14. Zhou X., He W.-z, Li G.-m, Zhang X.-j, Huang J.-w, Zhu S.-g., Recycling of electrode materials from spent lithium-ion batteries, 2010 4th International Conference on Bioinformatics and Biomedical Engineering, pp. 1–4 (2010).
15. Wang M., Tan Q., Liu L., Li J., A low-toxicity and high-efficiency deep eutectic solvent for the separation of aluminum foil and cathode active materials from spent lithium-ion batteries, Journal of Hazardous Materials, 380, 120846 (2019).
16. Zhang X., Xie Y., Cao H., Nawaz F., Zhang Y., A novel process for recycling and resynthesizing $LiNi_{1/3}Co_{1/3}Mn_{1/3}O_2$ from the cathode scraps intended for lithium-ion batteries, Waste Management, 34, 1715–1724 (2014).
17. He L., Sun S., Song X., Yu J., Recovery of cathode active materials and Al from spent lithium-ion batteries by ultrasonic cleaning, Waste Management, 46, 523–528 (2015).
18. Chow N., Jung J.C., Nacu A.M., Warkentin D.D. 'Processing of cobaltous sulfate/dithionate liquors derived from cobalt resource' Patent US 10,246,343 B2 (2019).
19. Genkinger S., Putnis A., Crystallisation of sodium sulfate: Supersaturation and metastable phases, Environmental Geology, 52(2), 329–337 (2007).

4 Closed-Loop Hydrometallurgical Processing Using LiOH for Coprecipitation and Electrodialysis for LiOH Regeneration

Joey Jung and Jiujun Zhang

CONTENTS

DOI: 10.1201/9781003269205-4

4.1　INTRODUCTION

Hydrometallurgical recycling of spent LIB cathode active materials generally runs below 100°C and involves leaching, impurity removal, metal extraction, and lithium (Li) recovery. During leaching, metal ions such as Ni^{2+}, Co^{2+}, Mn^{2+}, Al^{3+}, Fe^{2+}, and Li^+ in the spent cathode active materials dissolve into the leach solution. The impurity removal process removes the unwanted metal ions such as Al^{3+} and Fe^{2+}, as well as unleached solids like carbon and graphite powder. Metal extraction and lithium recovery can recover Ni^{2+}, Co^{2+}, Mn^{2+}, and Li^+ from the purified leach solution. The recovered Ni, Co, Mn, and Li can be used to produce new LIB cathode active materials [1]. Impurity removal and metal extraction are often conducted using sodium hydroxide (NaOH) [2–8] to precipitate metal ions as metal hydroxides or using solvent extraction [9–14] to extract individual metal ions. Lithium recovery is often conducted by adding sodium carbonate (Na_2CO_3) into the lithium-enriched solution to precipitate lithium as lithium carbonate (Li_2CO_3) [7, 15–18].

However, using NaOH for impurity removal and metal extraction together with Na_2CO_3 addition for lithium recovery can result in a solution enriched with Na^+ ions. The sodium-containing solution must then be treated before discharge or reuse, further complicating the hydrometallurgical recycling process and generating wastewater.

To minimize the impact of sodium ions on hydrometallurgical recycling of the spent LIB cathode active materials, lithium compounds such as lithium hydroxide (LiOH) or lithium carbonate (Li_2CO_3) can replace NaOH for impurity removal and metal extraction. The consumed lithium can be recovered using electrodialysis, creating a closed-loop process.

During impurity removal, impurities such as Al^{3+}, Cu^{2+} and Fe^{2+} are precipitated by using LiOH to adjust solution pH. Metal extractions of Ni, Co, and Mn are achieved by adding LiOH or Li_2CO_3 to further increase solution pH, leading to precipitation of the high-purity mixed metal hydroxide/carbonate. The decision to incorporate either LiOH or Li_2CO_3 for pH adjustment is nontrivial, as different metal recoveries can result. Using LiOH generates a superior yield of metal recovered with the drawback of worse water balance. On the other hand, when using Li_2CO_3, the hydrolyzing reaction of metal carbonate can result in the formation of free carbonic acid and limit the solution pH to around 8, regardless of the Li_2CO_3 dosage.

Figure 4.1 shows the Pourbaix (E-pH) diagrams for Al, Ni, Co, Mn, and Li [19]. At pH <4, Al^{3+}, Ni^{2+}, Co^{2+}, Mn^{2+} and Li^+ are soluble and remain in solution. As shown in Figure 4.1(a), Al^{3+} starts to precipitate as $Al(OH)_3$ once the solution pH is raised over 3.5. On the other hand, Ni^{2+}, Co^{2+}, Mn^{2+}, and Li^+ remain in the solution according to Figure 4.1(b), (c), (d), and (e). Furthermore, based on the Pourbaix diagrams, when the pH is gradually raised above 6.5, Co^{2+} will start to precipitate as cobalt hydroxide [$Co(OH)_2$]; above pH 8.0, Ni^{2+} will start to precipitate as nickel hydroxide [$Ni(OH)_2$]; above pH 8.3, Mn^{2+} will start to precipitate as manganese hydroxide [$Mn(OH)_2$]; and Li^+ will remain in solution in spite of the increasing pH. As such, carefully adjusting the solution pH allows Al, Ni, Co, Mn, and Li to be selectively separated. The chemical reactions of pH adjustment with LiOH are shown below, where aluminum requires 6 mol of LiOH (4.1) and this reaction produces lithium sulfate and aluminum hydroxide. Similarly, the hydrolysis reactions

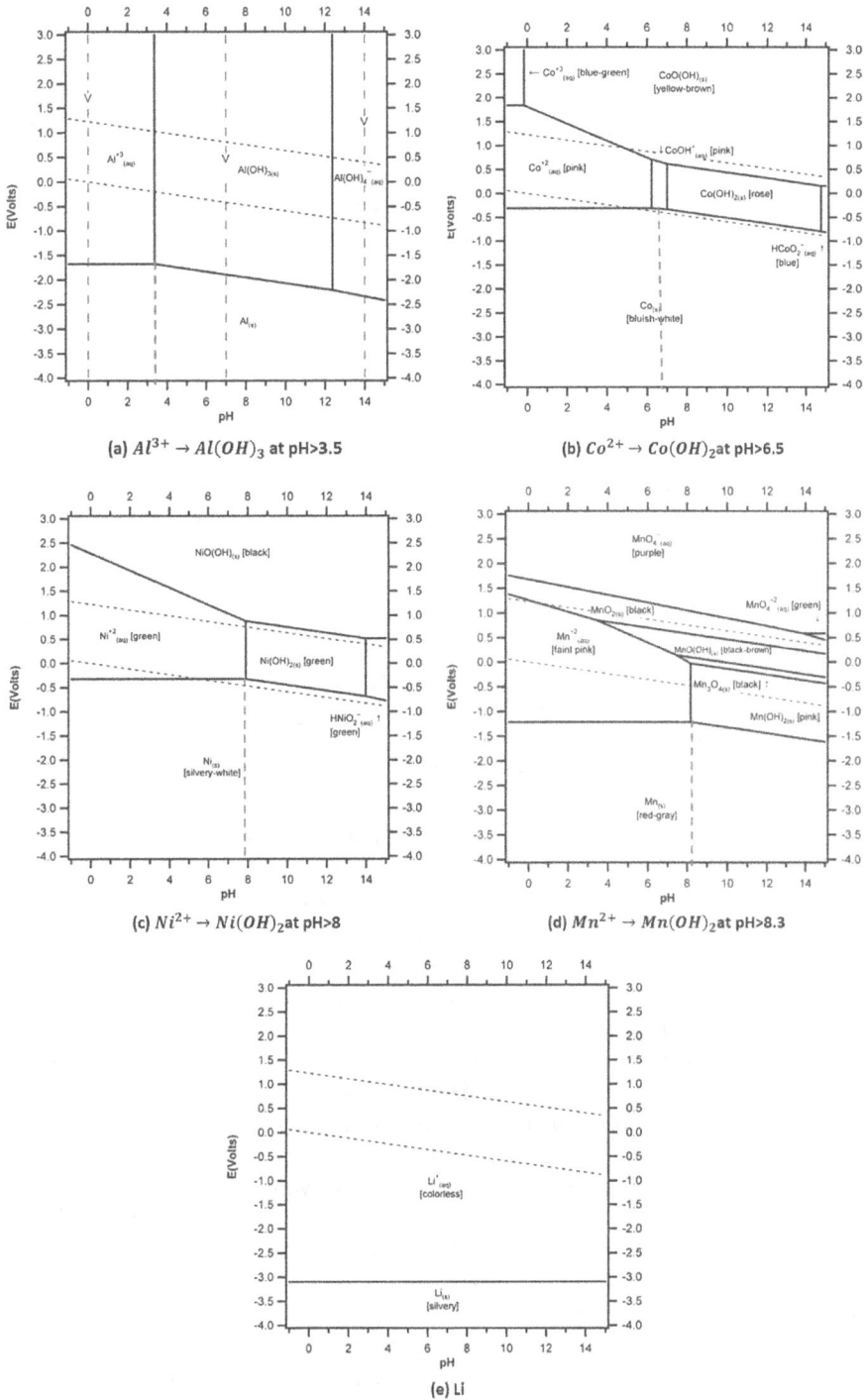

FIGURE 4.1 Pourbaix (E-pH) diagrams for (a) Al, (b) Ni, (c) Co, (d) Mn, (e) Li [19].

with other metals can produce metal hydroxides or oxides and lithium sulfate (4.2), (4.3), and (4.4):

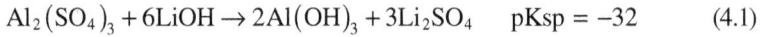

$$Al_2(SO_4)_3 + 6LiOH \rightarrow 2Al(OH)_3 + 3Li_2SO_4 \quad pKsp = -32 \quad (4.1)$$

$$NiSO_4 + 2LiOH \rightarrow Ni(OH)_2 + Li_2SO_4 \quad pKsp = -14.5 \quad (4.2)$$

$$CoSO_4 + 2LiOH \rightarrow Co(OH)_2 + Li_2SO_4 \quad pKsp = -15.3 \quad (4.3)$$

$$MnSO_4 + 2LiOH \rightarrow Mn(OH)_2 + Li_2SO_4 \quad pKsp = -12.7 \quad (4.4)$$

Figure 4.2 (a) shows the Al^{3+} concentration over time demonstrated in batch impurity removal experiments. Al^{3+} starts to precipitate out of solution at a relatively low pH of between 4.0–5.5 while the other metals remain in solution. Figure 4.2 (b) graphs the pKsp trends of Al^{3+}, Ni^{2+}, Co^{2+}, and Mn^{2+} together with the experimental data. Based on the pKsp, Al^{3+} should precipitate first, followed by Ni^{2+}, Co^{2+}, and Mn^{2+}. However, the experimental data shows that only Al^{3+} is precipitated out separately, whereas Ni^{2+}, Co^{2+}, and Mn^{2+} are coprecipitated.

Once the metals are precipitated as metal hydroxides, the filtrate contains mainly lithium sulfate (Li_2SO_4). LiOH and H_2SO_4 can be regenerated from Li_2SO_4 with electrodialysis through a three-compartment membrane electrodialysis cell. During electrodialysis, Li^+ migrates from central compartment towards the cathode compartment and SO_4^{2-} migrates from the central compartment toward the anode compartment, under the influence of an electric field. In the cathode compartment, water is reduced and OH^- and H_2 are produced. Li^+ migrates across the membrane into the cathode compartment to neutralize the charge, and then LiOH is formed. In the anode compartment, water is oxidized, O_2 and H^+ are produced. The SO_4^{2-} ions diffuse from the central compartment through the membrane to the anolyte and join with the protons generated at the anode and form H_2SO_4.

A schematic of the three-compartment electrodialysis cell is presented in Figure 4.3.

This approach enabled the recycling of a portion of the generated LiOH back to the impurity removal step while the majority of the LiOH crystallized to generate lithium hydroxide monohydrate crystal. The regenerated H_2SO_4 and the remaining Li_2SO_4 can be recycled back into the acidic leach, as depicted in Figure 4.4.

Hydrometallurgical recycling adapting (1) lithium hydroxide for impurity removal and coprecipitation, and (2) electrodialysis for regenerating H_2SO_4 and LiOH can create a closed-loop recycling process for spent/waste LIB cathode active materials, as shown in Figure 4.4. The regenerated LiOH, H_2SO_4, and process water can be reused in the recycling process. The recovered metal hydroxide and a portion of the lithium hydroxide can be used to generate new cathode active materials. The development of using LiOH for impurity removal, LiOH/Li_2CO_3 for metal coprecipitation, and electrodialysis to regenerate LiOH and H_2SO_4 is described below.

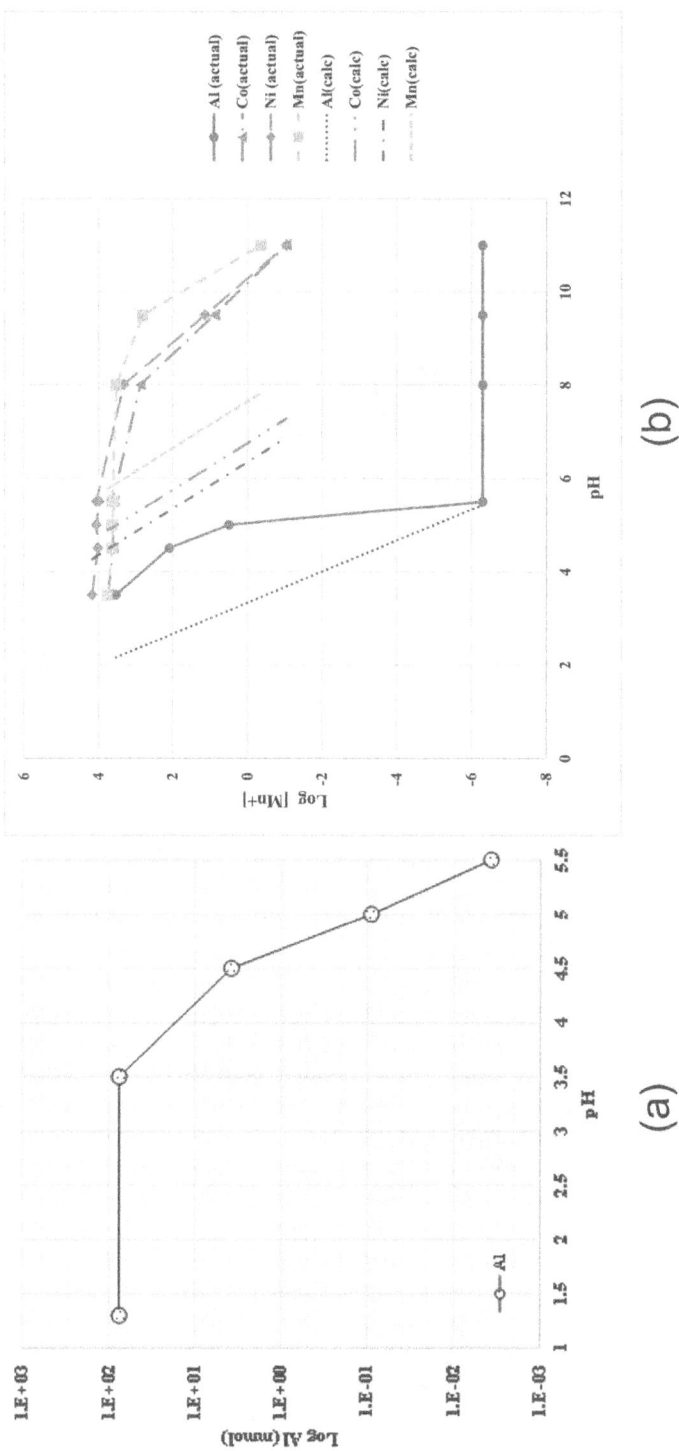

FIGURE 4.2 (a) Al stability versus pH and (b) pKa trend of Al^{3+}, Ni^{2+}, Co^{2+}, and Mn^{2+}.

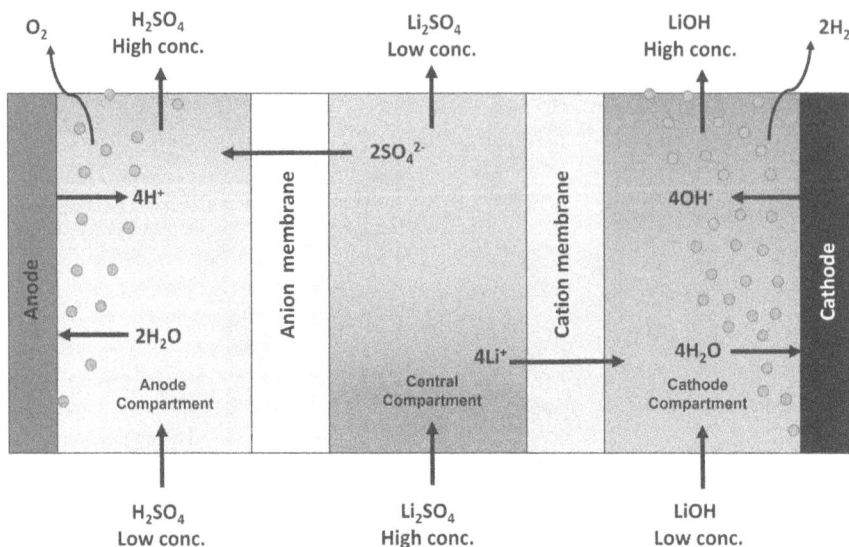

FIGURE 4.3 A three-compartment membrane electrodialysis cell for generating LiOH and H_2SO_4.

4.2   IMPURITY REMOVAL WITH LiOH

$LiNi_{1/3}Co_{1/3}Mn_{1/3}O_2$ (NMC) and $LiNi_{0.80}Co_{0.15}Al_{0.05}O_2$ (NCA) spent/waste cathodes were first soaked in 0.5 M H_2SO_4 to detach NMC and NCA cathode active materials from the aluminum foil by following the procedure shown in Chapter 3. Inductively Coupled Plasma (ICP) analysis data showed that when soaking NMC cathode in sulfuric acid, a small amount of aluminum foil could be leached when H_2SO_4 concentration was above 0.4 M, as shown in Table 4.1.

After removing the aluminum foil, the slurry containing detached NMC and NCA cathode active materials and aluminum ions (Al^{3+}) was further leached with H_2SO_4 and SO_2 to generate a leach solution containing Ni^{2+}, Co^{2+}, Mn^{2+}, Fe^{2+}, and Al^{3+} ions. The leach solution was then filtered to remove unleached solids and organics, such as residues, carbon, separator, and binder. Compared to leaching with H_2SO_4 and H_2O_2, leaching with H_2SO_4 and SO_2 resulted in a much smaller volume of leachate, which could reduce the degree of water management needed in impurity removal, metal extraction, and lithium recovery, thus providing a better opportunity to achieve a closed-loop process.

A series of impurity removal tests were carried out on the leachate in a 1-litre reactor using 1.0 M LiOH (prepared from analytical grade LiOH powder) to adjust the solution pH to 4.5 and 5.5 to precipitate impurities such as Al^{3+} and Fe^{2+}. Kinetic samples were taken hourly to observe changes in solution composition. At the end, the product was filtered and sampled for ICP analysis.

Since Al^{3+} was precipitated at a lower pH than the other metals, it should be easily separated from them. On the other hand, the analysis of precipitate obtained at pH 4.5

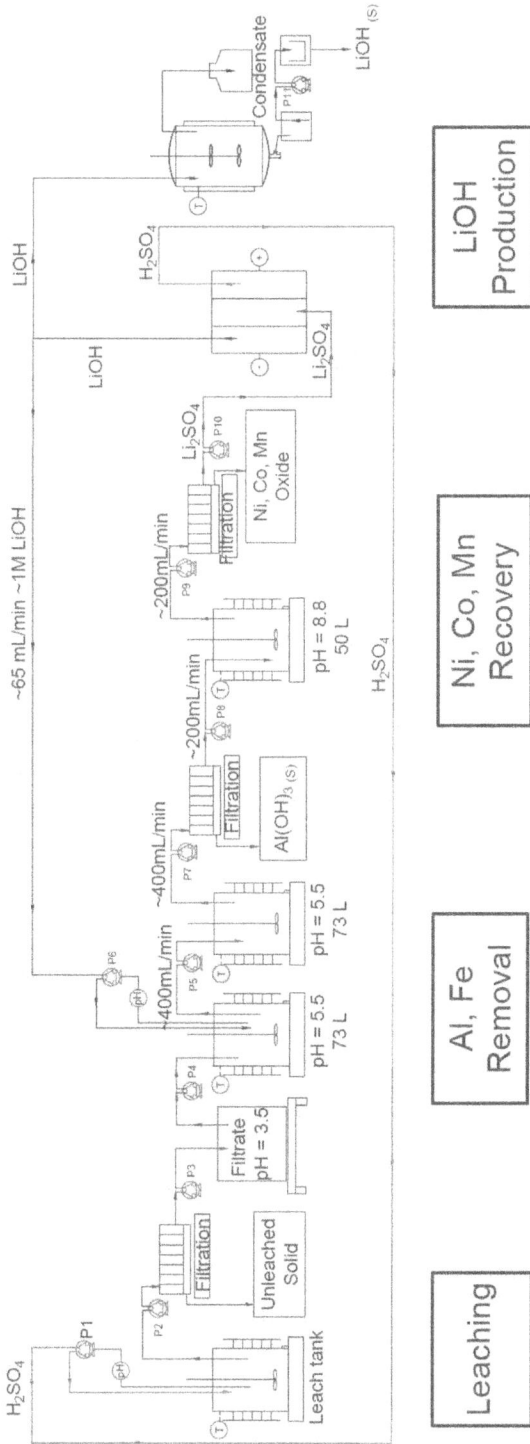

FIGURE 4.4 Conceptual closed-loop flowsheet and material flow employing LiOH and electrodialysis.

TABLE 4.1

Metal Concentrations in the Soaking H_2SO_4 Solution

Test#	H_2SO_4 Concentration (M)	Assay (mg/L)				
		Al	Li	Co	Mn	Ni
1	0.35	<2.0	2656	1744	2016	3266
2	0.41	44.3	2897	2231	2634	4497
3	0.60	84.7	4009	3677	4062	7164

identified that although the precipitate was mainly $Al(OH)_3$, it entrapped minor amount of Ni, Co, Mn, and Li. The presence of these metals was more pronounced at higher pHs. Table 4.2 (a) shows the LiOH dosages at different pHs and the resulting precipitate compositions.

As shown in Table 4.2 (a), very little precipitate can be collected at pH 3.5. The mass of precipitate collected from pH 4.5 to 5.5 varies between 17 to 20.8 g/L. The precipitate formed at pH 4.5 contains <1 g/kg Co and <8 g/kg Ni; at pH 5.5, the losses of these metals are increased to 16.1 g/kg Co and 69.8 g/kg Ni.

TABLE 4.2

(a) Summary of Al Precipitation Test with the pHs, (b) Semi-batch Reactor for Al Precipitation Test, (c) Estimated Composition of Al Precipitates Based on Elemental Analysis

(a)

Test ID	Time Hr	Reagents 1M LiOH		Final Solution						
		% Stoich	g/L	pH	Li mg/L	Co mg/L	Ni mg/L	Mn mg/L	Al mg/L	Fe mg/L
Al pH 3.5	22	39.1	222	3.6	4728	5525	14853	5681	3362	36
Al pH 4.5	22	118	574	4.5	5022	3989	10418	4242	125	1
Al pH 5.0	2.5	104	591	5.0	5362	4303	11307	4322	3	<1
Al pH 5.5	22	123	613	5.5	5254	3824	9700	4034	<1	<1

Test ID	Precipitate						
	Mass g/L Feed	Li g/Kg	Co g/Kg	Ni g/Kg	Mn g/Kg	Al g/Kg	Fe g/Kg
Al pH 3.5	<0.1						
Al pH 4.5	17.0	22.6	0.67	7.61	2.4	222	2.5
Al pH 5.0	18.4	1.02	3.29	13.0	2.5	233	2.6
Al pH 5.5	20.8	10.3	16.1	69.8	2.7	196	2.2

(Continued)

TABLE 4.2 *(Continued)*
(a) Summary of Al Precipitation Test with the pHs, (b) Semi-batch Reactor for Al Precipitation Test, (c) Estimated Composition of Al Precipitates Based on Elemental Analysis

(b)

Test ID	Feed L	Time hr	1M LiOH g/L	pH	Li mg/L	Co mg/L	Ni mg/L	Mn mg/L	Al mg/L	Fe mg/L
			NMC PP	**1.5**	**4.3**	**6.9**	**17.6**	**6.8**	**4.1**	**0.05**
Test 1 S-1	40	2	551	4.4	4.9	4.4	11.3	4.5	1.1	0.004
Test 1 S-2	54	2	62	5.5	5.0	3.8	9.9	4.1	0.2	<0.01
Test 2 S-1	40	4	561	4.5	4.8	4.1	10.5	4.3	0.5	0.002
Test 2 S-2	61	24	112	5.6	4.6	3.6	9.2	3.8	<0.01	<0.01
			NCA PP	**3.5**	**5.2**	**1.7**	**31.9**	**0.001**	**1.8**	**0.02**
Test 1 S-1	40	4	156	4.5	5.6	1.4	27.6	0.0002	0.23	<0.01
Test 1 S-2	43	4	45	5.5	5.3	1.3	23.6	0.0003	<0.01	<0.01

	Precipitate						
Test ID	Mass g/L Feed	Li g/Kg	Co g/Kg	Ni g/Kg	Mn g/Kg	Al g/Kg	Fe g/Kg
NMC PP							
Test 1 S-1	14.4	26.3	1.5	10.6	3.4	211	2.7
Test 1 S-2	3.2	22.0	2.1	20.8	3.3	215	1.0
Test 2 S-1	15.3	2talbe	2.1	10.8	3.8	194	2.6
Test 2 S-2	2.8	15.9	9.6	47.1	4.0	179	0.9
NCA PP							
Test 1 S-1	4.8	23.5	0.03	14.3	0.01	218	3.1
Test 1 S-2	1.6	6.38	10.2	200	0.01	144	0.7

(c)

Test ID/Solids		NMC Test 1 S-1	NMC Test 1 S-2	NMC Test 2 S-1	NMC Test 2 S-2	NCA Test 1 S-1	NCA Test 1 S-2
Feed	L	40	54	40	61	40	43
Solids Mass	g	575	171	613	175	193	70
$Al(OH)_3$	%	63.30	64.50	58.30	53.70	65.60	43.40
$Co(OH)_2$	%	0.24	0.33	0.34	1.51	0.004	1.60
$Ni(OH)_2$	%	1.67	3.29	1.71	7.45	2.26	31.60
$Mn(OH)_2$	%	0.56	0.53	0.62	0.65	0.002	0.002
$Fe(OH)_3$	%	0.51	0.20	0.50	0.17	0.60	0.13
LiOH	%	9.07	7.60	8.93	5.48	8.12	2.20
SO_4^{2-}	%	25.50	25.20	25.80	22.10	22.30	19.50
Sum	%	101	102	96.20	91.10	98.90	98.40

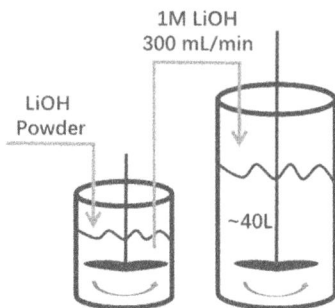

FIGURE 4.5   Semi-batch reactor.

Based on the bench test data, the conditions identified for impurity removal from the 1-litre bench test were validated using a larger-scale semi-batch type reactor setup in two stages, pH at 4.5 and pH at 5.5, respectively. A diagram of the reactor is shown in Figure 4.5.

The rationale for having two stages of impurity removal was to minimize Ni, Co, and Mn losses. Initially, the leach solution pH was raised to 4.5 by the addition of 1.0M LiOH solution, and most of the Al^{3+} precipitated as $Al(OH)_3$ and was removed by filtration. The remaining aqueous aluminum was removed in a second stage at pH 5.5. As shown in Table 4.2 (b), the precipitate collected after the second stage has a higher concentration of coprecipitated Co and Ni but is 20% by weight of the precipitate collected from the first stage. The precipitate collected after the second stage can be recycled back into the leaching step to redissolve and recover Co and Ni.

Furthermore, the precipitated $Al(OH)_3$ was observed to be instable when left unfiltered for an extended time. As shown in Table 4.2 (b), the solution pH of Test 1 S-1 is raised to 4.5 but is not filtered for 12 hours, causing the pH to drop from 4.5 to 4.0. Because of this pH change, the concentration of Al^{3+} in solution is increased from 70.4 mg/L to 1114 mg/L after 12 hours. A similar effect, to a lesser extent, is seen in Test 1 S-2. The pH drops from 5.5 to ~4.6 results in the Al^{3+} concentration risen from <2 mg/L to ~200 mg/L. According to Rubin et al. [20], this phenomenon was due to free sulfate ions (SO_4^{2-}) coagulating with $Al(OH)_3$ as it precipitated. Matijevic et al. [21] reported the formation of a polynuclear sulfatohydroxo-aluminum species such as $Al_8(SO_4)_5(OH)_{14}$ during aluminum hydrolysis in sulfate solution. These authors stated that the release of SO_4^{2-} contained in the aluminum hydroxide could produce acidity (pH drop). Hence, $Al(OH)_3$ precipitate containing sulfate should be filtered without delay to prevent redissolution.

ICP analysis of the $Al(OH)_3$ precipitate confirmed sulfur species present in the precipitate. Table 4.2 (c) shows the estimated chemical composition of $Al(OH)_3$ precipitate based on ICP analysis. The aluminum precipitate presumably contains two major phases: 43.4–65.6% $Al(OH)_3$ and 19.5–25.8% SO_4^{2-}. Precipitate obtained at pH 4.5 has minor inclusions of Co (<1%) and Ni (~2%). Precipitate obtained at pH 5.5 has up to 7.5% Ni (NMC leaching solution) and 31.6% Ni (NCA leaching solution). From a process design point of view, the precipitate generated at Stage 2 should be fed back into the leach to recover precipitated Ni.

FIGURE 4.6   SEM of $Al(OH)_3$ precipitate from NCA solution.

Figure 4.6 shows the micrograph of the $Al(OH)_3$ precipitate, where the particles appear very dense and have various sizes with sharp edges.

4.3   METAL COPRECIPITATION

The aluminum depleted, Co, Li, Mn, and Ni enriched leachate was treated with either 1 M LiOH, 4.6 M LiOH, or 30 wt% Li_2CO_3 slurry to examine the impact of using LiOH or Li_2CO_3 during metal coprecipitation. When treated with 1 M or 4.6 M LiOH, the solution pH was raised above 11 to coprecipitate Ni^{2+}, Co^{2+}, and Mn^{2+}. When treated with Li_2CO_3, a predetermined amount based on reaction stoichiometry was added.

As shown in Figure 4.7, when using LiOH to adjust solution pH, the pH could be adjusted to as high as 11, precipitating over 99.9% of Ni^{2+}, Co^{2+}, and Mn^{2+}. When adding Li_2CO_3 to coprecipitate, 97–98.5% of Ni^{2+}, Co^{2+}, and Mn^{2+} were coprecipitated at pH around 7.

The slightly lower metal recovery when using Li_2CO_3 might be due to using Li_2CO_3 can only adjust pH to ~8. Table 4.3 (a) presents reagent dosages and reaction product compositions. The carbonate method used 26.4 g Li_2CO_3 per litre of feed solution is compared to that used 533 g of 1 M LiOH per litre of feed solution. Compared to Li_2CO_3, using LiOH in the metal coprecipitation step will undoubtedly increase the total volume of downstream products needed to be treated at later stages of the recycling process. The amount of precipitate collected in Li_2CO_3 test is 36.7 grams per litre of feed solution and the precipitate contains 0.14% Li, ~10.9% Co, 27.9% Ni, and 11.5% Mn, respectively. The amount of precipitate collected in the

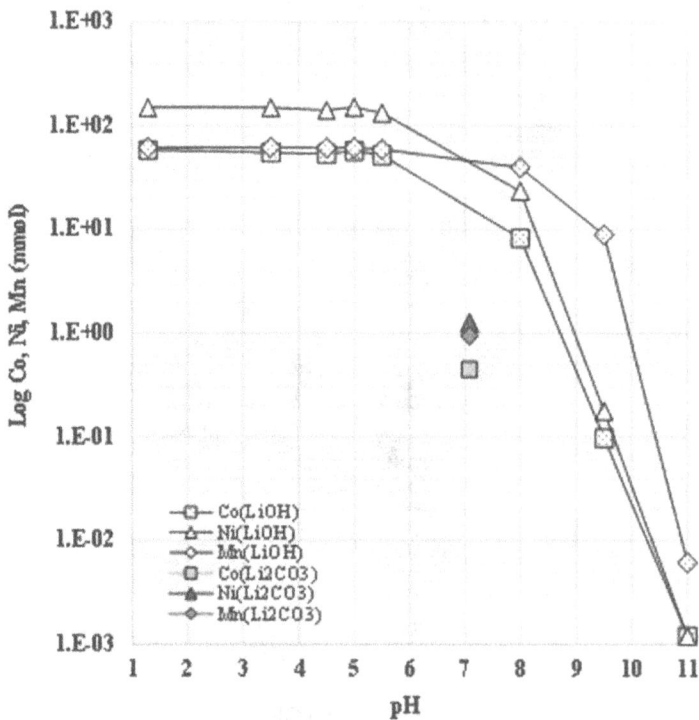

FIGURE 4.7   Co, Ni, and Mn stability versus pH.

LiOH test is 30.4 grams per litre of feed solution and contains ~0.05% Li, 12.2% Co, 29.9% Ni, and 13.3% Mn, respectively. Hence, metal coprecipitation using Li_2CO_3 can result in a relatively larger mass recovered but with a lower grade with respect to Co/Ni/Mn when compared to metal extraction using LiOH.

4.3.1   COPRECIPITATE THE METALS THROUGH Li_2CO_3 ADDITION

Li_2CO_3 is added to the impurities removed, Co, Li, Mn, and Ni enriched filtrate to coprecipitate metal carbonate/hydroxides. Comparing with LiOH, the system coprecipitate with Li_2CO_3 has less water in the process and simplifies the overall water balance. Table 4.3 (b) shows the coprecipitation results with various stoichiometric dosage of Li_2CO_3 based on the molar ratio of Co/Mn/Ni. T1 is 150% of stoichiometric requirements, T2 is 200%, and T3 is 280%.

The tests revealed that despite increasing dosages of lithium carbonate, the final pH of 8.8 remained unchanged. Mallya et al. [22] noticed a considerable buffering of the medium at pH 8.8. and explained the buffering was caused by the hydrolysis of precipitated nickel carbonate. In addition, they indicated that nearly 2.5 moles of alkali carbonate had to be added per mole of nickel salt to completely precipitate nickel. They estimated the formation of $NiCO_3.Ni(OH)_2$ composition at ~pH 8.

TABLE 4.3
(a) Batch Metal Extraction Experiment with LiOH or Li$_2$CO$_3$, (b) Semi-batch Metal Extraction Experiment with Li$_2$CO$_3$

(a)

	Reagents			Final Solution						
Test ID	Time hr	Li$_2$CO$_3$	1M LiOH g/L	pH	Li mg/L	Co mg/L	Ni mg/L	Mn mg/L	Al mg/L	Fe mg/L
BM1	2		533	11.0	5943	<0.5	<0.5	<0.1	<1	<1
BM2	4	26.4		7.08	10837	58.0	163.6	114.2	<1	<1

	Precipitate						
Test ID	Mass g/L Feed	Li g/Kg	Co g/Kg	Ni g/Kg	Mn g/Kg	Al g/Kg	Fe g/Kg
BM1	30.4	0.51	122.0	299.9	133.1	0.04	<0.01
BM2	36.7	1.37	109.1	279.4	114.8	0.12	<0.01

(b)

	Reagents				Final Solution						
Test ID	Feed L	Time hr	Li$_2$CO$_3$ g/L	% Stoich	pH	Li g/L	Co mg/L	Ni mg/L	Mn mg/L	Al mg/L	Fe mg/L
NMC T1 S-3	40.3	10	34.1	150	8.8	11.6	14.2	45.2	9.4	<1	<0.01
NMC T2 S-3	68.2	24	44.0	200	8.8	10.0	12.7	33.9	11.1	<1	<0.01
NCA T1 S-3	43.1	4	86.5	280	8.8	12.8	4.3	62.1	<1	<1	<0.01

	Precipitate						
Test ID	Solids g/L	Li g/Kg	Co g/Kg	Ni g/Kg	Mn g/Kg	Al g/Kg	Fe g/Kg
NMC T1 S-3	40	19	105	275	101	0.79	12.1
NMC T2 S-3	47	63	72	187	76	0.05	<0.01
NCA T1 S-3	91	86	13	258	0.005	0.04	<0.01

Davidson et al. [23] reported the formation of mineral zaratite NiCO$_3$.2Ni(OH)$_2$.4H$_2$O during nickel precipitation in carbonate medium. In our approach, to understand the composition of the precipitate obtained, samples from NMC T2-S3 and NCA T2-S3 precipitates were examined using X-ray diffraction and detected the presence of hydrated nickel carbonate such as Ni$_{12}$(CO$_3$)$_8$O(OH)$_6$(H$_2$O)$_6$, as shown in Figure 4.8.

The SEM micrograph of NMC T2-S3 sample is shown in Figure 4.9(a). The micrograph exhibits fine spherical precipitates which seem to aggregate on primary irregularly shaped particles. The primary irregularly particles appear to fill the internal

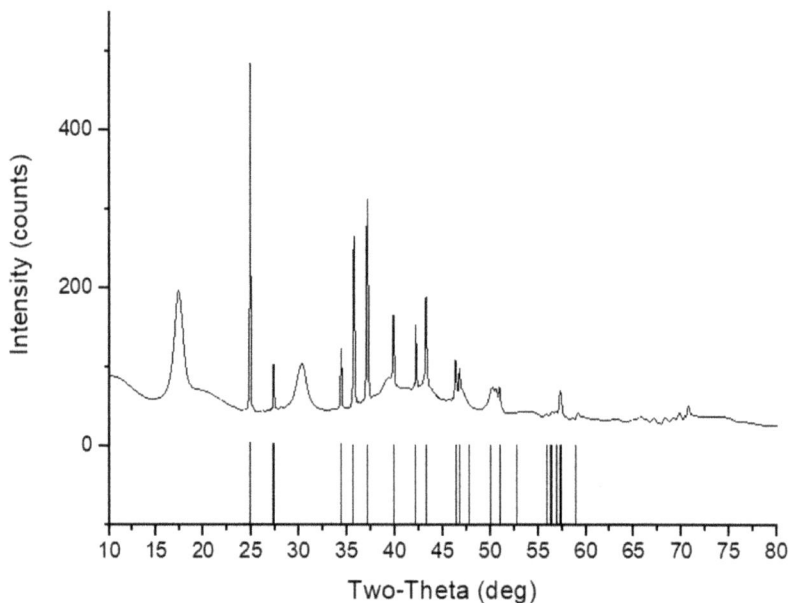

FIGURE 4.8   X-ray diffractogram of mixed metals carbonate precipitate.

void space within the secondary particle during the precipitation reaction, resulting in the growth of particles. However, the agglomerates seem loose and without definite shape, indicating that the particle size does not achieve a steady state. The agglomerates are thought to be caused by an inadequate reaction time for precipitation to reach equilibrium. The composition of the precipitates obtained from the metal extraction is given in Table 4.4.

After removing the metal carbonate/hydroxide, the filtrate mainly contains Li_2SO_4. Li_2CO_3 can be regenerated by adding Na_2CO_3 into Li_2SO_4 as shown in the following equation 4.5 thus a portion of the regenerated Li_2CO_3 can be circuited back.

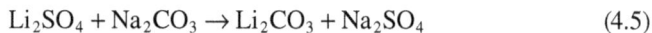

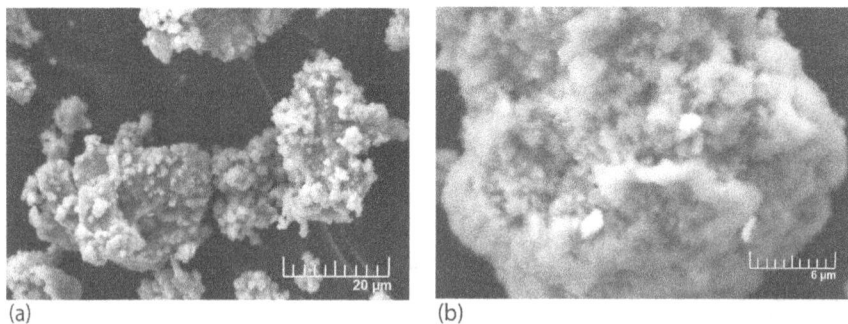

$$Li_2SO_4 + Na_2CO_3 \rightarrow Li_2CO_3 + Na_2SO_4 \tag{4.5}$$

(a)   (b)

FIGURE 4.9   (a) Micrograph of mixed metals carbonate precipitate from NMC solution, and (b) micrograph of mixed metal carbonate precipitate.

TABLE 4.4

Precipitates Compositions Produced from NMC Leach Solution

Test ID		NMC Test 1 S-3	NMC Test 2 S-3	NCA Test 1 S-3
Feed	L	40.3	68.2	43.1
Solid Mass	g	1622	3196	3942
Solid as	g/L Feed	40.3	46.8	91.4
$Al(OH)_3$	%	0.24	0.01	0.01
$CoCO_3$	%	21.2	14.5	2.65
$NiCO_3$	%	55.6	37.7	53.6
$MnCO_3$	%	21.1	15.9	0.0009
Li_2CO_3	%	10.1	33.8	43.6
SO_4^{2-}	%	0.98		0.09
Sum	%	109	102	99.9

Using Li_2CO_3 to coprecipitate the metals have the disadvantage of generating Na2SO4 that has very low market value and emitting CO_2 which contributes to the climate change and global warming.

4.3.2 COPRECIPITATE THE METALS THROUGH LiOH ADDITION

Using 1M LiOH to coprecipitate the metals although achieves 99.9% metal recovery but introduced large quantity of water into the process which challenges the water balance. The solubility of LiOH at 20°C is 128 g/L which is 5.3 M LiOH. To avoid reaching Li saturation, 4.6 M of LiOH can be used to remove impurities and coprecipitate the metals in order to improve the water balance of the overall recycling process.

Figure 4.10(a) shows the metal coprecipitation setup with pumping the Al^{3+} depleted, Ni, Co, Mn, Li enriched filtrate at pH 5.5 and 4.6 M LiOH solution concurrently into 2 reactors in series. The slurry was filtered immediately after the coprecipitation is finished. The ICP analysis of the filtrate shown in Table 4.5 shows that the filtrate contains majority Li with around 1 ppm of nickel and manganese. The recoveries of Co, Mn, and Ni from the filtrate are 99.9%.

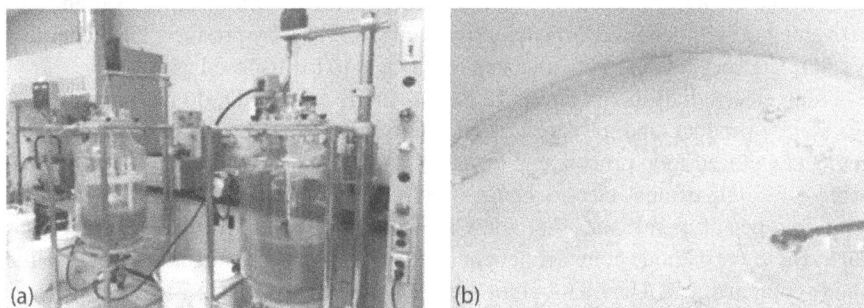

FIGURE 4.10 (a) Continuous pH adjustment experiment setup and (b) filtered Ni-Co carbonate.

TABLE 4.5
Summary of Precipitation Tests on NMC Leach Solution

			Reagents		Filtrate					
Test ID	Feed L	Time hr	4.6M LiOH L	pH	Li mg/L	Co mg/L	Ni mg/L	Mn mg/L	Al mg/L	Fe mg/L
T3	38	~1.5–2 hours	19.6	11.5	14869	<0.5	1.41	1.06	<1	<1

	Precipitate					
Test ID	Li mg/Kg	Co mg/Kg	Ni mg/Kg	Mn mg/Kg	Al mg/Kg	Fe mg/Kg
T3	16.4	81280	353271	105450	64.4	<5

The filtered precipitate is shown in Figure 4.10 (b) and the ICP analysis is listed in Table 4.5 which showed the precipitate contains some Li and Al impurities, which required further impurities removal. The precipitate appears to be a soft and fluffy aggregation with extremely small particles and is difficult to filter. The precipitate particle size is in the range of nanometers or a few micrometers. The small particle size is likely due to the fast chemical reaction, high nucleation rates, avoidance of agglomeration, and short growth periods [24]. To improve the particle size and density in order to improve filtration, chelating agents, such as ammonia and ammonium salts, can be used to assist the production of dense spherical particles [25].

4.4 ELECTRODIALYSIS TO REGENERATE H_2SO_4 AND LiOH

After filtration to separate the coprecipitate solid, the mother liquor is mainly Li_2SO_4 with minimum impurities. There are many methods to recover the Li from the Li_2SO_4 solution including (1) recover as Li_2SO_4 crystal via crystallization; (2) recover as Li_2CO_3 via adding Na_2CO_3; (3) recover as LiOH via adding NaOH; (4) recover as LiOH and regenerate H_2SO_4 via electrodialysis.

Li_2SO_4 crystal can be sold to Li producer for further processing to generate Li_2CO_3 or LiOH. Li_2SO_4 has a lower market value comparing to Li_2CO_3 and LiOH. Li_2CO_3 and LiOH generated by adding Na_2CO_3 and NaOH although has high market value but will generate large quantity of Na_2SO_4 as a by-product. The generated Na_2SO_4 has very low market value can often need to be disposed as waste.

Using electrodialysis to convert Li_2SO_4 to LiOH and H_2SO_4 do not generate any Na_2SO_4 by-product, and the regenerated LiOH and H_2SO_4 can be reused in the process and a closed-loop process can be established. The electrodialysis is carried out using a 3-compartment electrochemical cell. The compartments are separated by specific types of membranes that allow transportation of Li^+ from the central compartment to the cathode compartment and SO_4^{2-} from the central compartment to the anode compartment. The membranes are preconditioned with 0.5 M NaCl.

Electrodialysis of Li_2SO_4 is conducted by feeding three different solutions into the three-compartment cell. Li_2SO_4 solution is fed into the central compartment.

The central compartment is separated from the cathode compartment by a cation exchange membrane (Fumasep FKE-50). The central compartment is separated from the anode compartment by an anion exchange membrane (Fumasep FAA-3-PK-130). Dilute LiOH (0.1 M) is fed into the cathode compartment and dilute H_2SO_4 (0.05 M) is fed into the anode compartment.

During operation, hydrogen (H_2) is generated on the cathode and oxygen (O_2) is generated on the anode. H_2 and O_2 can be collected and used to power a fuel cell to provide electricity for the electrodialysis cell. H_2 can also be blended with natural gas to heat the facility. As H_2 is highly flammable, safety precautions are important and safety guidelines from the chlor-alkali and electrowinning industries, where H_2 is a common by-product, should be implemented.

To prove the concept, the electrodialysis experiments were conducted in a small cell with an electrode surface area of ~25 cm^2 at a controlled current of 2 A (750 A/m^2). The solutions were recirculated within each compartment. H_2SO_4 was circulated in the anode compartment, Li_2SO_4 was circulated in the central compartment, and LiOH is circulated in the cathode compartment. The current was fixed at 2 A current (750 A/m^2 current density). The test results showed that Li^+ and SO_4^{2-} ions were separated from central compartment and into relevant compartments.

After proofing the concept, further validation is conducted using a larger electrodialysis cell with an electrode surface area of 2455 cm^2, as shown in Figure 4.11.

With this large electrochemical cell, solutions are pumped in relevant compartments and run in the feed-and-bleed mode. Two different currents (34 Amp and 50 Amp) are applied in the tests corresponding to 138 A/m^2 and 204 A/m^2 of current densities.

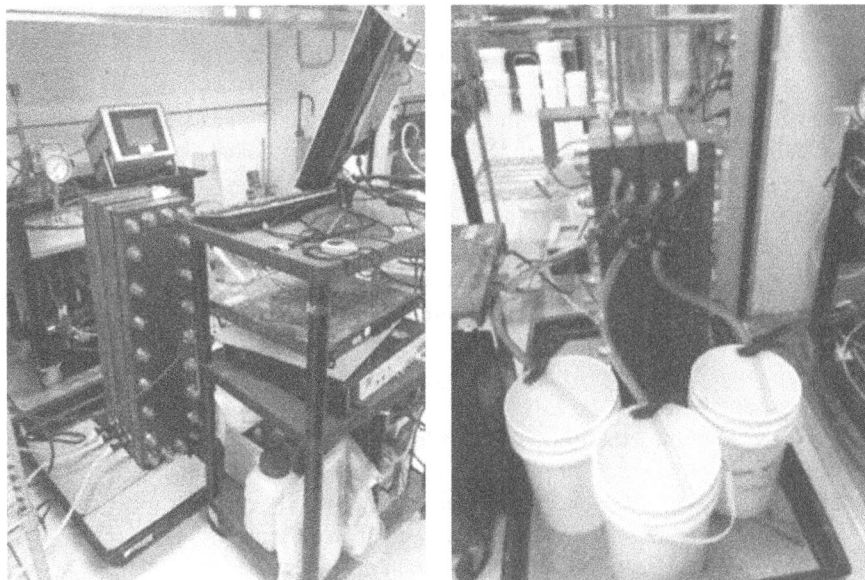

FIGURE 4.11   Large electrodialysis cell setup.

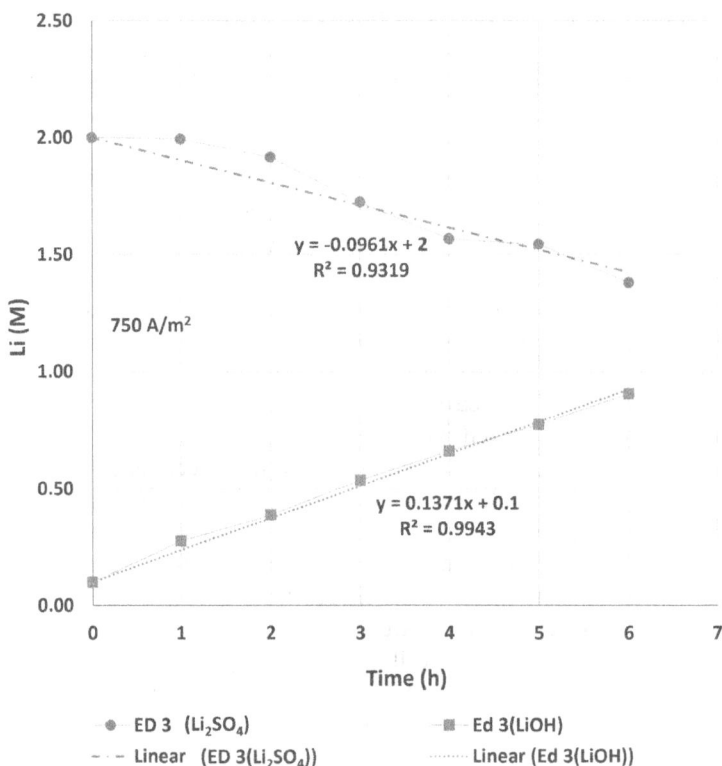

FIGURE 4.12   Li transport during electrodialysis based on synthetic Li_2SO_4 solution.

Figure 4.12 displays the rate of Li transfer with time. Ideally, the slopes for all lines are equal. In this case, the accumulation rate is ~0.127 M/h in catholyte and the depletion rate is 0.096 M/h in the central compartment. The difference is due to water being reduced at the electrodes, which affects the calculation. However, the rates are closer, and if the reaction time is extended to 7.5 hours, it can easily achieve the targeted 1 M LiOH, demonstrating that the three-compartment electrodialysis cell is capable of regenerating LiOH and H_2SO_4.

Operating conditions of LiOH and H_2SO_4 regeneration using a three-compartment electrodialysis cell with an electrode area of 2455 cm^2 (Figure 4.11) at two different current densities, 138 A/m^2 and 204 A/m^2 are shown in Table 4.6. The feed Li_2SO_4 solution is obtained from recycling waste NCA and NMC.

As shown in Table 4.6, regardless of the current applied, the voltage is decreased over the course of the experiments. The volume of the central compartment is about 7.46 liters and with the flow of ~2 L/h, resulting in a retention time of 3.5 hours.

Figure 4.13(a) and (b) show the Li^+ concentrations in the effluents (corresponding tests are ED-T6 and ED-T8) from the central compartment and the cathode compartment. In ED-T6, the test was run at a constant current density at 138 A/m^2 for six hours, which was approximately the first displacement of the cell volume. The concentration of Li^+ in the catholyte increased from 0.63 g/L to ~3.61 g/L. In ED-T8,

TABLE 4.6
Summary of Electrodialysis Tests

Test ID	Feed	Conditions Time hr	Current A	Voltage V
ED-T3	Synthetic	0	2.0	14.4
	Li$_2$SO$_4$ solution	6	2.0	10.7
ED-T6	NMC	0	34	19.7
	Composite	6	34	10.3
ED-T8 (6 hr)	NMC	0	50	14.7
	Composite	6	50	9.9
ED-T8 (12 hr)	ED-T8 (6)	7	50	16.0
		12	50	9.7
ED-T7 (6 hr)	NCA	0	34	13.1
	Composite	6	34	9.3
ED-T7 (12 hr)	ED-T7 (6)	7	50	15.3
		12	50	9.9

Test ID	Anode compartment Li mg/L	SO$_4^{2-}$ mg/L	S$_2$O$_6^{2-}$ mg/L	Central compartment Li mg/L	SO$_4^{2-}$ mg/L	S$_2$O$_6^{2-}$ mg/L	Cathode compartment Li mg/L	SO$_4^{2-}$ mg/L	S$_2$O$_6^{2-}$ mg/L
ED-T3		4909		13882	0.0	0.0	694		
		57974	366	9570	66120	188	6270	<10	<10
ED-T6	1.0	4909	<10	9354		3870	627	<10	<10
	18.3	26622	1047	6710	180859	1470	3606	62.6	<10
ED-T8 (6 hr)	0.0	4904	<10	10085	84623	3870	664	<10	<10
	0.0	40048	2924	5462	12715	912	5194	65	<10
ED-T8 (12 hr)	0.0	29833	1805	7028	58075	2698	4562	206	<10
	12.7	15143	2980	5924	14879	841	5648	111	<10
ED-T7 (6 hr)	1.2	4904	<10	8750	53190	8825	698	<10	<10
	12.5	28504	4036	5748	36086	1859	3928	<10	<10
ED-T7 (12 hr)	0.0	11358	1123	5504	49087	5648	2301	167	<10
	42.5	32186	5574	4290	11858	1442	5372	76.2	<10

a higher current density (204 A/m^2) was applied for six hours and the Li$^+$ increased from to 0.66 g/L to 5.19 g/L in the catholyte. The ED-T8 test was suspended after six hours for 12 hours and then continued for another six hours. As show in Table 4.7 and Figure 4.13(a), during suspension, the concentration of Li$^+$ in the central compartment increased from 5.46 g/L to 7.03 g/L, due to Li$^+$ back-diffusing to the central compartment.

Figure 4.13 (b) shows the difference between the initial and final Li$^+$ concentrations of Tests ED-T6 and ED-T8. The Li$^+$ concentration curves almost conver for both tests, except for during the ED-T8 test suspension.

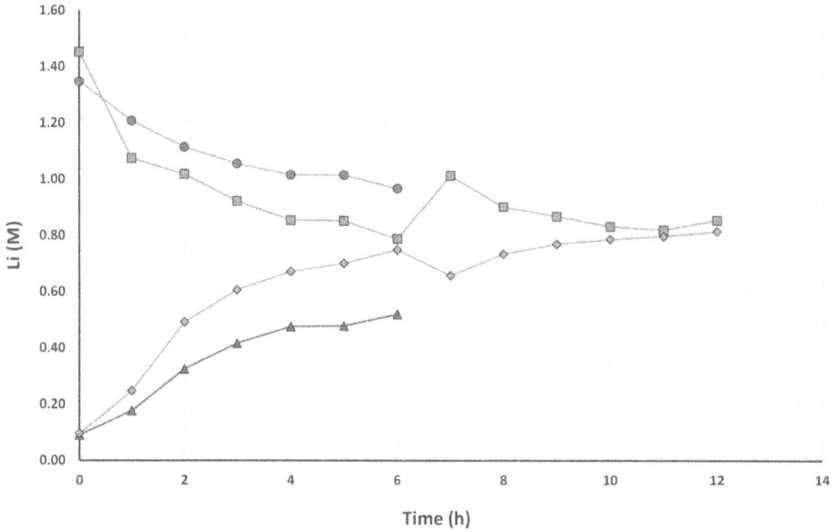

(a)

ED6 (Li2SO4) 34A   ED6 (LiOH) 34A   ED8(Li2SO4) 50A   ED8 (LiOH) 50A

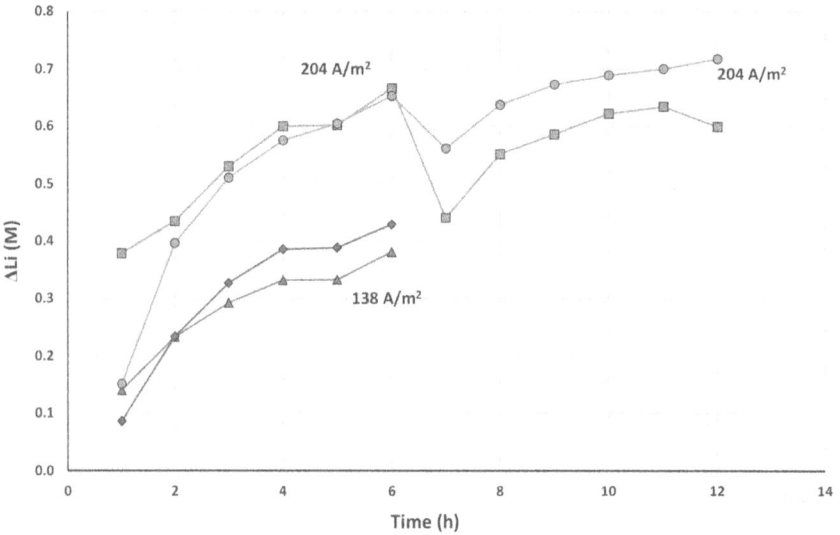

(b)

ED6 (Li2SO4) 34A   ED6 (LiOH) 34A   ED8(Li2SO4) 50A   ED8 (LiOH) 50A

FIGURE 4.13 (a) Salt-splitting experiments on NMC feed at two current densities, (b) the difference between the initial and final concentrations of NMC salt-splitting experiments, (c) salt-splitting experiments on NCA feed at two current densities, and (d) the difference between initial and final concentrations of NCA salt-splitting experiments. *(Continued)*

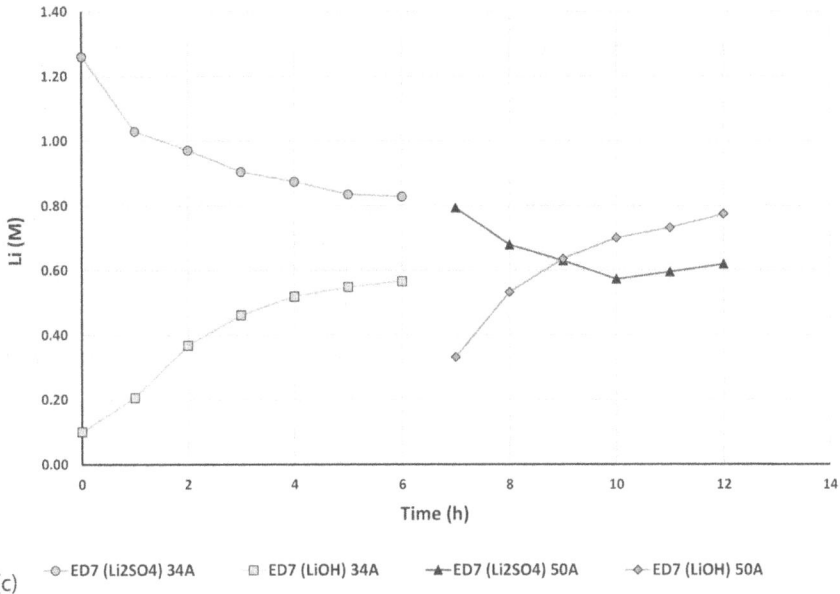

(c)   ⊙ ED7 (Li2SO4) 34A   □ ED7 (LiOH) 34A   ▲ ED7 (Li2SO4) 50A   ◇ ED7 (LiOH) 50A

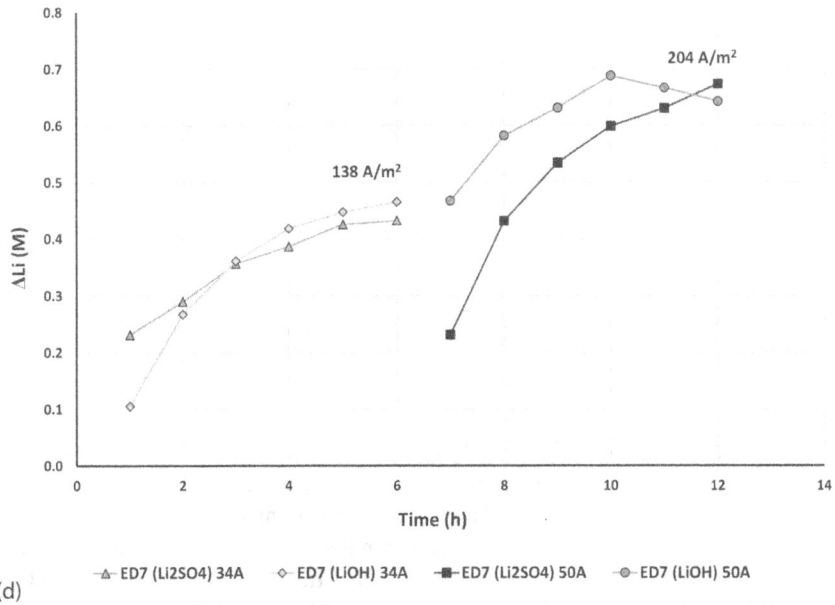

(d)   △ ED7 (Li2SO4) 34A   ◇ ED7 (LiOH) 34A   ■ ED7 (Li2SO4) 50A   ⊙ ED7 (LiOH) 50A

FIGURE 4.13   *(Continued)*

TABLE 4.7

Lithium Balance and Current Efficiency Based on Li Concentrations

			Lithium							Current Efficiency	
		Current Density A/m^2	Central			Cathode			Error %	Central %	Cathode %
Test ID	Feed		(M) 0 hr	(M) 6 hr	Difference	(M) 0 hr	(M) 6 hr	Difference			
ED-T6	NMC	138	1.35	0.97	0.38	0.09	0.52	0.43	11.6	81.5	101
ED-T8 (6 hr)	NMC	204	1.45	0.79	0.66	0.10	0.75	0.65	−1.5	99.1	94.9
ED-T8 (12 hr)	ED-T8(6)	204		0.85	0.60		0.81	0.71	15.5	75.2	84.2
ED-T7 (6 hr)	NCA	138	1.26	0.83	0.43	0.10	0.57	0.47	8.5	97.1	98.5
ED-T7 (12 hr)	ED-T7(6)	204		0.62	0.64		0.77	0.67	4.5	85.2	81.6

ED-T7 was run with a lower current density, 138 A/m^2, for 6 hours, suspended for 12 hours, and then run with higher current density, 204 A/m^2, for another 6 hours. Applying lower current density resulted in ~3.93 g/L LiOH. With a higher current density, ~5.73 g/L LiOH was generated. The Li profiles of ED-T7 [Figure 4.13 (c) and (d) are similar to the ED-T8].

In summary, to minimize the impact of sodium ions on hydrometallurgical recycling of the spent cathode material, LiOH can be used to remove impurities and to coprecipitate Co, Mn, and Ni. With electrodialysis, LiOH and H_2SO_4 can be regenerated which establish a closed loop hydrometallurgical recycling process with minimum reagent costs. The generated LiOH solution can be recircuilated to impurity removal stage as a reagent or distilled to produce LiOH crystals. The generated H_2SO_4 can be recirculated to the leach stage.

4.5 REGENERATE COMMERCIAL-GRADE NMC CATHODE ACTIVE MATERIAL FOR LITHIUM-ION BATTERIES

Current recycling of spent/waste LIBs mainly involve the separation of valuable metal ions, which includes selective precipitation [3], extraction [26, 27] and electrodeposition [28], as well as the re-synthesis or regeneration of cathode active materials via coprecipitation [29, 30] or sol-gel approach [31]. For instance, a common strategy to obtain $LiNi_xMn_yCo_zO_2$-type cathode active material is the calcination of the mixture of Li_2CO_3 as a typical Li precursor with $Ni_xMn_yCo_z (OH)_2$ or $Ni_xMn_yCo_zCO_3$ precursor that is acquired through coprecipitation of molarity-adjusted cations in the leachate [32–34]. Despite recent progress in recycling process of spent/waste LIBs, one of the challenges lies in the unsatisfactory purity of the regenerated cathode active materials; or more specifically, it remains difficult to achieve commercial-grade cathode active materials via a near-complete removal of impurity elements.

The hydrometallurgical recycling process of spent/waste Li cathode active materials using LiOH to coprecipitate the metals and generate high purity metal hydroxide precipitate and LiOH. The metal hydroxide and LiOH can be further processed to generate Li cathode precursors and Li cathode active materials. The Ni-Mn-Co hydroxide is first redissolved and then reprecipitated to generate $Ni_xMn_yCo_zCO_3$ precursors and NMC Li cathode active materials. The regenerated $LiNi_{0.5}Mn_{0.3}Co_{0.2}O_2$ cathode active materials possess remarkable discharge capacities, rate capabilities and cycling performances, which further demonstrates the success of the closed-loop recycling technology and the regeneration of Li cathode active materials based on the recycled materials.

4.5.1 Generation of $Ni_{0.5}Mn_{0.3}Co_{0.2}(OH)_2$ Precursor

To prepare $Ni_{0.5}Mn_{0.3}Co_{0.2}(OH)_2$, and $LiNi_{0.5}Mn_{0.3}Co_{0.2}O_2$ (NMC) cathode active material, the ternary Ni-Mn-Co precipitate was washed thoroughly to remove the entrained Li before redissolving with H_2SO_4 and H_2O_2 to produce a solution containing $NiSO_4$, $MnSO_4$, and $CoSO_4$ without other impurities. The content of Ni^{2+}, Mn^{2+}, and Co^{2+} in the solution was measured by ICP-OES in order to adjust the Ni^{2+}, Mn^{2+}, and Co^{2+} molarity ratio to 5: 3: 2 by adding $NiSO_4$, $CoSO_4$, and $MnSO_4$.

Precursor of $Ni_{0.5}Mn_{0.3}Co_{0.2}(OH)_2$ was generated by pumping the $NiSO_4$, $CoSO_4$, and $MnSO_4$ solution, 28 wt% ammonia (NH_4OH), and 20–50 wt% NaOH concurrently into a precipitation reactor under inner atmosphere (N_2) to adjust the solution pH to 10.9 at 50°C.

Figure 4.14 shows the experimental setup. The addition of 20–50 wt% NaOH to the solution can coprecipitate Ni, Co, and Mn ions and form $Ni_{0.5}Mn_{0.3}Co_{0.2}(OH)_2$ precursors. The purging of N_2 is to protect the metallic ions from being oxidized.

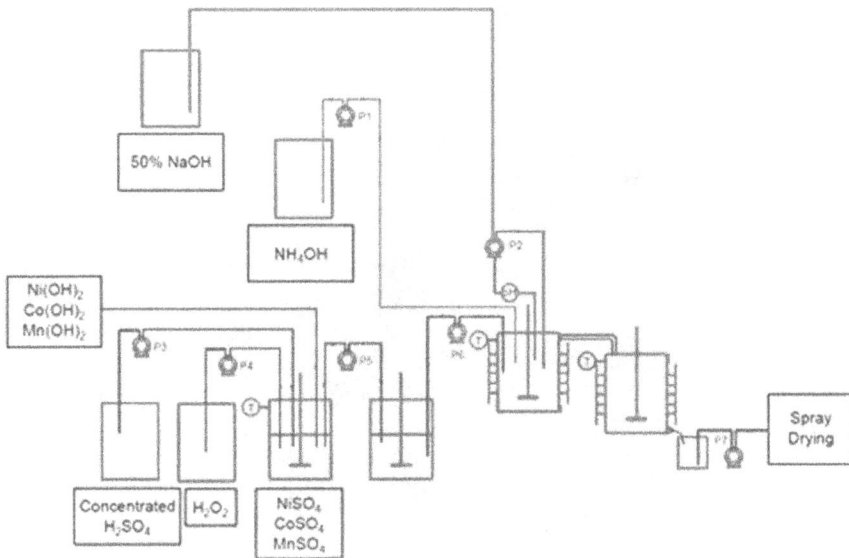

FIGURE 4.14 Experimental setup for NMC precursor generation.

Ammonia is used as a chelating agent to balance reduction-induced nucleation with crystal growth. The increasing content of NaOH breaks the stability of metal-ammonia complex to yield $Ni_{0.5}Mn_{0.3}Co_{0.2}(OH)_2$ precursors.

Inductively coupled plasma atomic emission spectroscopy (ICP-OES) was employed to determine the metal contents of $Ni_{0.5}Mn_{0.3}Co_{0.2}(OH)_2$ precursor. Laser diffraction was used to understand the particle size distribution of $Ni_{0.5}Mn_{0.3}Co_{0.2}(OH)_2$ precursor by measuring the angular variation in intensity of light scattered as a laser beam passes through a dispersed precursor sample. Scanning electron microscopy (SEM) of the $Ni_{0.5}Mn_{0.3}Co_{0.2}(OH)_2$ was taken to characterize the morphological structure.

The patterns of X-ray diffraction (XRD) of the materials were recorded to determine the crystallographic structure of the $Ni_{0.5}Mn_{0.3}Co_{0.2}(OH)_2$ precursor.

4.5.1.1 Precipitation of Ni, Co, and Mn Ternary Precipitates – Effect of Ammonia on Crystallographic Structure

During precipitation of Ni, Co, and Mn ternary precipitates, 4 M LiOH solution was used to raise the solution pH from 5.5 to 10.5. It was found that the obtained precipitates were difficult to dewater, which indicated that the precipitate may be amorphous. The amorphous precipitate might be due to the fast precipitation reaction kinetics, causing insufficient time for nucleation. Further precipitations were conducted by adding LiOH in the presence of ammonia. Ammonia was employed as a pH buffer to provide a stable environment for precipitation to from crystalline structure. The Ni, Co, and Mn ternary precipitates obtained with ammonia were easy to dewater and show good crystalline structure, as shown in the SEM photo of Figure 4.15.

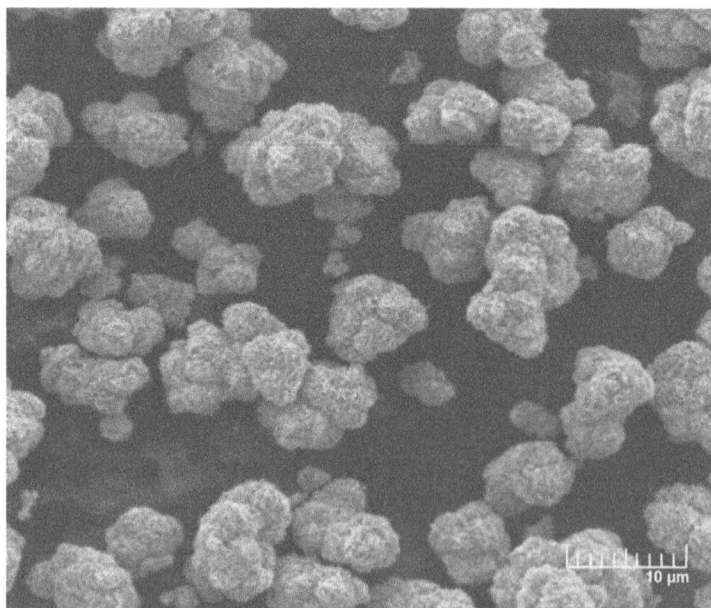

FIGURE 4.15 SEM photo of Ni, Co, and Mn ternary precipitates obtained with LiOH and ammonia.

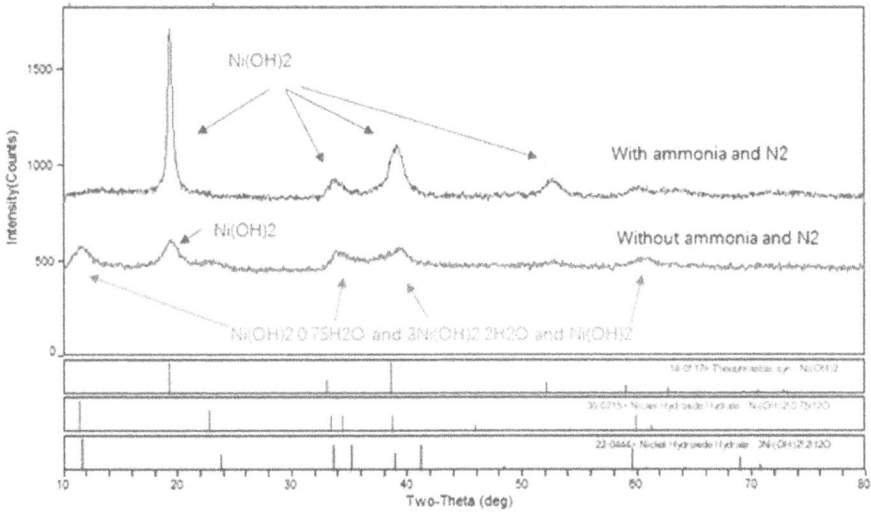

FIGURE 4.16 XRD pattern of Ni, Co, and Mn ternary precipitates with and without ammonia.

Figure 4.16 shows the XRD pattern comparison of the precipitate obtained with and without the presence of ammonia during precipitation. Comparing the two XRD patterns, the precipitate obtained with the presence of ammonia shows higher degree of crystallinity.

4.5.1.2 Factors Impacting the Precipitation of NMC Precursor

The ideal $Ni_{0.5}Mn_{0.3}Co_{0.2}(OH)_2$ precursors should have uniform spherical particles with narrow size distribution between 8 to 12 μm. Factors that will impact the morphology and particle size of the precursors include the mixing speed of the precipitation reactor, the retention time of the coprecipitation, precipitation pH, and the concentration of ammonia.

Figure 4.17 compares the effect of mixing speed on particle size distribution and particle morphology. With 5 hours of retention time, the particle size and tap density of the precursor increase with increasing the mixing speed from 600 rpm to 800 rpm. However, when the mixing speed increases to 900 rpm, the tap density of the precursor seems to decrease, and a more porous structure can be observed. As such, 800 rpm is determined to be the optimized mixing speed for precursor precipitation.

Figure 4.18 shows the SEM photos of the precursors generated at 5 hours, 16 hours, and 21 hours, respectively. With the prolonged retention time of the precipitation reaction at the constant stirring speed of 800 rpm, the primary grains grow larger and more interconnected with less voids among them.

Based on the experiments, the optimized experiment conditions to precipitate $Ni_{0.5}Mn_{0.3}Co_{0.2}(OH)_2$ precursors are mixing at 800 rpm for 21 hours at pH 10.9 while maintaining the temperature at 50°C.

On the other hand, ICP-OES analysis of the leftover solution showed the solution still contained Ni^{2+}, which was due to ammonia and nickel form nickel-ammonium sulfate complex $((NH_4)_2Ni(SO_4)_2)$.

FIGURE 4.17   Effect of mixing speed on particle size distribution and morphology of precursor.

FIGURE 4.18   Impact of retention time of coprecipitation to particle size.

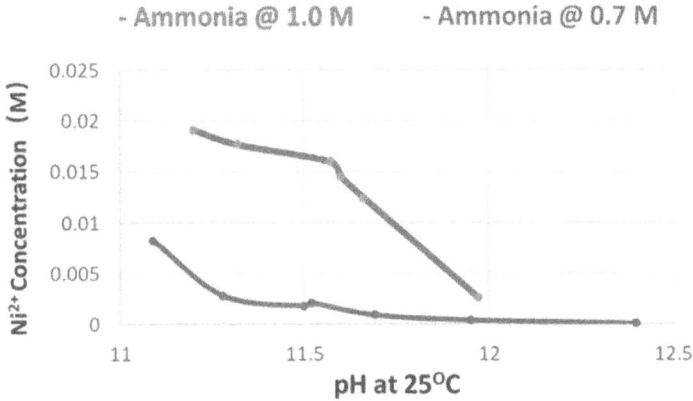

FIGURE 4.19 Effects of ammonia concentration and pH on Ni precipitation.

Figure 4.19 shows the effects of pH and ammonia concentration on the precipitation of Ni^{2+}. With 1.0 M ammonia, the remaining Ni^{2+} is precipitated at pH 12. When decreasing the ammonia concentration to 0.7 M, Ni^{2+} is more prone to precipitate out at pH >11 and complete precipitate at pH 12. The results suggest that a higher concentration of ammonia can result in a higher degree of $(NH_4)_2 Ni(SO_4)_2$ complex to be formed, and the complex can disintegrate to form $Ni(OH)_2$ precipitates under strong alkaline environment as ammonia ions at high pH will convert to ammonia gas [35].

4.5.1.3 Comparison between the Regenerated NMC Precursor and Commercial NMC Precursor

The SEM images of the regenerated NMC precursors are shown in Figure 4.20(a). The precursors are uniform spherical particles with narrow size distribution. They are composed of primary grains of cubic-like morphology with average volume of 587 μm^3.

Figure 4.20(b) and (c) shows the SEM photos of the regenerated NMC 523 precursors and commercial NMC 523 precursors respectively. Both precursors have similar spherical shape with narrow size distribution between 8 to 12 μm.

FIGURE 4.20 (a) Regenerated NMC precursor – 2 μm scale, (b) regenerated NMC precursor – 5 μm scale, and (c) commercial NMC precursor – 5 μm scale.

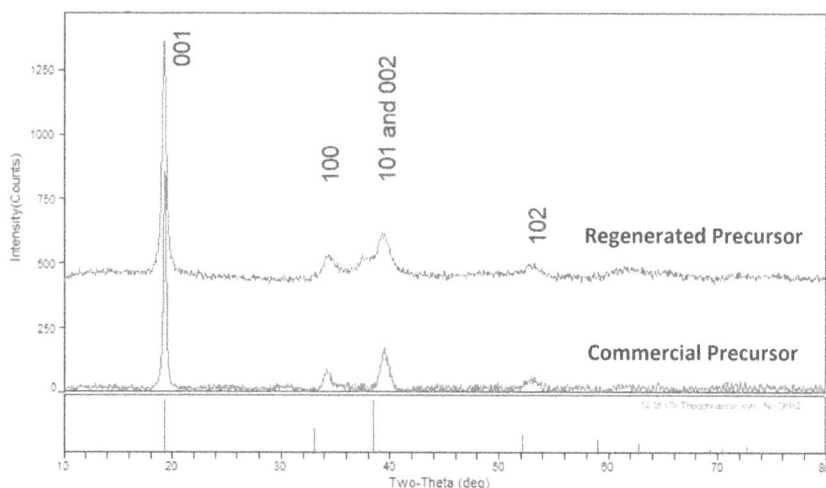

FIGURE 4.21   XRD patterns of regenerated NMC precursor and commercial NMC precursor.

Table 4.8 lists the ICP-OES analysis data of the regenerated NMC precursor and commercial NMC precursor. The regenerated NMC precursor in general has lower impurities than the commercial NMC precursor. The regenerated NMC precursor is 99.63% pure comparing to 99.31% pure of commercial NMC precursor.

The XRD patterns of the regenerated NMC precursor and the commercial NMC precursor are shown in Figure 4.21. The XRD patterns shows the regenerated NMC precursor is of hexagonal structure in the $R\bar{3}c$ space group. Although the peaks in the regenerated NMC precursor XRD are perfectly matched with the peaks in the commercial NMC precursor, the regenerated NMC precursor has lower intensity

TABLE 4.8
ICP-OES Analysis of Regenerated NMC Precursor and Commercial NMC Precursor

Material	Al (%)	Ca (%)	Cd (%)	Co (%)	Cr (%)	Cu (%)	Fe (%)	K (%)
Regenerated NMC 523 Precursor	0.0049	0.0510	0.0001	13.1	0.0006	0.0146	0.0105	0.0047
Commercial NMC 523 precursor	0.0142	0.0911	0.0001	11.7	0.0005	0.0181	0.0089	0.0152

Material	Mg (%)	Mn (%)	Na (%)	Ni (%)	Pb (%)	S (%)	Zn (%)
Regenerated NMC 523 Precursor	0.0862	19.0	0.0589	32.5	0.0000	0.1190	0.0152
Commercial NMC 523 precursor	0.0234	16.2	0.1830	28.8	0.1700	0.1640	0.0009

which suggested that although both share the same chemical compositions, but the regenerated NMC precursor might have lower crystallinity.

Figure 4.22 shows the particle size distribution of the regenerated NMC precursor and commercial NMC precursor. The particle size of commercial NMC precursor ranges from 6 µm to 20 µm with D10 at 6.154 µm, D50 at 11.76 µm, and D90 at 20.01 µm, while the particle size of the regenerated NMC precursor ranges from 6 µm to 12 µm with D10 at 6.097 µm, D50 at 8.773 µm, and D90 at 12.44 µm, respectively. The regenerated NMC precursor has a narrower particle size distribution and more uniform in size than commercial NMC precursor.

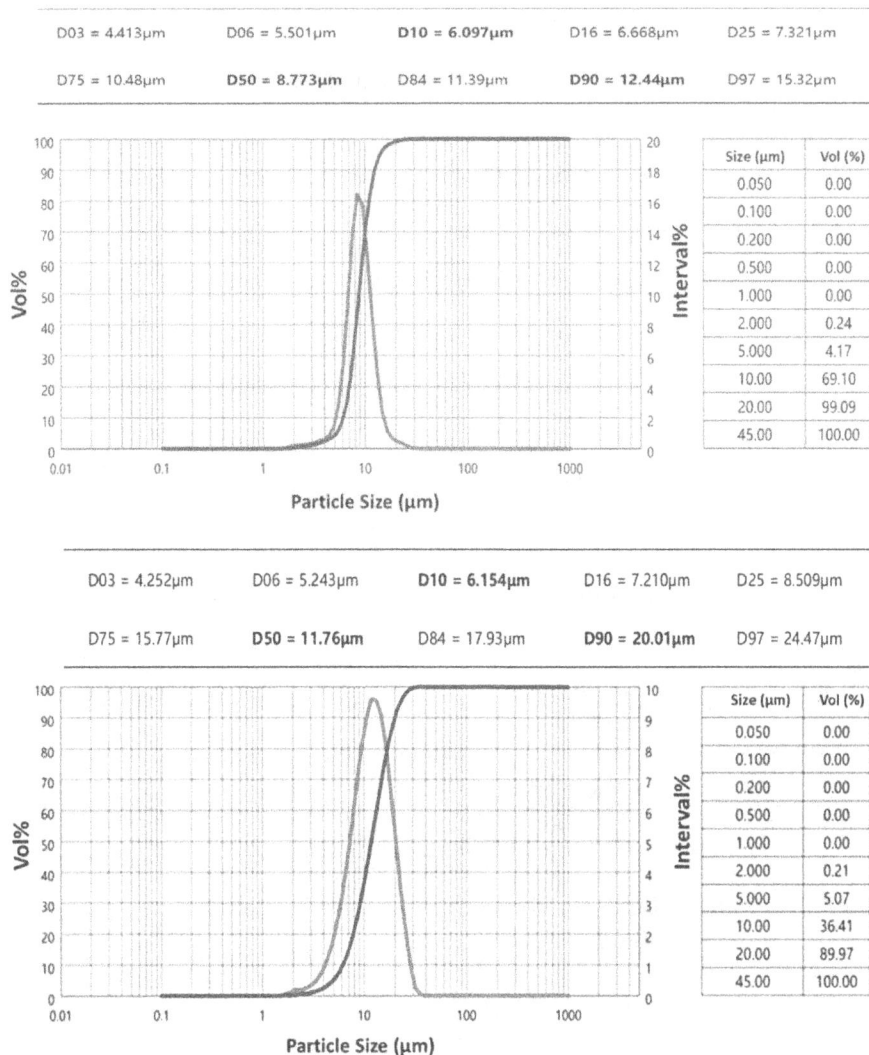

D03 = 4.413µm	D06 = 5.501µm	D10 = 6.097µm	D16 = 6.668µm	D25 = 7.321µm
D75 = 10.48µm	D50 = 8.773µm	D84 = 11.39µm	D90 = 12.44µm	D97 = 15.32µm

Size (µm)	Vol (%)
0.050	0.00
0.100	0.00
0.200	0.00
0.500	0.00
1.000	0.00
2.000	0.24
5.000	4.17
10.00	69.10
20.00	99.09
45.00	100.00

D03 = 4.252µm	D06 = 5.243µm	D10 = 6.154µm	D16 = 7.210µm	D25 = 8.509µm
D75 = 15.77µm	D50 = 11.76µm	D84 = 17.93µm	D90 = 20.01µm	D97 = 24.47µm

Size (µm)	Vol (%)
0.050	0.00
0.100	0.00
0.200	0.00
0.500	0.00
1.000	0.00
2.000	0.21
5.000	5.07
10.00	36.41
20.00	89.97
45.00	100.00

FIGURE 4.22 Particle size distribution of regenerated NMC precursor and commercial NMC precursor.

Based on analysis of particle size distribution, ICP-OES analysis, XRD pattern, and SEM photos comparison, the regenerated NMC precursor is matching the characteristics of commercial NMC precursor.

4.5.2   GENERATION OF $Ni_xMn_yCo_z(OH)_2$

Ternary Ni-Mn-Co precursor, $Ni_{0.6}Mn_{0.2}Co_{0.2}(OH)_2$ (NMC622), and $Ni_{0.8}Mn_{0.1}Co_{0.1}(OH)_2$ (NMC811) cathode precursors were generated from the high purity, coprecipitated Ni-Mn-Co precipitate. The Ni-Mn-Co precipitate was washed to remove any entrained Li and dissolved with H_2SO_4 and H_2O_2 to produce a solution containing $NiSO_4$, $MnSO_4$, and $CoSO_4$. The content of Ni^{2+}, Mn^{2+}, and Co^{2+} in the solution was measured by ICP-OES. The required amount of $NiSO_4$, $CoSO_4$, and $MnSO_4$ was then added into the solution to adjust the molar ratio to the required 6:2:2 and 8:1:1 ratio.

NMC622 and NMC811 precursors were generated by pumping the Ni, Co, and Mn enriched solution, 28 wt% ammonia (NH_4OH), and 20–50 wt% NaOH concurrently into a precipitation reactor under inner atmosphere (N_2) at pH 11.1 and 11.5 at 50°C.

Figure 4.23 shows the SEM and particle size distribution of the generated NMC 622. Figure 4.24 shows the SEM and particle size distribution of the generated NMC 811.

4.5.3   GENERATION OF $LiNi_{0.5}Mn_{0.3}Co_{0.2}O_2$ AND BATTERY CELL FABRICATION

The $Ni_{0.5}Mn_{0.3}Co_{0.2}(OH)_2$ precursor was mixed with the LiOH crystal in a molar ratio of 1:1.08 and sintered at 880°C for 10 hours to generate $LiNi_{0.5}Mn_{0.3}Co_{0.2}O_2$ cathode active material.

The regenerated $LiNi_{0.5}Mn_{0.3}Co_{0.2}O_2$ cathode active material was mixed with carbon black and polyvinylidene (PVDF) in a ratio of 8: 1: 1 by weight. The mixture was then mixed with n-methyl-2-pyrrolidone (NMP) to form an ink and coated on Al foil to form NMC cathode.

Graphite, carbon black, styrene-butadiene rubber (SBR) with weight ratio of 8:1:1 was mixed with 1.2% carboxymethyl cellulose (CMC) solution. The mixture was coated onto copper foil to make anode.

The electrolyte was 1.0M lithium hexafluorophosphate solution in ethylene carbonate, dimethyl carbonate, and diethyl carbonate. (EC/DMC/DEC=1:1:1(v/v/v))

The anode and cathode together with electrolyte were assembled under inner atmosphere into pouch cell battery for battery performance evaluation.

4.5.3.1   Impact of Calcining Temperature on the Crystallinity of NMC Cathode Active Material

The regenerated $LiNi_{0.5}Mn_{0.3}Co_{0.2}O_2$ cathode active material (NMC cathode active material) was produced by calcining the mixture of LiOH crystal and regenerated NMC precursor. Comparison experiments at various calcination temperatures (820°C, 860°C, 880°C, and 900°C) were carried out to explore the best temperature. Figure 4.25 shows the XRD patterns. The NMC cathode active material calcined at 880°C exhibits sharp peaks, a characteristic of high degree of crystallinity.

D03 = 5.557 um   D06 = 6.462 um   D10 = 7.105 um   D16 = 7.670 um   D25 = 8.375 um

D75 = 11.79 um   D50 = 9.966 um   D84 = 12.72 um   D90 = 13.65 um   D97 = 15.51 um

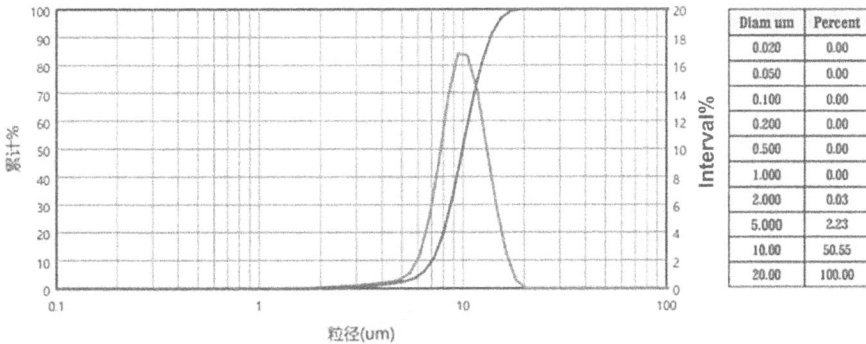

Diam um	Percent
0.020	0.00
0.050	0.00
0.100	0.00
0.200	0.00
0.500	0.00
1.000	0.00
2.000	0.03
5.000	2.23
10.00	50.55
20.00	100.00

FIGURE 4.23 SEM and particle size distribution of NMC 622 precursor.

Furthermore, the clean background of the XRD patterns suggests that the 880°C has higher phase purity. The optimum temperature is therefore determined to be 880°C. Moreover, the sample calcined for 4 hours exhibits better crystallinity and phase purity as compared to the 10-hour sample. In comparison with commercial $LiNi_{0.5}Mn_{0.3}Co_{0.2}O_2$ (NMC) cathode active material, the XRD patterns of the

Diam um	Percent
0.050	0.00
0.100	0.00
0.200	0.00
0.500	0.00
1.000	0.00
2.000	0.23
5.000	2.55
10.00	55.25
20.00	98.28
45.00	100.00

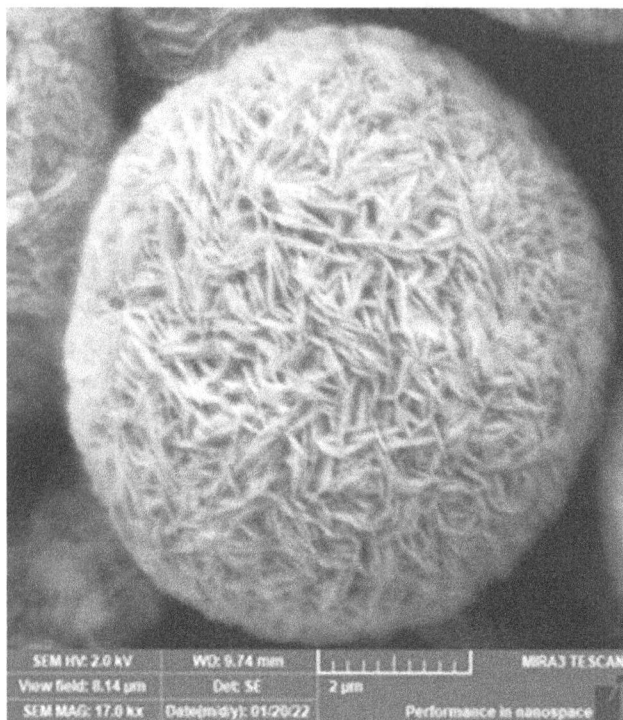

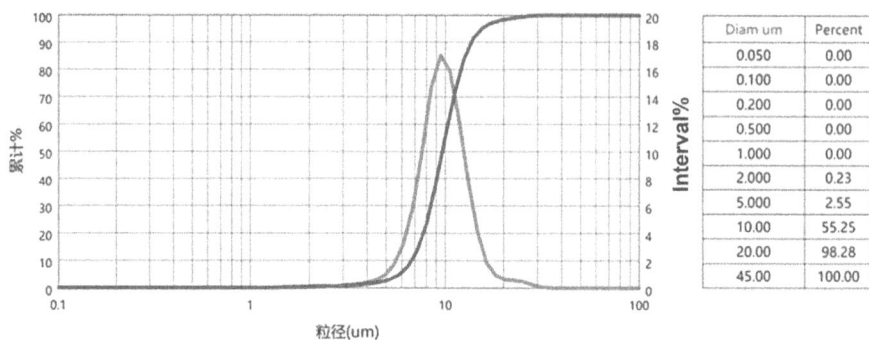

FIGURE 4.24   SEM and particle size distribution of NMC 811 precursor.

regenerated NMC cathode active material generated at 880°C matches the XRD patterns of commercial NMC cathode active material as shown in Figure 4.25.

Figure 4.26 and Table 4.9 display the Rietveld refinement results of regenerated $LiNi_{0.5}Mn_{0.3}Co_{0.2}O_2$. The precursor is of hexagonal structure in the R $\bar{3}$c space group, while the regenerated cathode active materials with α-NaFeO$_2$ structure belong to the R $\bar{3}$m space group, in which 3a positions are occupied by Ni/Co/Mn ions and mixing Li$^+$, 3b positions are occupied by Li$^+$ and mixing Ni^{2+}, and 6c positions are occupied by O^{2-}.

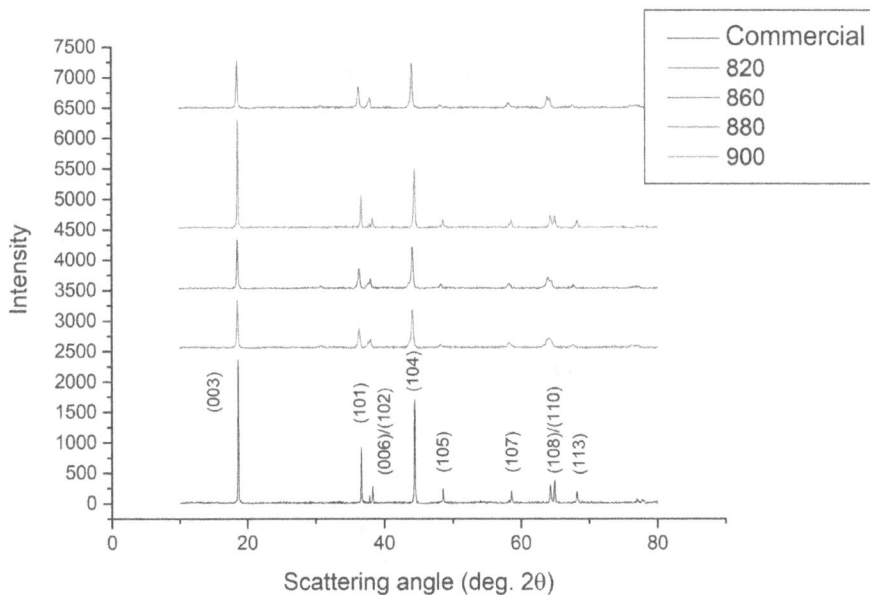

FIGURE 4.25   XRD pattern of regenerated NMC cathode active material calcined at 820°C, 860°C, 880°C, 900°C, and commercial NMC cathode active material.

The ordered layered structure can be manifested by the visible (006)/(012) and (018)/(110) peak splitting. The c/3a ratio of 1.6544 and the high value of I_{003}/I_{104} ratio imply that the regenerated cathode active material is featured by typical close-packed hexagonal lattices and well-ordered structure with weakened Li/Ni cation mixing (only 4.88%).

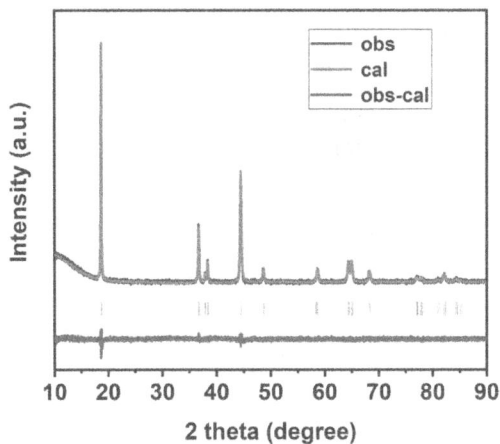

FIGURE 4.26   Rietveld refinement pattern of regenerated $LiNi_{0.5}Mn_{0.3}Co_{0.2}O_2$.

TABLE 4.9

Rietveld Refinement Results of Regenerated LiNi$_{0.5}$Mn$_{0.3}$Co$_{0.2}$O$_2$

Structural Parameters				Reliability Factors and Bragg R-factor			
a(Å)	c(Å)	c/3a	I$_{003}$/I$_{104}$	Rp(%)	Rwp(%)	Rexp(%)	Bragg R-factor(%)
2.8696	14.2421	1.6544	1.8320	1.93	2.47	0.80	9.34

Table 4.10 shows the ICP-OES analysis of the regenerated NMC cathode active material and commercial NMC cathode active material. The regenerated NMC cathode active material in general has the same level of impurities compared to the commercial NMC cathode active material. The regenerated NMC cathode active material is 99.54% pure comparing to 99.58% pure of commercial NMC cathode active material. On the other hand, the regenerated NMC has a % mole ratio of Ni:Mn:Co = 0.513:0.300:0.205 comparing to 0.526:0.288:0.202 of commercial NMC cathode active material. The mole ratio of the regenerated NMC cathode active material is closer to the required 0.5:0.3:0.2 ratio.

TABLE 4.10

ICP-OES Analysis of Regenerated NMC Cathode Active Material and Commercial NMC Cathode Active Material

Material	Al (%)	Ca (%)	Cd (%)	Co (%)	Cr (%)	Cu (%)	Fe (%)	K (%)
Regenerated NMC cathode active material	0.0141	0.0477	0.0001	12.1	0.0011	0.0141	0.0127	0.0072
Commercial NMC cathode active material	N/A	0.0351	N/A	11.9	0.0005	0.0136	0.0044	0.0063

Material	Li (%)	Mg (%)	Mn (%)	Na (%)	Ni (%)	Pb (%)	S (%)	Zn (%)
Regenerated NMC cathode active material	7.4	0.0718	16.5	0.0954	30.1	N/A	0.171	0.025
Commercial NMC cathode active material	7.8	0.0152	15.8	0.134	30.9	0.01	0.185	0.0157

Material	Li (% mole ratio)	Ni (% mole ratio)	Mn (% mole ratio)	Co (% mole ratio)
Regenerated NMC cathode active material	1.067	0.513	0.300	0.205
Commercial NMC cathode active material	1.12	0.526	0.288	0.202

4.5.3.2   Electrochemical Performance of the Regenerated NMC Cathode Active Material

The regenerated NMC cathode active material and the commercial NMC cathode active material were used to build pouch cell batteries to evaluate the power density and energy density of the regenerated cathode active material. Within the voltage range of 2.5–4.3 V, the regenerated NMC cathode active material exhibits initial discharge capacities of 162.15 mA·h/g and 153.95 mA·h/g at 0.2 C and 1 C, respectively. In contrast, the initial discharge capacities of commercial NMC cathode active materials are 151.64 mA·h/g and 141.30 mA·h/g at 0.2 C and 1 C, respectively. Figure 4.27 illustrates the 0.2 C and 1 C charge and discharge profiles of pouch cell batteries made with regenerated NMC cathode active material and commercial NMC cathode active material. The capacities of the pouch cell battery made with regenerated NMC cathode active material are 7% and 9% better at 0.2 C and 1 C discharge rate than the pouch cell battery made with commercial NMC cathode active material.

Moreover, Figure 4.28 shows the 1 C cycling performance of pouch cell batteries made with regenerated NMC cathode active materials and commercial NMC cathode active material. The batteries were subject to cycling test between 2.5–4.3 V. The capacity of the pouch cell battery made with commercial NMC cathode active material decays 9.3% after 100 cycles. The capacity of the pouch cell battery made with regenerated NMC cathode active material decays 6.5% after 100 cycles and 11.3% after 300 cycles. The electrochemical performances suggest that the regenerated NMC cathode active material is superior to the commercial NMC cathode active material. The faster performance degradation of the battery equipped with commercial NMC cathode active material might be due to the % mole ratio of Ni, Mn, and Co being 0.526:0.288:0.202 instead of 0.5:0.3:0.2, where the higher content of Ni causing the cathode active material become less stable.

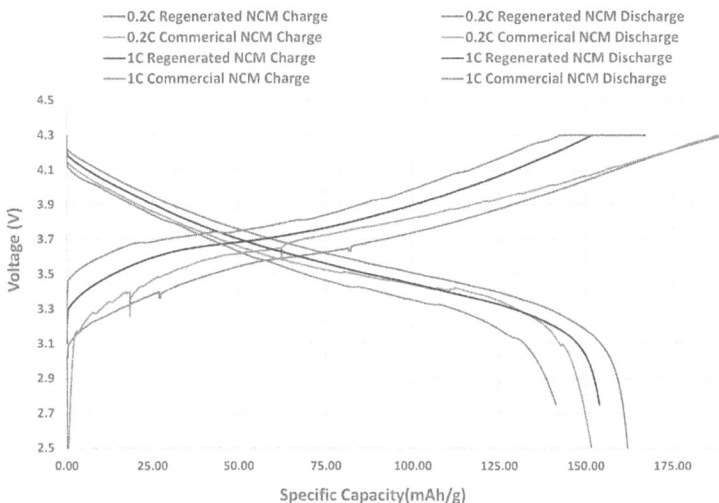

FIGURE 4.27   0.2 C and 1 C charge and discharge profile of pouch cell batteries with regenerated NMC cathode active material and commercial NMC cathode active material.

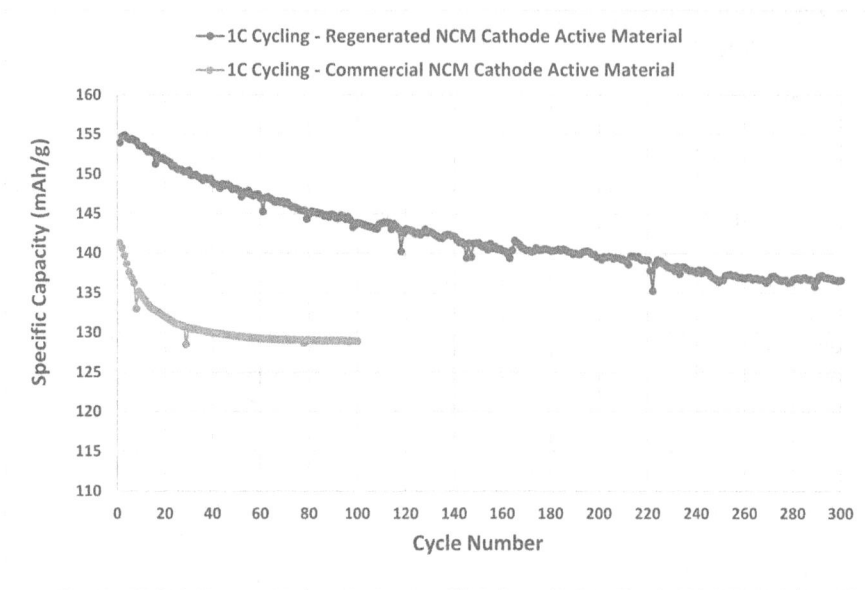

FIGURE 4.28 1 C cycling profile pouch cell batteries made with regenerated NMC cathode active material and commercial NMC cathode active material.

REFERENCES

1. Jung J.C.Y., Sui P.C., Zhang J., A review of recycling spent lithium-ion battery cathode active materials using hydrometallurgical treatments, Journal of Energy Storage, 35, 102217 (2021).
2. Takacova Z., Havlik T., Kukurugya F., Orac D., Cobalt and lithium recovery from active mass of spent Li-ion batteries: Theoretical and experimental approach, Hydrometallurgy, 163, 9–17 (2016).
3. Nayl A.A., Elkhashab R.A., Badawy S.M., El-Khateeb M.A., Acid leaching of mixed spent Li-ion batteries, Arabian Journal of Chemistry, 10, S3632–S3639 (2017).
4. Chen L., Tang X., Zhang Y., Li L., Zhen Z., Zhang Y., Process for the recovery of cobalt oxalate from lithium-ion batteries, Hydrometallurgy, 108, 80–86 (2011).
5. Kang J., Senanayake G., Sohn J.S., Shin S.M., Recovery of cobalt sulfate from spent lithium ion batteries by reductive leaching and solvent extraction with Cyanex 272, Hydrometallurgy, 100, 168–171 (2010).
6. Shin S.M., Kim N.H., Sohn J.S., Yang D.H., Kim Y.H., Development of a metal recovery process from Li-ion battery wastes, Hydrometallurgy, 79, 172–181 (2005).
7. Zhu S., He W., Li G., Zhou X., Zhang X., Huang J., Recovery of Co and Li from spent lithium-ion batteries by combination method of acid leaching and chemical precipitation, Transactions of Nonferrous Metals Society of China, 22, 2274–2281 (2012).
8. Swain B., Jeong J., Lee J., Lee G.H., Sohn J., Hydrometallurgical process for recovery of cobalt from waste cathodic active material generated during manufacturing of lithium-ion batteries, Journal of Power Sources, 167, 536–544 (2007).
9. Dorella G., Mansur M.B., A study of the separation of cobalt from spent Li-ion battery residues, Journal of Power Sources, 170, 210–215 (2007).

10. Nan J., Han D., Yang M., Cui M., Hou X., Recovery of metal values from a mixture of spent lithium-ion batteries and nickel-metal hydride batteries, Hydrometallurgy, 84, 75–80 (2006).

11. Xu J., Thomas H.R., Francis R.W., Lum K., Wang J., Liang B., A review of processes and technologies for the recycling of lithium-ion secondary batteries, Journal of Power Sources, 177, 512–527 (2008).

12. Swain B., Jeong J., Lee J.C., Lee G.H., Development of process flow sheet for recovery of high pure cobalt from sulfate leach liquor of lib industry waste: A mathematical model correlation to predict optimum operational conditions, Separation and Purification Technology, 63, 360–369 (2008).

13. Lupi C., Pasquali M., Electrolytic nickel recovery from lithium-ion batteries, Minerals Engineering, 16, 537–542 (2003).

14. Zante G., Braun A., Masmoudi A., Barillon R., Trebouet D., Boltoeva M., Solvent extraction fractionation of manganese, cobalt, nickel and lithium using ionic liquids and deep eutectic solvents, Minerals Engineering, 156, 106512 (2020).

15. Zhang P., Yokoyama T., Itabashi O., Suzuki T., Inoue K., Hydrometallurgical process for recovery of metal values from spent lithium-ion secondary batteries, Hydrometallurgy, 47, 259–271 (1998).

16. Nan J., Han D., Zuo X., Recovery of metal values from spent lithium-ion batteries with chemical deposition and solvent extraction, Journal of Power Sources, 152, 278–284 (2005).

17. Granata G., Moscardini E., Pagnanelli F., Trabucco F., Toro L., Product recovery from Li-ion battery wastes coming from an industrial pretreatment plant: Lab scale tests and process simulations, Journal of Power Sources, 206, 393–401 (2012).

18. Gao W., Zhang X., Zheng X., Lin X., Cao H., Zhang Y., Sun Z., Lithium carbonate recovery from cathode scrap of spent lithium-ion battery: A closed-loop process, Environmental Science & Technology, 51, 1662–1669 (2017).

19. Schweitzer G.K., Pesterfield L.L., The Aqueous Chemistry of the Elements. Oxford University Press (2010).

20. Rubin A.J., Hayden P.L., Studies on the hydrolysis and precipitation of aluminium (III). Project Rep., Water Resources Ctr., Ohio State University (1973).

21. Matijevic' E., Stryker L.J., Counterion complexing and sol stability. II. Coagulation effects of aluminum sulfate in acidic solutions, The Journal of Physical Chemistry, 73(5), 1484–1487, May (1969).

22. Mallya R.M., Vasudeva Murthy A.R., Studies on the Basic Carbonates on Nickel Part 111: Potentiometric Study of Precipitation. Department of inorganic and Physical Chemistry, Indian Institute of Science, Bawalure-12 (1960).

23. Davidson J.M., Glass D.H., Nucleation kinetics in the reactions of nickel basic carbonates with hydrogen sulphide: The carbonate precipitation reactions of divalent nickel, Industrial & Engineering Chemistry Research, 46(14), 4772–4777 (2007).

24. Mersmann A., Löffelmann M., Review crystallization and precipitation: The optimal supersaturation, Chemical Engineering & Technology, 23(1) (2000).

25. Bommel A., Dahn J.R., Analysis of the growth mechanism of coprecipitated spherical and dense nickel, manganese, and cobalt containing hydroxides in the presence of aqueous ammonia, Chemistry of Materials, 21(8), 1500–1503 (2009).

26. Wang R.-C., Lin Y.-C., Wu S.-H., A novel recovery process of metal values from the cathode active materials of the lithium-ion secondary batteries, Hydrometallurgy, 99(3–4), 194 (2009).

27. Jha A.K., Jha M.K., Kumari A., Sahu S.K., Kumar V., Pandey B.D., Selective separation and recovery of cobalt from leach liquor of discarded Li-ion batteries using thiophosphinic extractant, Separation and Purification Technology, 104, 160 (2013).

28. Barbieri E.M.S., Lima E.P.C., Cantarino S.J., Lelis M.F.F., Freitas M.B.J.G., Recycling of spent ion-lithium batteries as cobalt hydroxide, and cobalt oxide films formed under a conductive glass substrate, and their electrochemical properties, Journal of Power Sources, 269, 158 (2014).

29. Gao R., Sun C., Xu L., Zhou T., Zhuang L., Xie H. Recycling LiNi0.5Co0.2Mn0.3O2 material from spent lithium-ion batteries by oxalate co-precipitation, Vacuum, 173 (2020).

30. Yang L., Xi G., Preparation and electrochemical performance of LiNi1/3Co1/3Mn1/3O2 cathode active materials for lithium-ion batteries from spent mixed alkaline batteries, Journal of Electronic Materials, 45(1), 301, (2016).

31. Zhao T., Chen S., Li L., Zhang X., Chen R., Belharouak I., Wu F., Amine K., Synthesis, characterization, and electrochemistry of cathode active material Li Li0.2Co0.13Ni0.13Mn0.54O2 using organic chelating agents for lithium-ion batteries, Journal of Power Sources, 228, 206 (2013).

32. Gratz E., Sa Q., Apelian D., Wang Y., A closed loop process for recycling spent lithium ion batteries, Journal of Power Sources, 262, 255 (2014).

33. He L.-P., Sun S.-Y., Yu J.-G., Performance of LiNi1/3Co1/3Mn1/3O2 prepared from spent lithium-ion batteries by a carbonate co-precipitation method, Ceramics International, 44(1), 351 (2018).

34. Sa Q., Gratz E., He M., Lu W., Apelian D., Wang Y., Synthesis of high performance LiNi1/3Mn1/3Co1/3O2 from lithium-ion battery recovery stream, Journal of Power Sources, 282, 140 (2015).

35. Mohammed-Nour A., Al-Sewailem M., El-Naggar A.H., The influence of alkalization and temperature on ammonia recovery from cow manure and the chemical properties of the effluents, Sustainability, 11, 2441 (2019). doi:10.3390/su11082441

5 Analysis of Mass Balance, Energy Consumption, and Economics of the Closed-Loop Hydrometallurgical Recycling Waste/Spent Lithium-Ion Battery Cathode Active Materials

Joey Jung and Jiujun Zhang

CONTENTS

5.1 SUMMARY OF THE DEVELOPED CLOSED-LOOP HYDROMETALLURGICAL RECYCLING WASTE/SPENT LITHIUM-ION CATHODE ACTIVE MATERIALS

A closed-loop hydrometallurgical recycling process can be achieved by adapting lithium hydroxide (LiOH) to remove impurity, LiOH to co-precipitate metal hydroxide, and employing electrodialysis to regenerate LiOH and sulfuric acid (H_2SO_4) [1]. The co-precipitated metal hydroxide can be used to regenerate lithium nickel manganese cobalt oxide (NMC) cathode precursor, the resultant H_2SO_4 mainly participates

DOI: 10.1201/9781003269205-5

in the leaching process, while LiOH serves as the precipitating agent as well as the lithium source to calcine with the NMC precursors to regenerate NMC cathode active material.

The detail flowsheets are shown in the following figures: Figure 5.1: separation of lithium-ion battery (LIB) cathode active material and aluminum foil and leach of cathode active material; Figure 5.2: impurities removal; Figure 5.3: metal recovery; Figure 5.4: LIB cathode precursor generation; Figure 5.5: sodium sulfate crystallization; and Figure 5.6: lithium hydroxide and sulfuric acid regeneration and electrodialysis.

As shown in Figure 5.1, the concentrations of sulfuric acid and the diluted lithium sulfate (Li_2SO_4) solution generated from the electrodialysis cell are pumped into a mixing tank to prepare weak sulfuric acid. The spent/waste cathode is fed into a shredder to reduce the size to 1 cm x 4 cm. The shredded cathode and the weak acid are then pumped into two separation tanks alternatively and mixed until the cathode active material separated from aluminum foil [2].

The obtained slurry containing aluminum foil and cathode active material is then discharged into a sieve shaker where the aluminum foil is screened and removed, while the slurry containing cathode active material passes through. The collected aluminum foil is washed to remove any carried-over cathode active material before passing through a second sieve shaker. The two slurries containing cathode active material are then joined together and pumped into four cascade leach tanks. The leach of cathode active material is conducted by controlling the pH at 1.5 using concentrated sulfuric acid with a reductive agent such as sulfur dioxide (SO_2) [3] or hydrogen peroxide (H_2O_2). The washed aluminum foil can be sent to a metal recycler.

As shown in Figure 5.2, impurities removal is conducted by pumping the leachate into three cascade reactors continuously to precipitate aluminum as aluminum hydroxide ($Al(OH)_3$).

The pH and the temperature of the three cascade reactors is controlled by pumping lithium hydroxide to control at 5.0 and 80°C. [4] The leachate containing aluminum hydroxide precipitate is then collected and pumped into a filter press to filter out the aluminum hydroxide and the unleached solid. The aluminum hydroxide precipitate and the unleached solid are then washed inside the filter press to remove any entrain leachate. The metal-enriched pregnant leach solution (PLS) and the wash water are then collected. The aluminum hydroxide and the unleached solid are sent for waste collection.

Figure 5.3 shows the metal recovery process. Metal recovery is conducted by pumping the leachate into 4 reactors concurrently while controlling the pH of the cascade reactors at 10.9 using lithium hydroxide and buffered with ammonia [5, 6]. The cobalt, nickel, and manganese in the solution are precipitated as hydroxide slurry which is collected at the end and then filtered out by a filter press. The precipitate-free lithium sulfate solution is then collected. The metal hydroxide can then be used to produce LIB cathode precursor.

Figure 5.4 shows the developed flowsheet for LIB cathode precursor generation. The Ni-Co-Mn hydroxide generated from metal recovery process is redissolved using sulfuric acid and hydrogen peroxide to produce a solution containing cobalt sulfate, manganese sulfate, and nickel sulfate. The solution is then pumped into

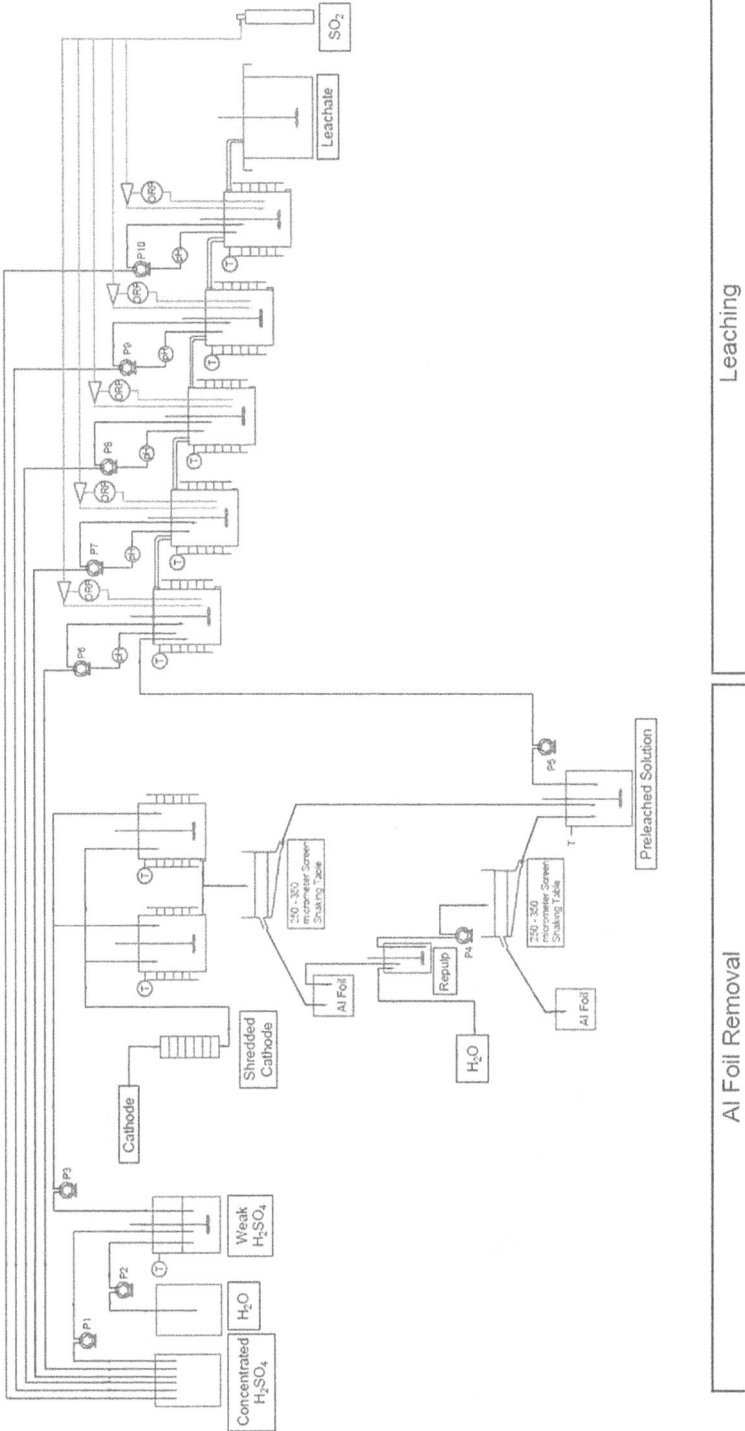

FIGURE 5.1 Separation of lithium-ion battery cathode active material and aluminum foil and leach of cathode active material.

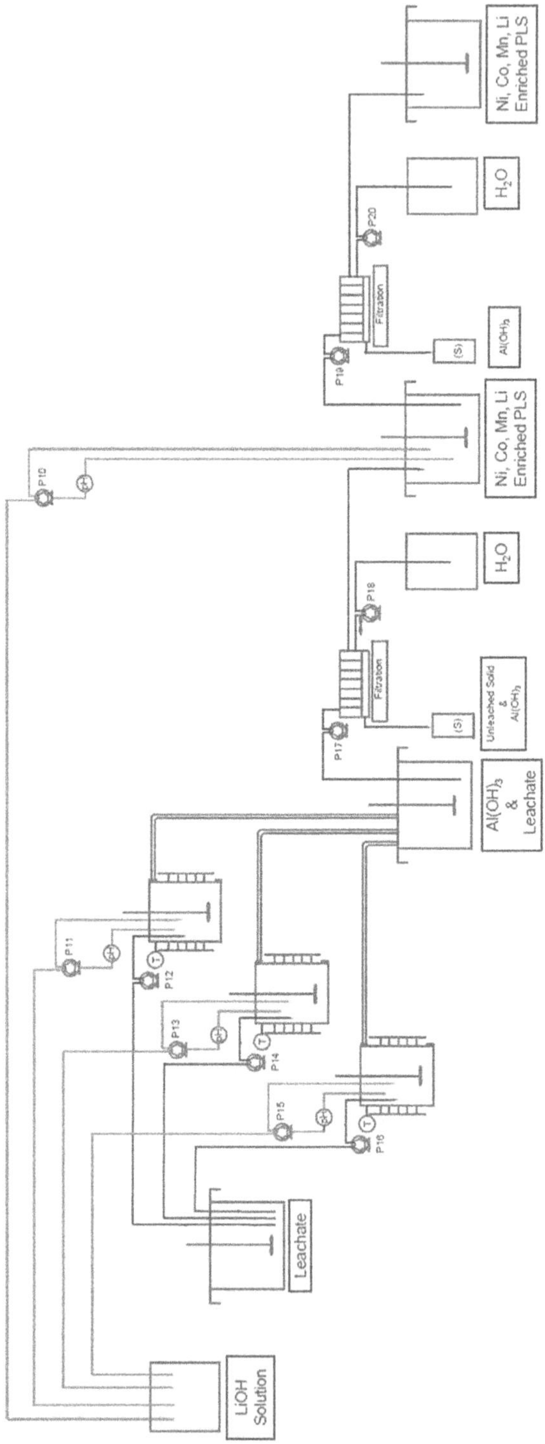

FIGURE 5.2 Impurities removal.

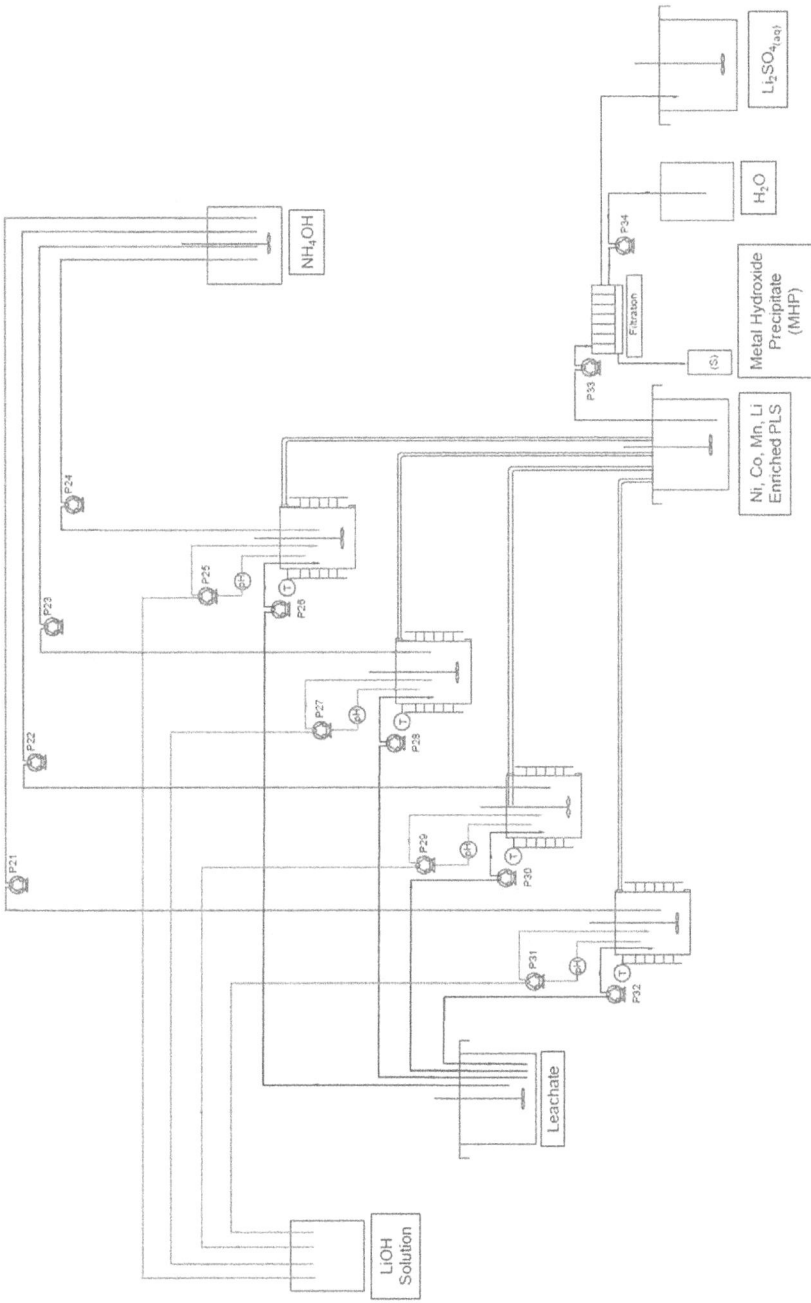

FIGURE 5.3   The recovery of nickel, cobalt, and manganese.

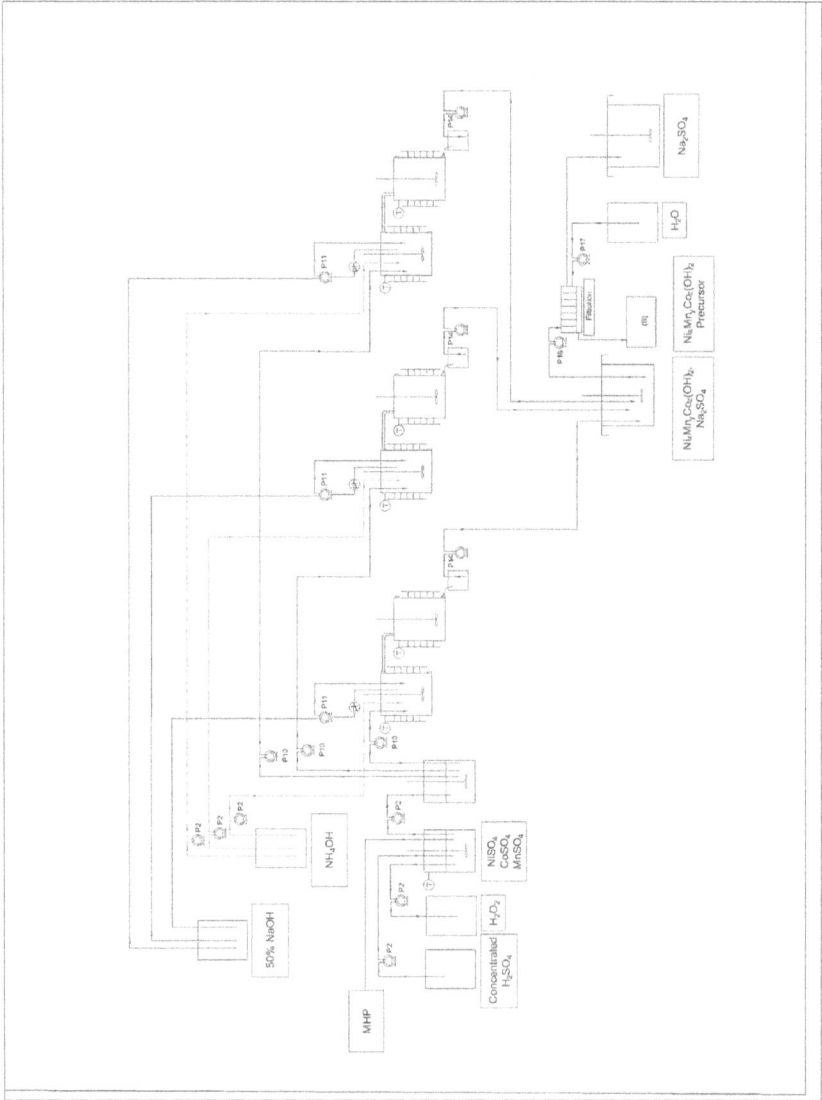

FIGURE 5.4 LIB cathode precursor generation.

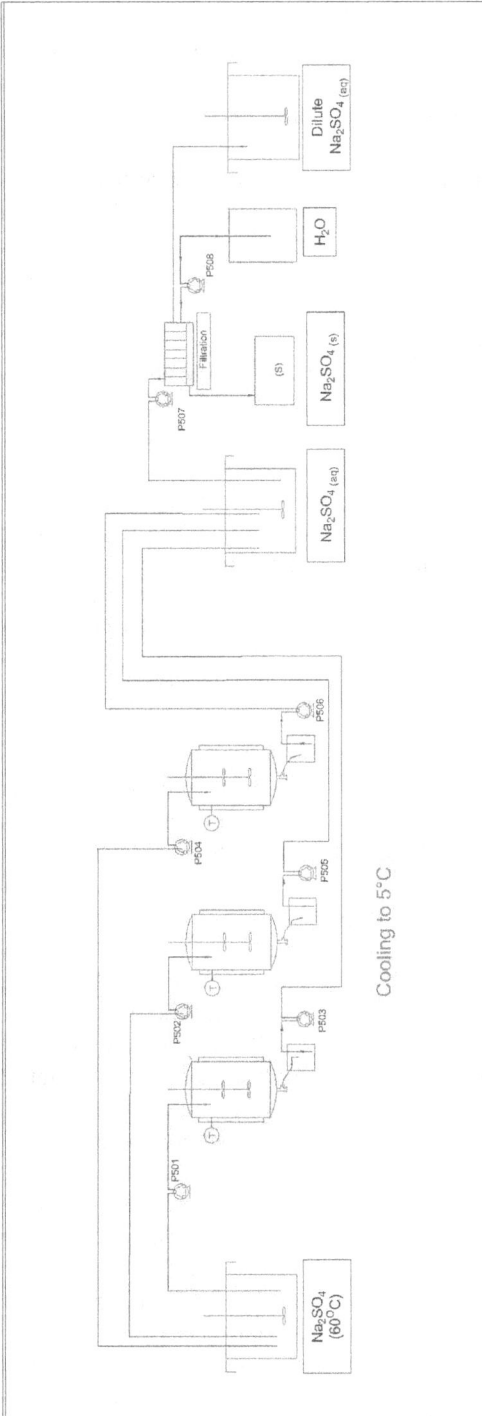

FIGURE 5.5   Sodium sulfate crystal generation.

3 two-cascade reactors simultaneously with a fixed flow rate. The pH of the 3 two-cascade reactors is controlled by 20–50% sodium hydroxide (NaOH) and buffered by ammonia at 10.9–11.5 to allow LIB cathode precursor to precipitate and to grow to the required particle size. The slurry containing the LIB cathode precursor is then pumped to a pressure filter to separate the LIB cathode precursor and the remaining sodium sulfate (Na_2SO_4) solution.

Figure 5.5 shows the process to recover sodium sulfate through crystallization. The sodium sulfate solution is pumped into three crystallizers simultaneously. The temperature of the crystallizers is controlled at 5°C as sodium sulfate has a low solubility at low temperature as shown in Figure 5.7 [7]. The solubility of sodium sulfate decreases more than tenfold between 32.4°C to 0°C, where it reaches a minimum of 4.76 g Na_2SO_4 per 100 g water at 0°C.

The sodium sulfate crystals can then be filtered out by a filter press. The sodium sulfate can be sold as a product to improve the economy of the recycling process.

Figure 5.6 shows the process of using three electrodialysis cells connected in series to recover lithium and to regenerate sulfuric acid. The lithium sulfate solution is first preheated to 80°C and then pumped into an evaporator to evaporate excess water and concentrate the lithium sulfate solution. The evaporated water is condensed and collected into a tote, which can be used as wash water in the filtration processes. The concentrated lithium sulfate is then pumped into the central compartment of the first electrodialysis cell. Dilute sulfuric acid is pumped into the anode compartment and dilute lithium hydroxide solution is pumped into the cathode compartment of the first electrodialysis cell. The lithium sulfate will be split into lithium ions and sulfate ions where lithium ions will penetrate into the cathode compartment and sulfate ions will penetrate into the anode compartment. In the anode compartment, water is split into oxygen and hydrogen ions. The hydrogen ions join with sulfate ions to form sulfuric acid. In the cathode department, water is split into hydrogen and hydroxide ions. The hydroxide ions join with lithium ions to form lithium hydroxide. With electrodialysis, the lithium sulfate concentration is decreased, the sulfuric acid concentration is increased, and the lithium hydroxide concentration is increased. The diluted lithium sulfate solution is fed from the first electrodialysis cell into the central compartment of the second electrodialysis cell and then the third electrodialysis cell where the lithium sulfate concentration is further decreased. The concentrated sulfuric acid and the concentrated lithium hydroxide solution from the first electrodialysis cell are fed into the anode compartment and the cathode compartment of the second electrodialysis cell and then the third electrodialysis cell, where the concentration of sulfuric acid and lithium hydroxide are further increased. The diluted lithium sulfate and the concentrated sulfuric acid can be reused in the process to prepare weak sulfuric acid to separate aluminum foil and cathode active material. The concentrated lithium hydroxide solution is fed into crystallizers to form lithium hydroxide crystals [1]. The lithium hydroxide crystals can then be combined with the lithium battery cathode precursor to produce new LIB cathode active material. Therefore, the closed-loop recycling process has achieved two perspectives; (1) commercial grade NMC cathode active materials can be generated from recycling spent/waste LIB cathode active material,

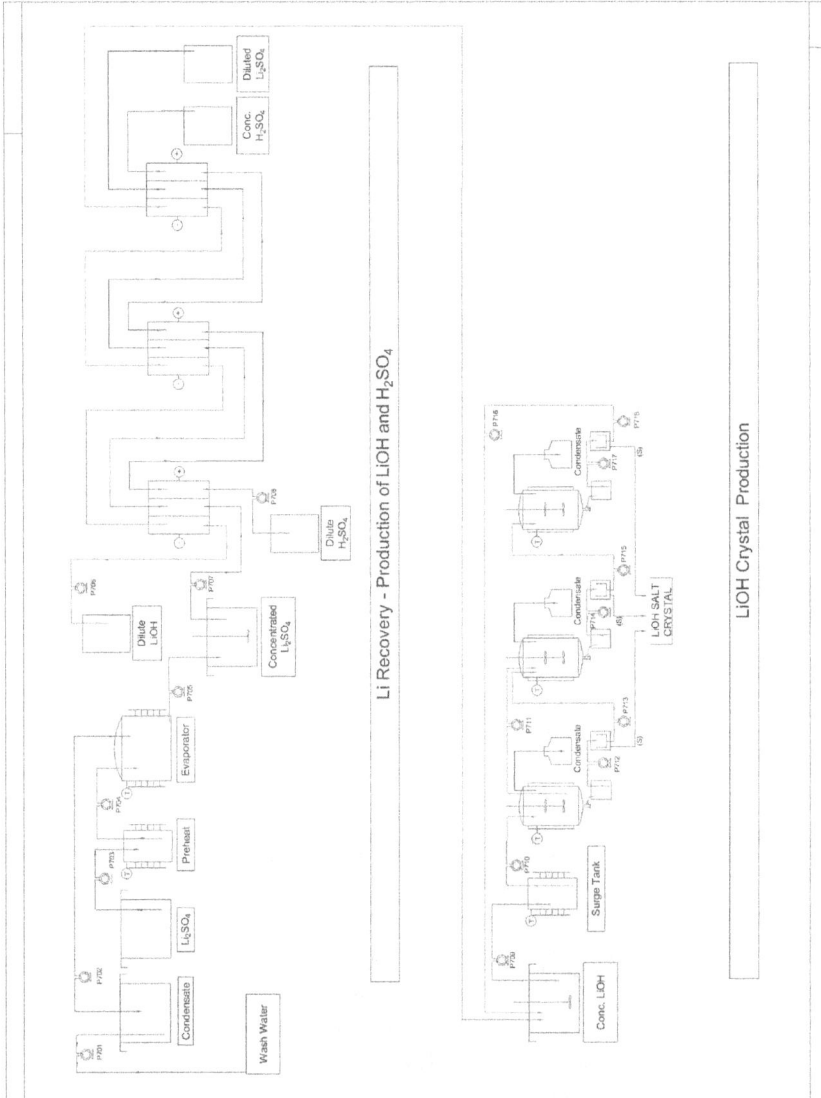

FIGURE 5.6 Lithium recovery and sulfuric acid regeneration via electrodialysis and lithium hydroxide production.

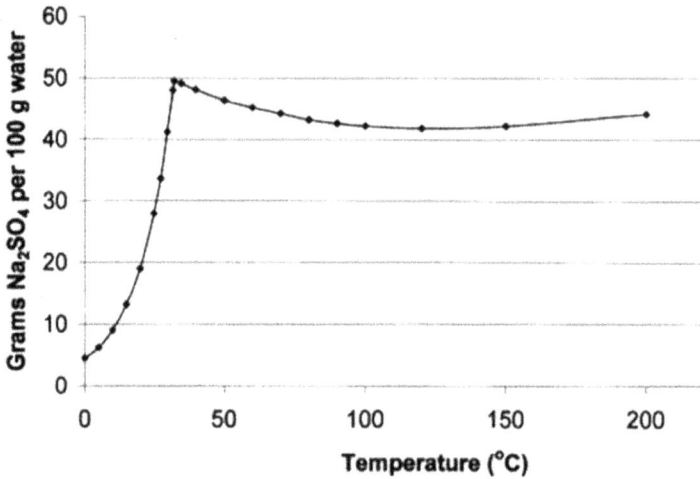

FIGURE 5.7   Solubility of sodium sulfate vs temperature [7].

and (2) H_2SO_4 and LiOH deriving from mother liquor are re-engaged in the whole process to achieve the mass balance without waste discharge.

5.2   MASS BALANCE ANALYSIS OF SEPARATION, LEACH, AND METAL RECOVERY

As shown in Figures 5.1–5.6, the hydrometallurgical regeneration of cathode precursor process from spent LIB cathode active material includes 6 stages, which are separation of aluminum foil and cathode active material, leaching of cathode active material, aluminum recovery, nickel, cobalt, and manganese recovery, lithium recovery, and LIB cathode precursor generation. During lithium recovery, sulfuric acid and lithium hydroxide are generated. The mass balance of lithium can be expressed as follows:

$$W_{SpentCathode}C_{Li_{spent}} + W_{LiOH_{pH5.5}}\,C_{LiOH_{pH5.5}} + W_{LiOH_{pH11}}C_{LiOH_{pH11}}$$
$$= W_{D\_Li_2SO_4}C_{D\_Li_2SO_4} + W_{C\_LiOH}C_{C\_LiOH} + W_{LiOH\_Crystal}$$

where $W_{SpentCathode}$ is the weight of spent LIB cathode active material in Kg, $C_{Li_{spent}}$ is the concentration of Li in the spent LIB cathode active material in mg/Kg, $W_{LiOH_{pH5.5}}$ is the weight of LiOH solution consumed to increase the leachate pH to 5.5 in Kg, $C_{LiOH_{pH5.5}}$ is the concentration of Li in the LiOH solution in mg/Kg, $W_{LiOH_{pH11}}$ is the weight of LiOH solution consumed to increase the leachate pH to 11 in Kg, $C_{LiOH_{pH11}}$ is the concentration of Li in the LiOH solution in mg/Kg, $W_{D\_Li_2SO_4}$ is the weight of the dilute Li_2SO_4 solution from the output of the central compartment of the electrodialysis cell in Kg, $C_{D\_Li_2SO_4}$ is the concentration of Li in the dilute Li_2SO_4 solution in mg/Kg, W_{C_LiOH} is the weight of a portion of the concentrated LiOH solution from the output of the cathode compartment that redirect back to the recycling process in

Kg, C_{C_LiOH} is the concentration of Li in the concentration LiOH solution in mg/Kg, and $W_{LiOH\_Crystal}$ is the weight of LiOH crystal obtained from the crystallization process in Kg.

The mass balance of Al in the leachate can be expressed as follows:

$$W_{SpentCathode}C_{Al_{spent}} = W_{Al(OH)_3}C_{Al(OH)_3}$$

where $W_{SpentCathode}$ is the weight of spent LIB cathode active material in Kg, $C_{Al_{spent}}$ is the concentration of Al in the spent LIB cathode active material in mg/Kg, $W_{Al(OH)_3}$ is the weight of $Al(OH)_3$ precipitate obtained at pH 5.5 in Kg, and $C_{Al(OH)_3}$ is the concentration of Al in the $Al(OH)_3$ in mg/Kg.

The mass balance of Ni, Co, or Mn can be expressed as follows:

$$W_{SpentCathode}C_{Metal_{spent}} = W_{Al(OH)_3}C_{Metal} + W_{Metal(OH)_2}C_{Metal(OH)_2}$$

where $C_{Metal_{spent}}$ is the concentration of Ni, Co, or Mn in the spent LIB cathode active material in mg/Kg, C_{Metal} is the concentration of Ni, Co, or Mn in the $Al(OH)_3$ precipitate in mg/Kg, $W_{Metal(OH)_2}$ is the weight of the metal hydroxide precipitate obtained at pH 11 in Kg, and $C_{Metal(OH)_2}$ is the concentration of Ni, Co, or Mn in the metal hydroxide precipitate in mg/Kg.

The total consumption of sulfuric acid for the recycling process can be expressed as follows:

$$W_{H_2SO_4Consumption} = W_{H_2SO_4Separation} + W_{H_2SO_4Leach} + W_{H_2SO_4Redisolve} - W_{C\_H_2SO_4}C_{C\_H_2SO_4}$$

Where $W_{H_2SO_4Separation}$ is the weight of 98 wt% sulfuric acid consumed in the separation of aluminum foil and LIB cathode active material in Kg, $W_{H_2SO_4Leach}$ is the weight of 98 wt% sulfuric acid consumed during leaching of LIB cathode active material in Kg, $W_{H_2SO_4Redisolve}$ is the weight of 98% sulfuric acid consumed during redissolve metal hydroxide to make LIB cathode precursor in Kg, $W_{C\_H_2SO_4}$ is the weight of concentrated sulfuric acid from the anode compartment of electrodialysis cell in Kg, and $C_{C\_H_2SO_4}$ is the concentration of concentrated sulfuric acid from the anode compartment of electrodialysis cell in weight percentage (wt%).

Table 5.1 shows the mass balance data of lithium, nickel, manganese, and cobalt of recycling from waste and spent NMC cathode obtained from one of the recycling experiments. The concentration of aluminum, cobalt, lithium, nickel, and manganese were track by ICP in separation and leach step, in aluminum, nickel, cobalt, and manganese recovery step, and in LIB cathode precursor generation step. As shown in Figure 5.1, the accountability of cobalt, manganese, and nickel from separation to precursor generation is 94%, 95%, and 93% respectively. The accountability of lithium including the lithium in the waste and spent cathode, LiOH added to recover aluminum, and LiOH added to recover cobalt, manganese, and nickel is about 99%.

TABLE 5.1

Mass Balance of Aluminum, Cobalt, Lithium, Manganese, and Nickel Form Separation to Precursor Generation

Precursor NMC-T5		Wt/Vol	Assay					Overall Extraction/Recovery (%)				
			Al (mg/L) or (mg/kg)	Li (mg/L) or (mg/kg)	Co (mg/L) or (mg/kg)	Mn (mg/L) or (mg/kg)	Ni (mg/L) or (mg/kg)	Al	Li	Co	Mn	Ni
Separation and leach	T5 NMC cathode Feed (kg)	1.55	97534.1	63207.2	130476.2	140392.9	244200.2					
	Al Foil residue (kg)	0.24	486070.7	53596.6	20889.0	59941.4	103597.5	0.7238	0.0873	0.0264	0.0549	0.0640
	Leaching Slurry (Kg)	9.52	4586.2	13603.7	19063.1	25669.7	37357.1	0.2663	0.8642	0.9386	0.9163	0.8997
	Pre-leach Wash1 (L)	1.94	754.8	3529.8	3338.5	3683.9	7016.3	0.0090	0.0458	0.0336	0.0269	0.0345
	Pre-leach Wash2(L)	2.23	65.3	178.2	126.3	236.2	324.0	0.0009	0.0027	0.0015	0.0020	0.0018
	Calculated Head							1.0000	1.0000	1.0000	1.0000	1.0000
Al removal, bulk precipitation, and re-dissolution	LiOH consumption adjust pH to 5.5 (ml)	2393.0		29780.0								
	pH5.5 Al Cake (g)							0.9962	0.1379	0.3712	0.3024	0.4308
	pH5.5 Al removal Filtrate (ml)	6940.0	5.0	21107.0	16515.5	25112.4	29376.4	0.0008	0.3725	0.6373	0.7534	0.5795
	LiOH consumption (bulk precipitation) (ml)	2393.0		29780.0								
	Bulk precipitation filtrate (ml)	3840.0	1.0	23860.3	0.5	0.2	0.5	0.0001	0.2330	0.0000	0.0000	0.0000
	Bulk precipitation Wash 1 (ml)	7470.0	1.0	9289.9	0.5	0.2	0.5	0.0002	0.1765	0.0000	0.0000	0.0000

(Continued)

TABLE 5.1 (Continued)
Mass Balance of Aluminum, Cobalt, Lithium, Manganese, and Nickel Form Separation to Precursor Generation

Precursor NMC-T5	Wt/Vol	Assay					Overall Extraction/Recovery (%)				
		Al (mg/L) or (mg/kg)	Li (mg/L) or (mg/kg)	Co (mg/L) or (mg/kg)	Mn (mg/L) or (mg/kg)	Ni (mg/L) or (mg/kg)	Al	Li	Co	Mn	Ni
Bulk precipitation Wash 2 (ml)	6925.0	1.0	3391.2	0.5	0.2	0.5	0.0002	0.0597	0.0000	0.0000	0.0000
Bulk precipitation Wet cake (g)	4455.0										
Bulk precipitation Dry cake (g)	952.0	150.2	14159.5	122998.4	177964.4	217546.3	0.0033	0.0343	0.6511	0.7324	0.5887
H2SO4 used for redissolution (g)	675.7						0.0000	0.0000	0.0000	0.0000	0.0000
Bulk precipitation redissolution filtrate (ml)	9383.4	1.0	1217.9	10693.8	12654.6	18954.1	0.0002	0.0291	0.5579	0.5133	0.5055
Bulk precipitation redissolution dry solid (g)	159.1	868.4	3226.8	80035.5	267909.3	140695.2	0.0032	0.0013	0.0708	0.1843	0.0636
Calculated Head							1.0000	1.0000	1.0000	1.0000	1.0000
Precursor test Precursor Reactor Feed (ml)	9383.4	1.0	1217.9	10693.8	12654.6	18954.1					
25% NaOH Consumption (ml)	1334.2										
NH4OH Consumption (L)	903.1										
O/F Reactor Filtrate (ml)	9569.6	5.0	588.0	52.7	1.0	3011.9	0.4455	0.4804	0.0053	0.0001	0.1739
O/F Reactor Repulp wash 1 (ml)	4948.6	1.0	37.1	0.9	0.2	3.9	0.0461	0.0157	0.0000	0.0000	0.0001

(Continued)

TABLE 5.1 (Continued)
Mass Balance of Aluminum, Cobalt, Lithium, Manganese, and Nickel Form Separation to Precursor Generation

Precursor NMC-T5	Wt/Vol	Assay					Overall Extraction/Recovery (%)				
		Al (mg/L) or (mg/kg)	Li (mg/L) or (mg/kg)	Co (mg/L) or (mg/kg)	Mn (mg/L) or (mg/kg)	Ni (mg/L) or (mg/kg)	Al	Li	Co	Mn	Ni
O/F Reactor Repulp wash 2 (ml)	4806.0	1.0	2.8	0.5	0.2	0.5	0.0447	0.0012	0.0000	0.0000	0.0000
O/F Reactor Wet cake (g)	519.9										
O/F Reactor Dry cake (g)	218.4	16.3	57.7	179158.0	213031.1	206483.6	0.0330	0.0011	0.4127	0.4129	0.2720
O/F Reactor Dry-cake Moisture%	0.6										
Precursor Reactor Filtrate (ml)	6945.6	5.0	798.8	254.5	1.0	4037.8	0.3233	0.4737	0.0186	0.0001	0.1692
Precursor Reactor Repulp wash 1 (ml)	4375.9	1.0	60.1	19.0	0.2	42.9	0.0407	0.0225	0.0009	0.0000	0.0011
Precursor Reactor Repulp wash 2 (ml)	4547.9	1.0	5.4	2.0	0.2	0.5	0.0423	0.0021	0.0001	0.0000	0.0000
Precursor Reactor Wet cake (g)	626.1										
Precursor Reactor Dry cake (g)	306.0	8.5	128.8	174197.0	216102.0	207855.5	0.0242	0.0034	0.5623	0.5870	0.3837
Precursor Reactor Dry cake Moisture%	0.5										
Precursor Stage Calculated Head											
Accountability								99%	94%	95%	93%

5.3   ENERGY CONSUMPTION AND ECONOMICS ANALYSIS OF THE CLOSED-LOOP HYDROMETALLURGICAL PROCESS TO RECYCLE WASTE/SPENT LITHIUM-ION BATTERY CATHODE ACTIVE MATERIAL

Energy consumption can be calculated as follows:

$$E_{consumption} = \sum_{i=1}^{n} E_{p_i} + \sum_{i=1}^{n} E_{qh_i} + \sum_{i=1}^{n} E_{qc_i} + \sum_{i=1}^{n} E_{m_i} + \sum_{i=1}^{n} E_{c_i} + \sum_{i=1}^{n} E_{ED_i}$$

where E_p is the energy consumed by the pumps, E_{qh} is the energy consumed by the heater, E_{qc} is the energy consumed by the chiller, E_{mn} is the energy consumed by the mixer, E_c is the energy consumed by the controller such as pH controllers and ORP controllers and EED is the energy consumed by the electrodialysis cells.

The energy consumed by each pump is related to the head rise of the solution through a pump, which can be expressed as follows [8]:

$$h_a = (p_2 - p_1)/\gamma + (h_2 - h_1) + (v_2^2 - v_1^2)/2g \qquad (5.1)$$

where h_a is the actual head rise in m, p is the pressure in N/m^2 or Pa, h is the elevation in m, γ is equal to ρg which is the specific weight of fluid in N/m^3, v is velocity in m/s, ρ is the density of fluid in kg/m^3, and g is the acceleration of gravity (9.81 m/s^2).

The actual head rise can be calculated using the following Eq. 5.2:

$$h_a = h_{shaft} - h_{loss} \qquad (5.2)$$

where h_{shaft} is the shaft work in the pump and h_{loss} is the head loss through the pump.

Normally, the inlet velocity and the outlet velocity of a pump are the same. Assuming the inlet and outlet elevation are equal, the head rise can be expressed as

$$h_a = (p_2 - p_1)/\gamma \qquad (5.3)$$

The energy consumed by a pump is calculated trough Eq. 5.4:

$$Ep = h_a \cdot g \cdot T_h \qquad (5.4)$$

where Ep = specific work in Nm/kg or J/kg, and T_h is operation time in hours.

The energy consumed by the heater or chiller is calculated using Eq. 5.5 or Eq. 5.6:

$$Eq_h = m \cdot Cp \cdot \Delta T \cdot T_h \qquad (5.5)$$

$$Eq_c = m \cdot Cp \cdot \Delta T \cdot T_h \qquad (5.6)$$

where Eq_h and Eq_c is the heat variable, m is the mass of the object, Cp is the specific heat, and ΔT is the temperature change.

The hydrometallurgical recycling process uses mixer to mix acid, waste and spent cathode active material, SO_2 gas, and lithium hydroxide in various stage. The horsepower required by a mixer can be calculated by the following Eq. 5.7 [9, 10]:

$$HP_{theoretical} = 6.54 \times 10^{-14} \cdot g_{slurry} \cdot p_i \cdot D \cdot T_h \cdot 1.5 \times 10^3 \qquad (5.7)$$

where g_{slurry} is the specific gravity of the slurry in Kg/L, p_i is the power factor for the specific impeller type, and D is the diameter of impeller in inches.

The energy consumed by the mixer in kW is calculated by Eq. 5.8 and the energy consumed by the electrodialysis cell in kW is calculated by Eq. 5.9:

$$E_m = HP_{theoretical} \times 0.7355 \qquad (5.8)$$

$$E_{ED} = V_{ED} \times C_{ED} \qquad (5.9)$$

where V_{ED} is the voltage of the electrodialysis cell in Volt and C_{ED} is the current of electrodialysis cell in Amp.

Based on the developed flowsheets, experimental data, and mass balance data, the electricity consumed by the pumps, mixers, heaters, chillers, pH controllers, ORP controllers, centrifuge, power suppliers, crystallizers, and distillators are calculated. The total cost of electricity based on USD 0.12 per kW is estimated. The regeneration of sulfuric acid and lithium hydroxide as well as the consumption of sulfuric acid, sulfur dioxide, lithium hydroxide, ammonia, sodium hydroxide, manganese sulfate, cobalt sulfate, and distilled water are recorded. Based on the available chemical cost, the economics of the closed-loop hydrometallurgical process to recycle waste/spent LIB cathode active material is analyzed as shown in Table 5.2.

As shown in Table 5.2, the cost of chemical and energy to treat 1000 kg of spent/waste cathode active materials is USD 14,268.99, where the cost of spent/waste LIB cathode is 46.9% of the total chemical and energy cost.

Table 5.2 shows that processing 1000 Kg of waste/spent NMC 532 cathode can generate 1148 Kg of NMC 532 LIB cathode precursors, 178 kg of lithium hydroxide crystals, and 3200 kg of sodium sulfate crystals, and a total revenue of USD 21,062.25 can be obtained based on the market data in June 2021 with the selling price for NMC 532 cathode precursor being USD 15.10 to USD 16.00 per kg, the selling price for lithium hydroxide being USD 13.35 to USD 13.82 per kg, and the selling price for sodium sulfate crystals being USD 0.22 per kg.

5.4   SUMMARY

The economic analysis shows that the cost to obtain waste/spent cathode will greatly impact the economy of the developed closed-loop hydrometallurgical process. As such, how to secure the supply and the cost of the waste/spent cathode will define if the process can be commercialized.

Environmentally friendly electrical vehicles powered by LIBs have been rapidly developed and commercialized with a trend to replace vehicles powered by combustion engines. For example, sales totaled 1.367 million electric vehicles in 2020 and

TABLE 5.2

Economic Analysis of the Closed-Loop Hydrometallurgical Process to Recycle Waste/Spent Lithium-Ion Battery Cathode Active Material

	Chemitcal	Requirement Per Day (Kg)	Unit Cost (USD)	Total Cost (USD)
Chemical cost	Spent NMC Cathode	1000.00	$6.69	$6686.48
	98 wt% H_2SO_4	498.38	$0.45	$222.16
	SO_2	431.29	$0.23	$100.61
	NaOH	2283.93	$0.16	$373.30
	NH_4OH	688.99	$0.19	$819.00
	$MnSO_4 \cdot H_2O$	443.70	$0.48	$210.97
	$CoSO_4 \cdot 7H_2O$	318.96	$7.13	$2274.90
	H_2O	3446.43	$0.01	$51.21
			Subtotal	$10738.64

	Process	Equipment	kWh	Energy Cost (USD)
Energy cost @ USD $0.12 per kW	Separation & Leach	Pumps & Mixers	1953.20	$232.18
	Al, Ni, Co, Mn Recovery	Heaters, Pumps, Mixers, Controllers	2295.90	$272.92
	Redissolve	Pumps & Mixers	666.67	$79.25
	Co-precipitate	Heaters, Pumps, Mixers, Controllers	16800.00	$1997.03
	Filtration & Dry	Filter presses & centrifuges	1000.00	$118.87
	Electrodialysis	Power suppliers	278.16	$33.07
		Pumps & Mixers	111.76	$13.29
	Crystallization and distillation	Crystallizers and distillators	6593.33	$783.75
			Subtotal	$3530.34
			Chemical + Energy cost	$14268.99

	Product	Production per Day (Kg)	Unit Cost (USD)	Total Cost (USD)
Revenue	$Ni_{0.5}Mn_{0.3}Co_{0.2}(OH)_2$	1148.81	$15.60	$17923.46
	LiOH	178.39	$13.60	$2425.36
	$Na_2SO_4 \cdot 10H_2O$	3200.88	$0.22	$713.42
			Revenue	$21062.25

are expected to reach 245 million by 2030 [11]. With the booming growth of electric vehicles, the number of LIB gigafactory has increased drastically in the past five years, which significantly increased the consumption of lithium, nickel, cobalt, and manganese. Furthermore, with some of these LIB reach end of life, the disposal of LIBs needs to be addressed. In China, the weight of spent or discarded LIBs reached 500,000 tons in 2020, 591,000 tons in 2021, and is expected to reach 2,312,000 ton by 2026. [11, 12, 13]

The high demand of lithium, nickel, cobalt, and manganese and the piling up spent/waste LIBs raises the awareness to recycle them to recover metals for reuse in new LIB cathodes. As recognized, the spent/waste LIBs contain a high content of valuable metals in the range of 5~7% Li, 55~10% Ni, and 5~20% Co.

Methods to recycle spent/waste LIB cathode active materials can categorize into pyrometallurgical, hydrometallurgical, and biometallurgical. Hydrometallurgical recycling of spent/waste LIB cathode active materials, compared to other methods, has many advantages, including high recovery of metals, high purity of the recovered metals, and low-energy consumption. The most common hydrometallurgical recycling method involves using organic solvent to separate cathode active material from aluminum foil; using sulfuric acid and hydrogen peroxide to leach valuable metals such as lithium, cobalt, manganese, nickel, aluminum, and iron from the cathode active material; using sodium hydroxide to remove impurities; using solvent extraction to recover nickel, cobalt, and manganese; and using sodium carbonate to recover lithium. Using sodium hydroxide in the impurity removal step and sodium carbonate in the lithium recover step can result in the final solution enriched with sodium ions. The sodium containing solution needs to be treated before discharge or reuse, which complicates the hydrometallurgical recycling process, makes the management of overall process mass-balance difficult and generates wastewater [11, 12].

An alternative hydrometallurgical treatment to recycle spent/waste LIB cathode active materials is proposed in this book. Separation of cathode active material and aluminum foil is successfully carried out by using weak sulfuric acid without any organic solvent; leach of cathode active material by using sulfuric acid and hydrogen peroxide or leaching with sulfuric acid and sulfur dioxide; impurity removal by using lithium hydroxide instead of sodium hydroxide to eliminate sodium ions contaminating the solution; metal recovery by using lithium hydroxide or lithium carbonate to co-precipitate nickel, cobalt, and manganese into metal hydroxide instead of recovering by solvent extraction; lithium recovery by using electrodialysis to recover lithium as lithium hydroxide and regenerate sulfuric acid. The precipitated metal hydroxide and lithium hydroxide crystals are reprocessed to regenerate LIB cathode active materials. The developed closed-loop hydrometallurgical process to recycle LIB cathode active material recycles and reutilizes all materials without environmental pollution.

REFERENCES

1. Jung C.Y.J., Zhang J., Dai. L., Sui P., A method to recycle lithium-ion battery material, CN Patent ZL201911411831.7, (March 2021).
2. Chow N., Jung C.Y.J., Nacu A., Warkentin D.D., Processing of cobaltous sulphated/dithionate liquors derived from cobalt resource, US Patent 10,308,523 B1, (June 2019).
3. Chow N., Jung C.Y.J., Nacu A., Warkentin D.D., Processing of cobaltous sulpha/dithionate liquors derived from cobalt resource, US Patent 10,246,343 B1, (April 2019).
4. Jung C.Y.J., Zhang J., Dai. L., Sui P., An impurity removal and treatment method in the process of recycling of lithium battery scrap cathode material, CN Patent ZL202011347197.
5. Jung C.Y.J., Zhang J., Dai. L., Sui P., An impurity removal and treatment method in the process of recycling of lithium battery scrap cathode material, CN Patent ZL202110215526.1.

6. Jung C.Y.J., Zhang J., Dai. L., Sui P., An impurity removal and treatment method in the process of recycling of lithium battery scrap cathode material, CN Patent ZL202110212725.7.
7. Sodium sulfate, McGill University, https://www.cs.mcgill.ca/~rwest/wikispeedia/wpcd/wp/s/Sodium_sulfate.htm
8. Engineering ToolBox, Head Rise and the Energy Equation – for Pump or Fan, (2004).
9. Chang R., Physical Chemistry for the Biosciences. University Science Books, 132–133, (2005).
10. Meyer E.F., Thermodynamics of "mixing" of ideal gases: A persistent pitfall, Journal of Chemical Education, 64, 676–677, (1987).
11. Jung J., Sui P.C., Zhang J., A review of recycling spent lithium-ion battery cathode active materials using hydrometallurgical treatments, Journal of Energy Storage, 35, 102217, (2021).
12. Zeng X., Li J., Singh N., Recycling of spent lithium-ion battery: A critical review, Critical Reviews in Environmental Science and Technology, 44(10), 1129, (2014).
13. White Paper on the Development of China's Waste Lithium-ion Battery Recycling, Dismantling and Echelon Utilization Industry, EV Tank and China YiWei Institute of Economics, 2022

6 Modeling and Simulation on the Recycling Process of Spent Lithium-Ion Battery Cathode Materials

Pang-Chieh Sui

CONTENTS

DOI: 10.1201/9781003269205-6

6.1 INTRODUCTION AND OBJECTIVES

The methodology for closed-loop recycling of spent lithium-ion battery (LIB) materials has been introduced in the previous chapters. In essence, such recycling procedures consist of three major processes: leaching of valuable elements from the slurry of spent LIB material, precipitation of metal hydroxides from the leachate, and a salt-splitting process to recycle the lithium-rich solution, which produces sulfuric acid and lithium hydroxide that can be used for the leaching and the precipitation processes, respectively. Figure 6.1 depicts a representative recycling system for spent LIB materials. Black mass enters the system in powder form through leaching, precipitation, and salt-splitting reactors. Valuable metal hydroxides are extracted and eventually made into a new cathode. Experiments can be done to find the optimal operating conditions of these processes, e.g., the temperature, pH value, and retention time through each reactor. These tests are often carried out on a bench scale and then scaled up to a pilot plant scale before mass production. The scale-up of any chemical process is no doubt time-consuming and costly. Proper simulation tools can be employed to accelerate the scale-up and reduce the overall cost of experiments and time.

This chapter aims to review the state-of-the-art modeling and simulation methods that can help expedite the research and development of LIB material recycling. Various simulation methods that have been developed for chemical engineering applications are introduced. The strengths and weaknesses of these methods are discussed. Some examples are provided following the discussion to help the readers understand these methods.

6.1.1 CLASSIFICATION OF SIMULATION METHODS

The modeling for chemical processes can be roughly divided into three categories i.e., simplistic, system dynamics, and phenomenological models. These models are developed with different levels of complexity and principles and can be employed to fulfill different purposes as needed. A simplistic model is developed based on a minimum information requirement and simple principles. The calculation can be carried out by office software

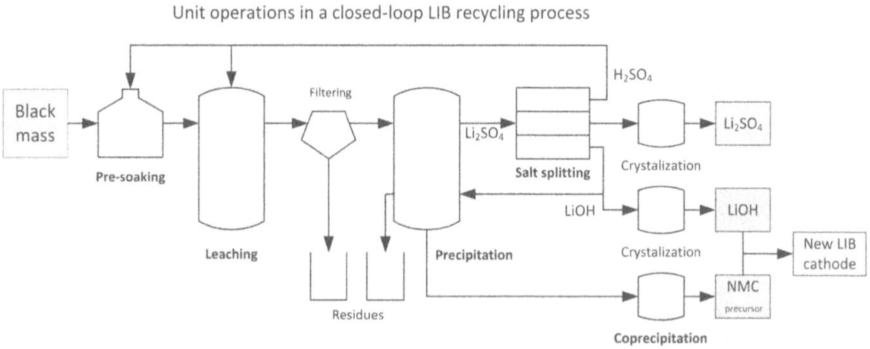

FIGURE 6.1 A closed-loop hydrometallurgical recycling system for spent lithium-ion battery cathode materials.

such as Microsoft Excel. This model category is ideal for the first-cut estimate at the initial design stage. A system dynamic model considers the entire system as a whole. In the simulation, each component of the recycling process is represented by a unit block with input and output ports connected to other components. The whole system's mass and energy balance are solved in a coupled fashion to obtain the physical conditions, e.g., flow rate, temperature, and composition in each component. Information obtained from this software is crucial and helpful for plant design. A phenomenological model considers the physics of relevant transport phenomena and the chemical/electrochemical reactions involved in recycling. Conservation equations of mass, momentum, species, and energy are solved in actual reactor geometry to obtain spatial distributions of physical fields such as flow velocity, composition, and temperature. The phenomenological model can provide predictions with high fidelity, and it is thus ideal for reactor design and detailed investigation of the transport processes at the device level. An excellent review of different levels of modeling for chemical engineering applications can be found in [1].

6.1.2 Organization of This Chapter

Most models described above can be considered simplified models derived from a comprehensive set of conservation equations. Appropriate assumptions can be made to reduce computational time. Reduced dimensionality is often required to simplify the problem. Nevertheless, each model still captures specific, significant characteristics of the processes. Thus, it can simulate the features of the recycling system. The remaining sections of this chapter are organized as follows. In Section 6.2, the mathematical framework that describes the physical phenomena of the recycling processes is first introduced. Conservation equations of mass, species, momentum, energy, and reaction kinetics for various reactions or phase changes are discussed. Representative simulation models and results are presented in Sections 6.3 (system modeling) and 6.4 (computational fluid dynamics) to demonstrate how these models are solved. Finally, a summary of the capabilities of these models is provided.

6.2 MATHEMATICAL FRAMEWORK

A recycling process of spent LIB materials involves a series of chemical/electrochemical processes that treat the raw materials and convert them into final products such as ternary precursor materials and lithium hydroxide. In the present hydrometallurgical approach, spent LIB cathode material in powder form (black mass) is first prepared and fed into a series of leaching reactors. Unwanted materials such as iron, carbon particles, copper, and aluminum are filtered out from the dissolved liquid. The slurry goes through a precipitation stage to form NMC hydroxides, which are collected and used to produce ternary precursor materials from a separate co-precipitation process. The remaining solution, consisting mainly of lithium sulfate, goes through a salt-splitting reactor to produce sulfuric acid and lithium hydroxide, which are reused in the leaching and precipitation processes, respectively, to close the loop of liquid streams of the system.

This section describes the mathematical framework for the transport of the ionic species and liquid flows and the transfer of heat and water in the entire system.

6.2.1 GOVERNING EQUATIONS

From a thermo-fluid discipline's perspective, a recycling process can be described mathematically by a set of conservation equations. The conservation of mass, species, momentum, and energy can be expressed in the following generalized form:

$$\frac{\partial}{\partial t}(\rho\psi) + \nabla \cdot (\rho\vec{v}\psi) + \nabla \cdot (\vec{j}) + \rho\phi + \vec{\gamma} = 0 \tag{6.1}$$

where ρ is the fluid density, ψ is the primary variable, $\vec{v}$ is the fluid velocity, j is the diffusional flux, ϕ is the volume source term, and $\vec{\gamma}$ is the contribution due to driving forces. The primary variables for mass, species, momentum, charge, and energy are 1, x_i (concentration), $\vec{v}$, and h (enthalpy), respectively. The corresponding parameters for these conservation equations are summarized in Table 6.1. The source terms in the species are due to chemical reactions, see Section 6.2.2. The transport of charged species, i.e., positive and negative ions, can be written to recover the Nernst-Plank equation:

$$J_i = \frac{z_i F D_i c_i^m}{RT} \frac{\Delta\varphi_m}{l_m} + D_i^m \left(\frac{dc_i^m}{l_m} \right) + u_m c_i^m \tag{6.2}$$

where superscript i refers to the i-th species, F is the Faraday's constant = 96485 A·s·mol^{-1}; φ is the electrolyte potential, u is the velocity of electrolyte solution, and c_i, D_i, M_i, and z_i are the concentration, diffusion coefficient, mobility, and valence of species i, respectively.

6.2.2 REACTION KINETICS

The source terms in Table 6.1 account for the mass and species transfer due to chemical or electrochemical reactions and the heat generation/absorption associated with these reactions. In hydrometallurgical recycling, multiple types of reactions occur. Homogeneous chemical or electrochemical reactions occur for species in the same phase, mainly in aqueous solutions, where ions and molecules form different

---

TABLE 6.1
Variables and Parameters of Conservation Equations

Conservation / Equation	Primary Variable, ψ	Diffusion, $\vec{j}$	External Force, $\vec{\gamma}$	Source/sink, ϕ
Mass	1	0	0	$\sum \dot{m}_i$
Momentum	$\vec{u}$	$-\mu\nabla|\vec{u}|$	$\vec{g}$	0
Species	x_i	$-D_{ij}\nabla x_i$	0	$\dot{m}_i$
Charge	c_i	$-D_i^m\nabla c_i$	$\dfrac{z_i F D_i c_i^m}{RT}\nabla\varphi$	$\sum \dot{c}_i$
Energy	T	$-k\nabla T$	0	$\sum \dot{q}_i$

molecules, e.g., proton and sulfate ions become sulfuric acid during the electrodialysis process. On the other hand, heterogeneous reactions involve reactions taking place on phase boundaries, e.g., leaching and precipitation.

A general form for a chemical reaction can be expressed as

$$\sum_{i=1}^{N} \upsilon' M_i \left(\xrightarrow{k} \sum_{i=1}^{N} \upsilon'' M_i \right) \tag{6.3}$$

where k is the reaction rate constant, M is the molecular weight of species, and υ' and υ'' are the mole number of the i-th species before and after the reaction.

The rate of reaction for a particular species, r_i, can be written as:

$$r_i = k(\upsilon'' - \upsilon') \prod_{i=1}^{N} M_i^{\upsilon'} \tag{6.4}$$

The homogeneous reactions can be described by the Arrhenius form:

$$k = A e^{-E_a/RT} \tag{6.5}$$

where k is the rate constant of the reaction, A is the pre-exponential factor, E_a is the activation energy of the chemical reaction, R is the universal gas constant, and T is the absolute temperature. Heterogeneous reactions generally follow the same expression as the Arrhenius with an additional dependence on several factors, such as the specific surface area for reaction, phase change rates such as nucleation rate and absorption/desorption rates, etc. Nucleation rate and site density play an essential role in the precipitation processes, especially for the co-precipitation of multiple molecules to form cathode precursors. It is noted that the activation energy for electrochemical reactions is more often expressed in terms of the so-called overpotential, which represents the potential difference between two reactant phases, i.e., the Butler-Volmer equation [2]:

$$I = i_0 \left(e^{-\alpha n F \eta / RT} - e^{(1-\alpha) n F \eta / RT} \right) \tag{6.6}$$

where I is the external current density, n is the number of electrons in the anodic half-reaction, F is Faraday's constant, i_0 is the exchange current density, α is a coefficient with values ranging from 0 to 1, and η is the electrode overpotential.

6.2.3 Phase Equilibria

Equations (6.1) to (6.6) with the variables and terms defined in Table 6.1, along with appropriate boundary conditions (BCs), and initial conditions, constitute the governing equations needed to describe the transport and reactions in a hydrometallurgical system. However, many model parameters and thermodynamic factors must be determined for model closure. Phase equilibria conditions are often required to reflect the relationships among physical and thermodynamic variables properly. The phase equilibrium of a multi-component system cannot be easily described mathematically. It is generally described in phase diagrams, where phase boundaries define the thermodynamic states of the involved components. A widely used tool is the E-pH diagram (potential-pH diagram, or Pourbaix diagram), which shows boundary lines that separate the domains of the predominance of the same species. A recent article by Li et al. [3] reporting on a Li-Ni-Co-Mn-H2O system is an excellent example of thermodynamic analysis on this topic.

6.3 UNIT OPERATION AND SYSTEM DYNAMICS MODELING

The comprehensive mathematical framework discussed in the previous section is a mathematical description of the transport phenomena in a chemical system. However, solving the entire governing equations is not practical for real-world applications. Solving these equations numerically can be daunting and demands substantial computational resources. Simplified models, ideally at the unit-operation level that can yield rapid results, are desirable for general system design and implementation. Processes such as leaching, filtration, precipitation, crystallization, evaporation, and electrodialysis are likely involved in hydrometallurgical recycling. System modeling is then built upon all relevant unit operations. Appropriate assumptions can be made to substantially reduce the complex equations to simple ordinary algebraic equations or ordinary differential equations. For general unit operations, we are primarily interested in the relationship between the inlets, outlets, and unit boundaries. Take a continuous stirred-tank reactor (CSTR), for example, the analysis of mass balance and energy can be significantly simplified if (1) perfect mixing occurs within the reactor, (2) the reactor reaches a steady state, (3) the reactions are fast, or they occur at prescribed reaction rates. With these assumptions, most system devices can be represented by nodal points where simple conservation law holds between the inlets and the outlets. Table 6.2 lists

TABLE 6.2
Transport Processes Taking Place in Relevant Unit Operations of Litium-Ion Battery Recycling

Transport Unit Operation	Heat Transfer	Mass Transfer	Ion Transfer	Reaction	Fluid Dynamics	Phase Change
Leaching	√	√		√	√	√
Precipitation	√	√			√	√
Filtration					√	
Evaporation	√	√			√	√
Crystallization	√	√			√	√
Membrane separation	√	√	√	√	√	√
Sintering	√	√				√

the transport processes taking place in relevant unit operations of a LIB recycling system. Each transport process has been well-documented in common chemical engineering or fluid dynamics literature. However, the complexity of the problem is the coupling among some of the transport processes, particularly the temperature dependence that affects both mass transfer and heat transfer.

The derivation of mass balance and energy balance for each nodal point is described as follows. By applying the Reynolds' transport theorem [4], one can integrate the conservation equations over the entire control volume and convert the differential equations into the flows in and out of the control volume:

$$\frac{d}{dt}\int_{\Omega(t)} f\, dV = \int_{\Omega(t)} \frac{\partial f}{\partial t}\, dV + \int_{\partial\Omega(t)} (\vec{v}_b \cdot \vec{n}) f\, dA \tag{6.7}$$

where $\Omega(t)$ is the domain, t is time, f is the variable of interest, V is the system volume, A is the outer area of the system, $\vec{v}_b$ is the velocity at the boundary, and $\vec{n}$ is the normal vector on the boundary surface. Consider a control volume with multiple inlets and outlets, each with an unknown or known value of mass flow rate $m_{i,IN}$ and $m_{i,OUT}$ respectively, where i is the i-th species and enthalpy per unit mass $h_{i,IN}$ and $h_{i,OUT}$. Conservation of species for the control volume can be written as:

$$\underset{\text{Inlet}}{\sum \dot{m}_{i,IN}} + \underset{\text{Outlet}}{\sum \dot{m}_{i,OUT}} = 0 \tag{6.8}$$

When chemical reactions are involved in a system, the full enthalpy h of each species is used to handle the transfer of energy better:

$$h = \Delta h_f^0 + \int_{T_0}^{T} Cp\, dT \tag{6.9}$$

where Δh_f^0 is the enthalpy of formation at reference temperature T_0 and Cp is specific heat. In this way, the heat of reaction of any given reaction can be computed directly by summing up the energy conservation equation:

$$\underset{\text{Inlet}}{\sum \dot{m}_{i,IN} \cdot h_{i,IN}} + \underset{\text{Outlet}}{\sum \dot{m}_{i,OUT} \cdot h_{i,OUT}} + \Gamma_{ht} = 0 \tag{6.10}$$

where Γ_{ht} is heat transfer across the system boundary.

Each component in a unit operation satisfies species conservation and energy conservation law. Figure 6.2 illustrates the reactors' mass/energy balance in a leaching process. Lines between them represent the streams of liquid and gas flow between reactors. Each reactor has its ports of inlet and outlet, and one can derive the balance equations with the reactor's volume. By writing all conservation

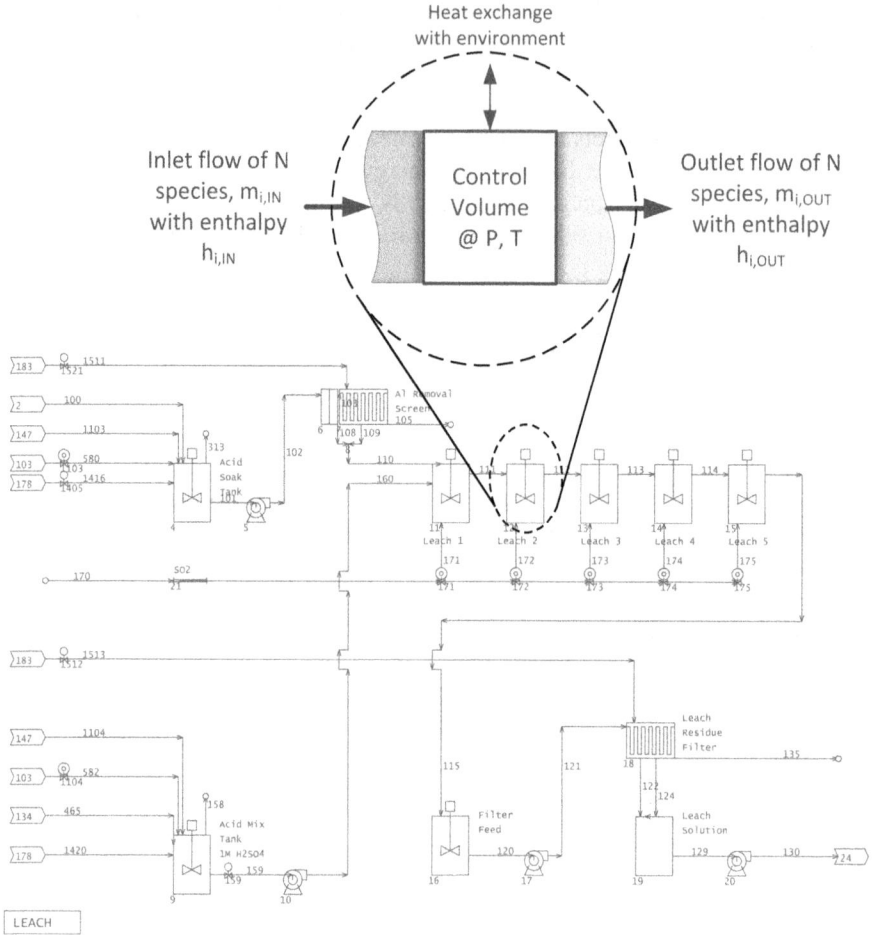

FIGURE 6.2 Schematic diagram of the unit operation in a recycling process.

equations for the system components, one can translate a flow sheet into a flow network, where each component is a nodal point, and the interconnection among components constitutes a network, similar to an electric circuit. The unknowns are each species' mass flow rate and the device's temperature. It should be noted that for a recycling procedure, unlike a general flow network with flows that can go either way of a pipe, the direction of flow is naturally in one direction. This feature makes the analysis much more straightforward. To solve the system of equations, model closure is needed to make the problem well-posed, which can be found through the reactions taking place in each reactor. For instance, for the CSTR model, the relationships of species flow rates between the inlet and outlet flow can be established through reaction kinetics and reactor dimensions. Since most reaction rates are temperature-dependent, the solution procedure of these system equations is likely iterative.

6.3.1 METSIM Examples

Currently, several commercial software tools can perform system-level simulations for hydrometallurgy. Two notable tools are METSIM [5] and Aspen Plus [6]. The former is aimed at hydrometallurgical and inorganic-chemical process applications, whereas the latter is general-purpose software for chemical engineering applications and is not limited to metallic species. This book uses METSIM [7] to demonstrate how system modeling can be set up and employed to study mass/energy balance in a recycling system. A case implemented with Aspen Plus is also provided as a comparison.

6.3.1.1 Variations of Reagents for Acid Dissolution and Precipitation

We first look at several METSIM cases for systems with different reagents and investigate their results. Figures 6.3(a)–(d) show four different configurations for the recycling process. Each case is fed by a mixed mixture consisting of six different types of cathode materials; see the compositions shown in Table 6.3. These cases consider two reduction reagents (SO_2 and H_2O_2) and two bases for precipitation (sodium, NaOH, and lithium, LiOH). An electrodialyzer (ED) is added to the cases with lithium as the base for precipitation. Figure 6.3(e) summarizes the reagents and products for these cases. When NaOH is used for precipitation, Na_2CO_3 is added to extract lithium in the form of Li_2CO_3 and sodium in the form of Na_2SO_4, the former being a chemical product ready to make a new cathode. For cases with an ED, H_2SO_4, and LiOH are produced, which can be introduced back to the leaching and precipitation processes.

Figure 6.4 illustrates representative flowcharts of the procedure for Case 2, which is a recycling system using SO_2 and LiOH, along with an ED at the end. The procedure starts at the leaching of black mass using H_2SO_4 and SO_2 as the reagents, see Figure 6.4(a). Five CSTR reactors are employed in this process to achieve complete leaching. The temperature, pH value, and flow rate are controlled to extract valuable elements, i.e., Li, Ni, Mn, and Co. All other metals, e.g., Al, Fe, Cu, etc., are filtered out from the solution. The liquid solution then undergoes Ni, Mn, and Co precipitation, Figure 6.4(b), forming hydroxides that will be subsequently used in a co-precipitation process to produce NMC precursor material. The flowsheet of Al precipitation (not shown) is similar to the NCM precipitation and separation. The remaining solution contains mainly Li_2SO_4, which is treated by an ED to obtain H_2SO_4 and LiOH, Figure 6.4(c). Excessive Li_2SO_4 effluent from the ED can be crystallized to become a by-product or circulated to the recycling process. The H_2SO_4 is reused in the leaching, whereas the LiOH can be recirculated to the precipitation processes or simply crystallized to become a by-product, as depicted in Figure 6.4(d). The flowsheet of Li_2SO_4 crystallization (not shown) is similar to LiOH crystallization. Water distribution and mass balance are a crucial part of the entire recycling process, throughout which water is the primary carrier of all liquid solutions. Ideally, all water can be balanced by collecting all excessive water from some processes, e.g., distillation, crystallization, and ED. Water can be treated with a reverse-osmosis membrane and redistributed to reactors where water is needed, see Figure 6.4(e). It should be noted that when H_2O_2 is used in leaching, water also enters the system

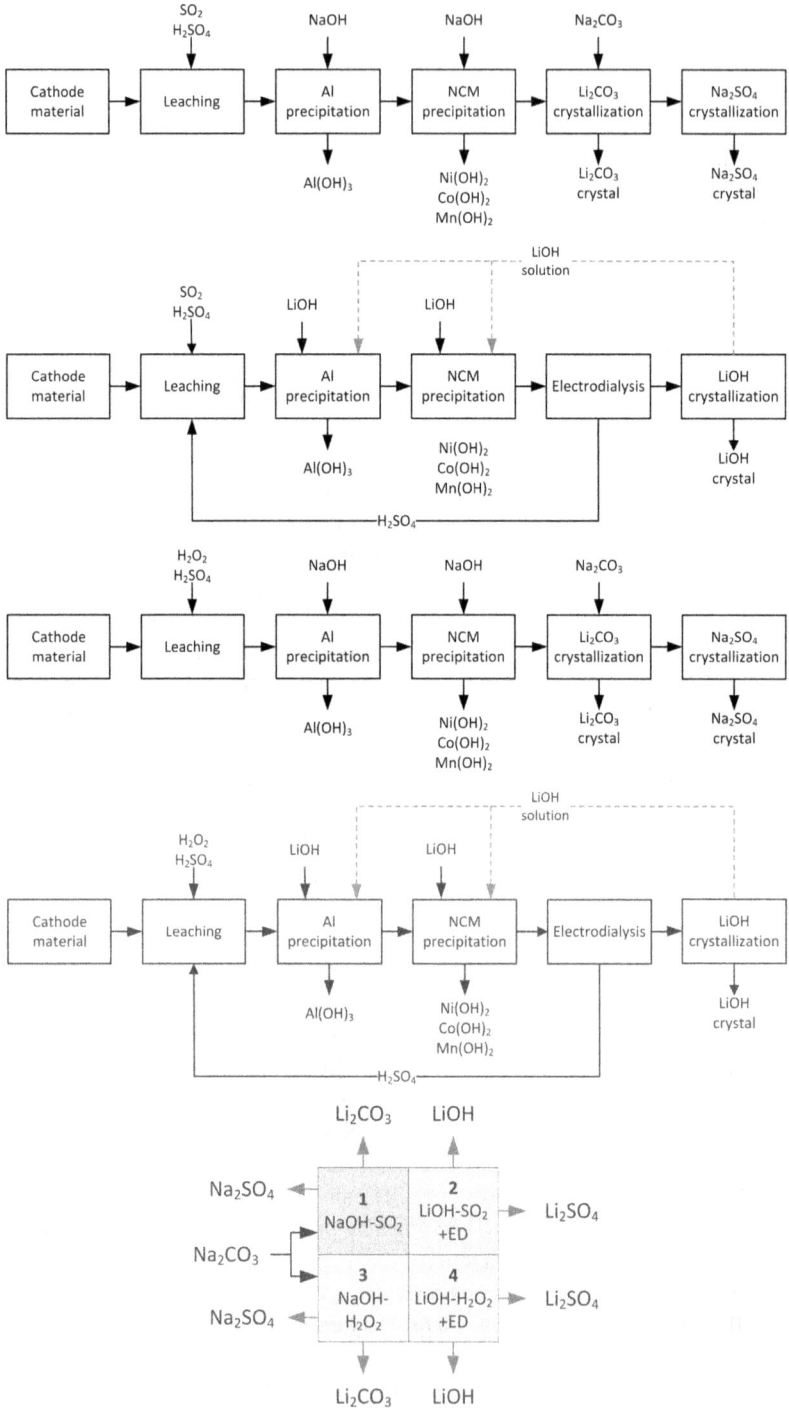

FIGURE 6.3 Flowcharts of four recycling procedures.

TABLE 6.3

Composition of the Mixture Feed for the METSIM Simulation (Cases 1–4)

Flow Rate (g/h) \ Species	Li	Al	Mn	Co	Ni
NCM Type 1	1207.3	4200.0	2867.5	2050.6	5107.2
NCM Type 2	1202.7	4200.0	1904.3	2042.6	6104.8
NCM Type 3	1198.3	4200.0	948.7	1017.6	8110.1
NCA Type 1	1209.3	4388.1	0.0	1129.7	8696.5
NCA Type 2	1209.2	4388.1	0.0	1437.7	8389.0
NCA Type 3	1213.2	4435.9	0.0	1545.4	8211.2

after oxygen dissociates from H_2O_2. Not all water can be recovered, and freshwater will be supplied to compensate for the water lost during certain processes.

Table 6.4 summarizes the mass balance for Cases 1–4. Ni, Co, and Mn yields for these cases are very similar. Since the efficiencies of leaching and precipitation are high for all these cases, most of the valuable elements from the black mass feed are extracted. The yield of lithium depends on the base used in the precipitation process. For the sodium cases (Cases 1 & 3), lithium appears as Li_2CO_3 since Na_2CO_3 is added to replace the sulfate from leaching. When LiOH is used in the precipitation, an ED is included to split the lithium and sulfate ions. Part of the LiOH extracted from the ED returns to the system, and the remaining LiOH is crystallized. Similarly, the Li_2SO_4 after the ED can be either recirculated back to the system or crystallized. The proportions of recirculated LiOH and Li_2SO_4 can be adjusted to maximize the value of end products from the recycling process.

Adding an ED to the system provides benefits, e.g., the consumption of H_2SO_4 for leaching is significantly reduced (Case 4) or becomes zero (Case 2). For the same reason, the NaOH needed for precipitation can be eliminated. The water balance results from these calculations are interesting in that the net change of water for the cases with an ED is relatively lower than those without.

6.3.2 MASS BALANCE STUDY FOR CASES USING DIFFERENT BASES FOR PRECIPITATION

Two additional cases using SO_2 as the reductant are created and analyzed with METSIM. Both cases use 622 NMC black mass at the inlet at a 21 kg/h supply rate. Case 5, see Figure 6.5(a) has a similar configuration as Case 2, which uses sodium as the base for precipitation. Case 6 has a similar configuration to Case 1, see Figure 6.5(b), uses lithium as the base, and is equipped with an ED for salt splitting. The mass flow rates in kg/h are also shown in both figures for significant components. As discussed earlier, the LiOH recycled via the ED cell is used as the base to precipitate metal ions. Thus, lithium from the black mass is reused in the cycle or converted into lithium hydroxide. Similarly, the net input of sulfuric acid to the system can be maintained at zero except for the initial filling before the recycling process starts.

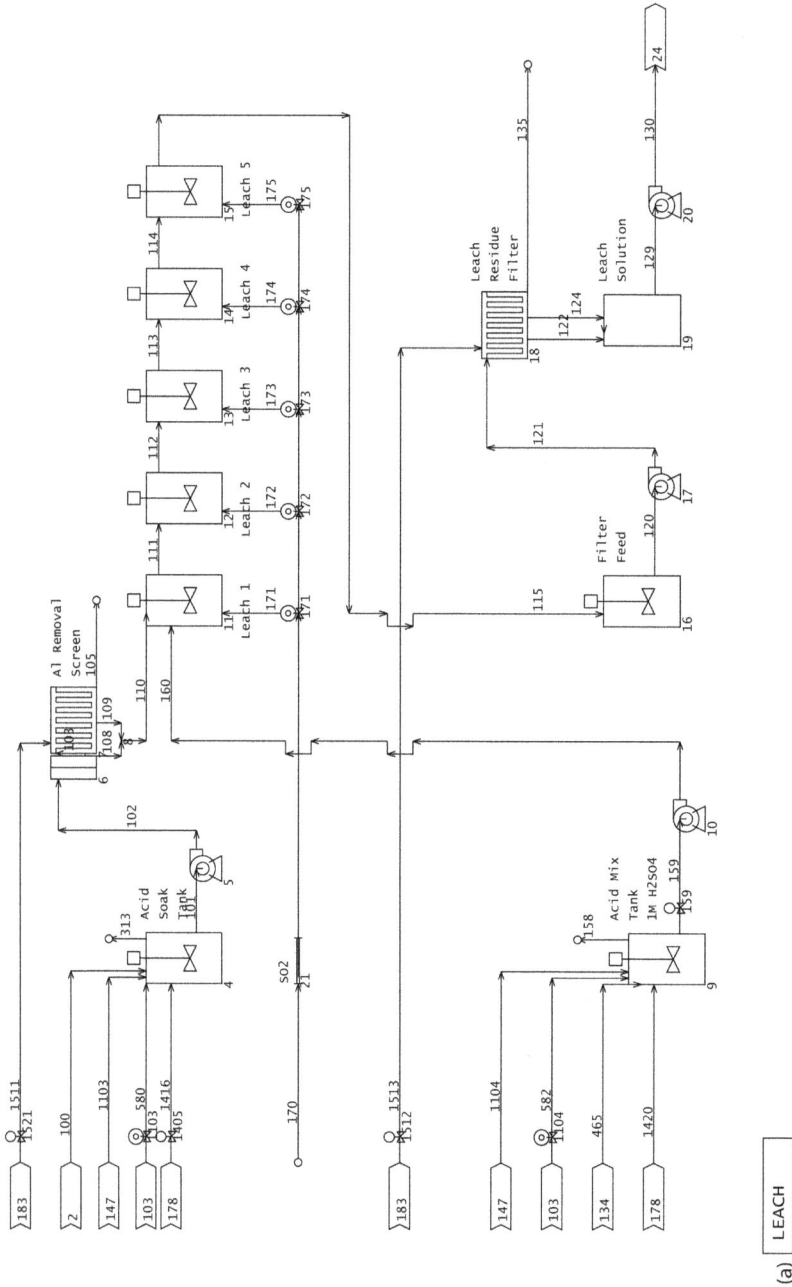

FIGURE 6.4 Flowsheets created in METSIM. (a) Leaching, (b) NCM precipitation and separation, (c) Electrodialysis of Li_2SO_4, (d) Crystallization of LiOH, and (e) Water distribution. (*Continued*)

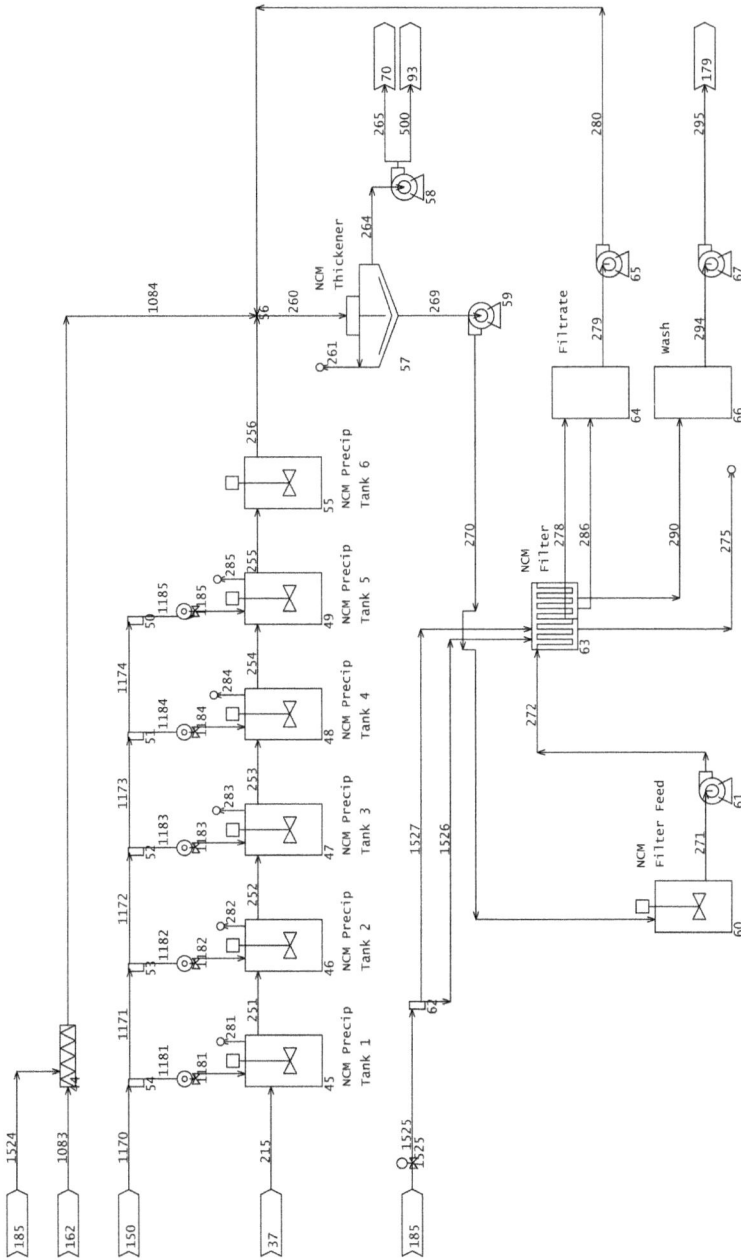

(b) PLS NEUT – Ni,CO,Mn

FIGURE 6.4 *(Continued)*

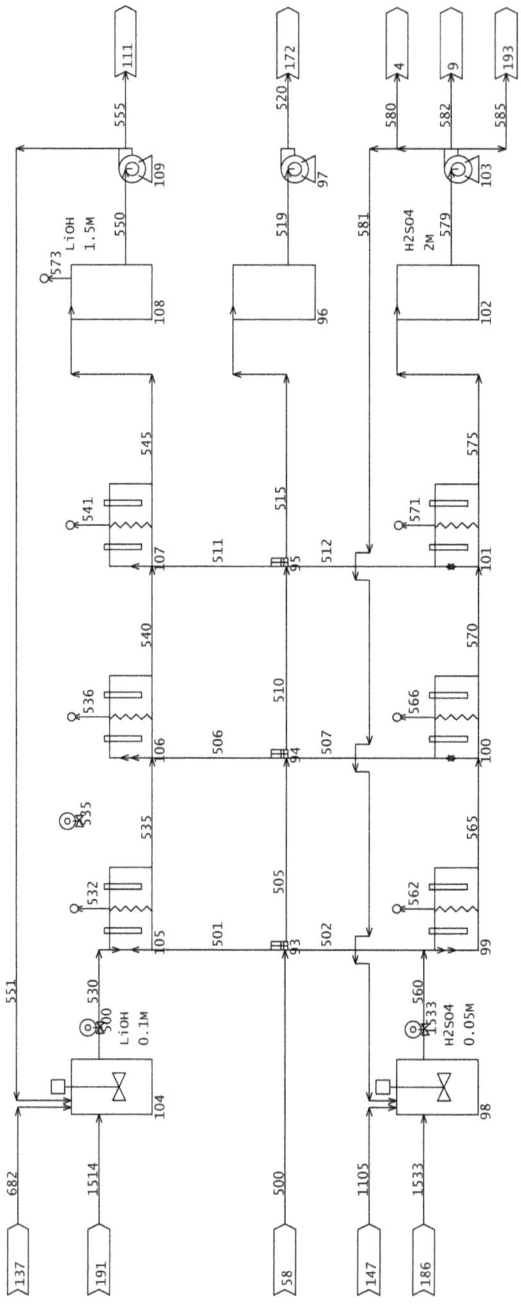

(c) ELECTRODIALYSIS

FIGURE 6.4 *(Continued)*

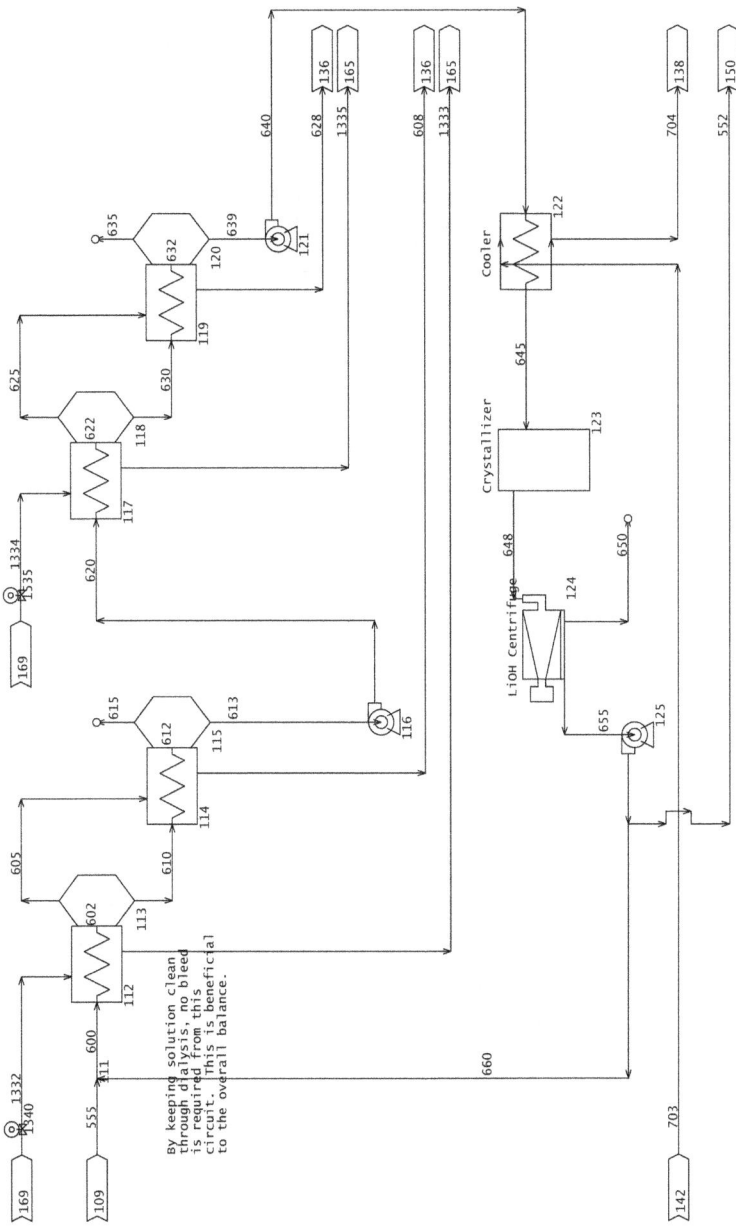

By keeping solution clean through dialysis, no bleed is required from this circuit. This is beneficial to the overall balance.

(d) CRYSTALLIZER – LIOH

FIGURE 6.4 (*Continued*)

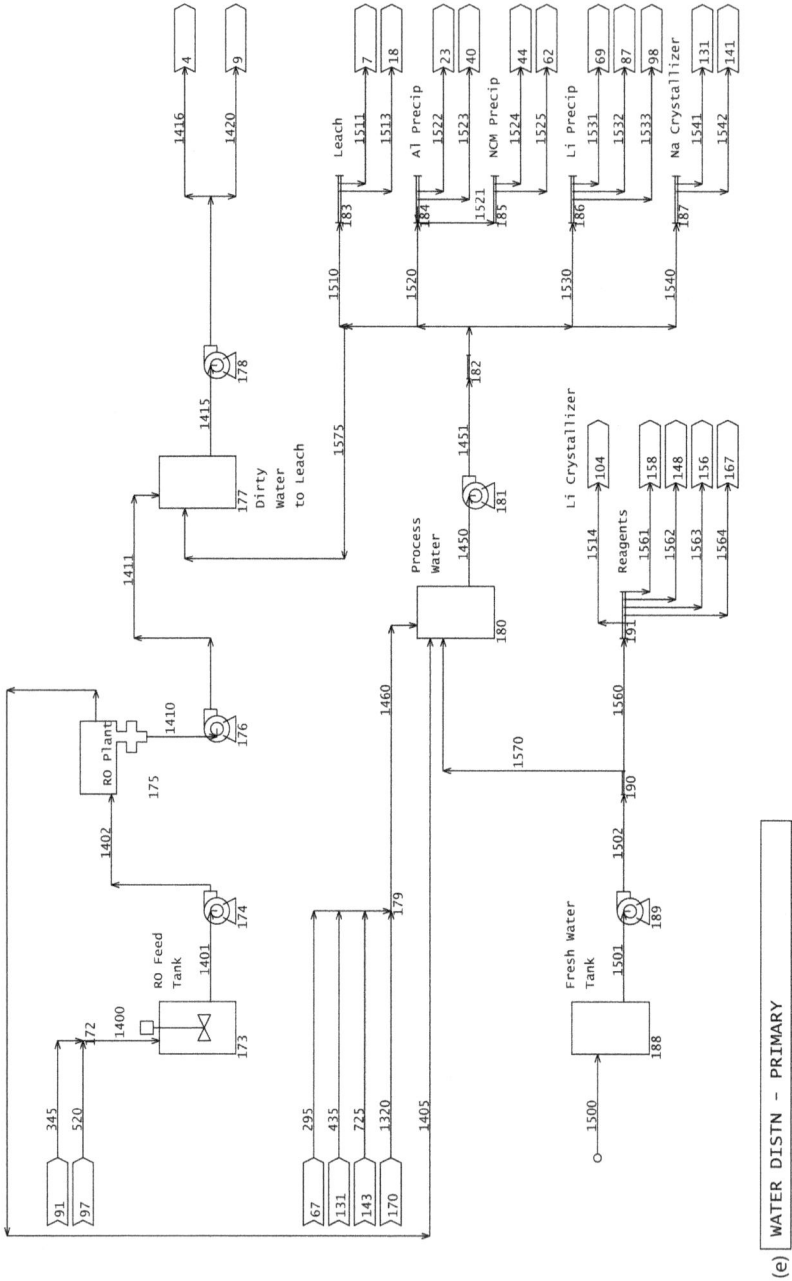

(e) WATER DISTN – PRIMARY

FIGURE 6.4 (Continued)

TABLE 6.4
Summary of Results from METSIM

Case	Reagents							Products							
	SO₂	H₂O₂	NaOH	Na₂CO₃	H₂SO₄	LiOH	Water In/Out	Na₂SO₄	Al(OH)₃	LiOH	Co(OH)₂	Mn(OH)₂	Ni(OH)₂	Li₂SO₄	Li₂CO₃
1	33.34		97.26	66.01	136.69		777.16 337.04	562.35	1.699		14.082	8.964	68.250		35.33
2	33.81					0	1291.21 267.10		1.704	39.02	14.101	8.978	68.298		
3		153.62	114.53	82.41	213.35		958.71 337.80	588.00	1.683		14.096	8.974	68.276		33.32
4		124.99			34.53	0	897.14 363.53		1.702	12.4	14.100	8.980	68.294	39.25	

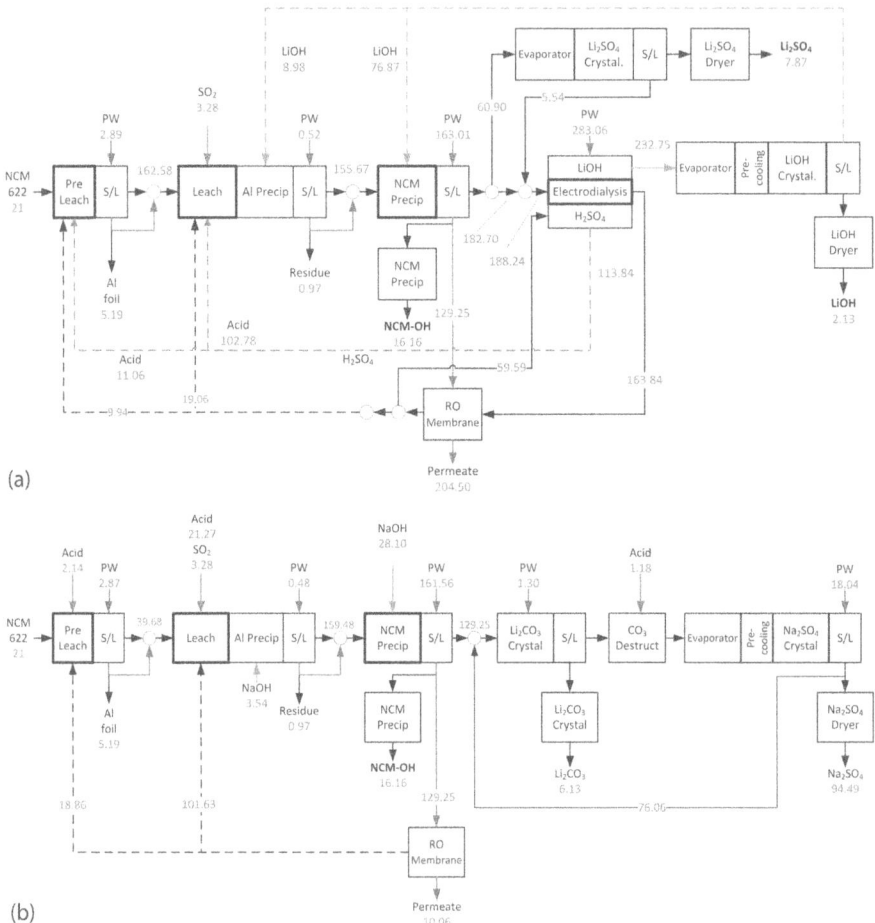

(a)

(b)

FIGURE 6.5 Flowcharts and METSIM results of (a) Case 5 and (b) Case 6.

This arrangement is an elegant design that minimizes the consumption of reagents. It should be noted that the present models do not consider subsequent treatment on the NMC materials to produce precursors at desired NMC proportions, which requires leaching of the NCM-hydroxides and precipitation again in a co-precipitation reactor. If the co-precipitation of NMC materials is also included in the model, then part of the acid and base derived from the ED cell can be redirected to this process.

6.3.3 ASPEN PLUS

General-purpose system software that deals with standard chemical engineering unit operations are desired for design, construction, and operation. Unlike METSIM, which has been widely employed for hydrometallurgy applications, software such as Aspen Plus can also be employed to perform similar tasks. Aspen Plus provides a graphical user interface environment, which can integrate with its proprietary programing language for extended capabilities for user-defined functionality. Live links to communicate with software such as Microsoft Excel™ are also available. This feature makes it a versatile simulation tool for hydrometallurgy. Another advantage of Aspen Plus over METSIM is that it can assist the user in doing some sensitivity analysis, thus optimizing the process design.

In this section, a leaching case built with Aspen Plus is demonstrated. The results are compared with those obtained from METSIM, i.e., Figure 6.4(a). A leaching process is built with Aspen Plus, see Figure 6.6. The system consists of five CSTRs along with two filters. The black mass mixture is fed to the pre-soak tank with acid, and its overflow enters the first CSTR, followed by four other tanks. The outflow concentrations for each tank are prescribed as 50% of its inflow, same as set in the METSIM model. This example gives the readers some flavor of system modeling with general-purpose software. There are abundant studies in the literature on hydrometallurgical applications with Aspen Plus.

Thus far, we have looked at system modeling that can be employed to assist the design and optimal operation of recycling processes for spent LIB materials. The accuracy of these models heavily depends on the empirical or semi-empirical correlations prescribed in the unit operations. To mitigate this drawback, advanced simulation tools based on the computational fluid dynamics method can be employed. This topic will be discussed in detail in the next section with two models on the ED process.

6.4 CFD SIMULATION

Most of the processes in the recycling procedure for spent LIB materials, such as leaching, precipitation, crystallization, and ED, involve multiple chemical components interacting with the varying flow and temperature fields. To better understand the actual change of these materials within each reactor and throughout the entire cycle, computational fluid dynamics (CFD) technology [8] can be employed to numerically investigate the change of all physical fields, e.g., velocity, concentration, and temperature, in actual reactor geometry. CFD is a well-established technique for industrial design, and it has been utilized to assist research and development in a wide variety of areas, such as automotive and aerospace industries, among many functional areas. CFD is an ideal tool to study the transport during hydrometallurgy processes, e.g., McBride et al. [9].

Current mainstream CFD software is built upon the finite-volume method [10, 11], and OpenFOAM [12], and the finite-element method [13], although few are developed using methods such as the lattice Boltzmann method [14]. In a nutshell, in CFD, we solve the governing Eqs. (6.1) to (6.6) in discretized form over the computational domain (CD), meshed at a reasonable density to resolve the geometry under investigation. A sparse, large matrix consisting of the terms of the discretized equations, along with appropriate BCs of the problem, is then solved directly or iteratively to obtain the approximated solution of the governing equations. One of the advantages of CFD over the methods mentioned earlier is its capability to solve fully coupled transport processes, making it ideal for the detailed study of complicated systems. However, this advantage comes at a price – CFD is CPU and RAM intensive and often requires substantial computational resources and time.

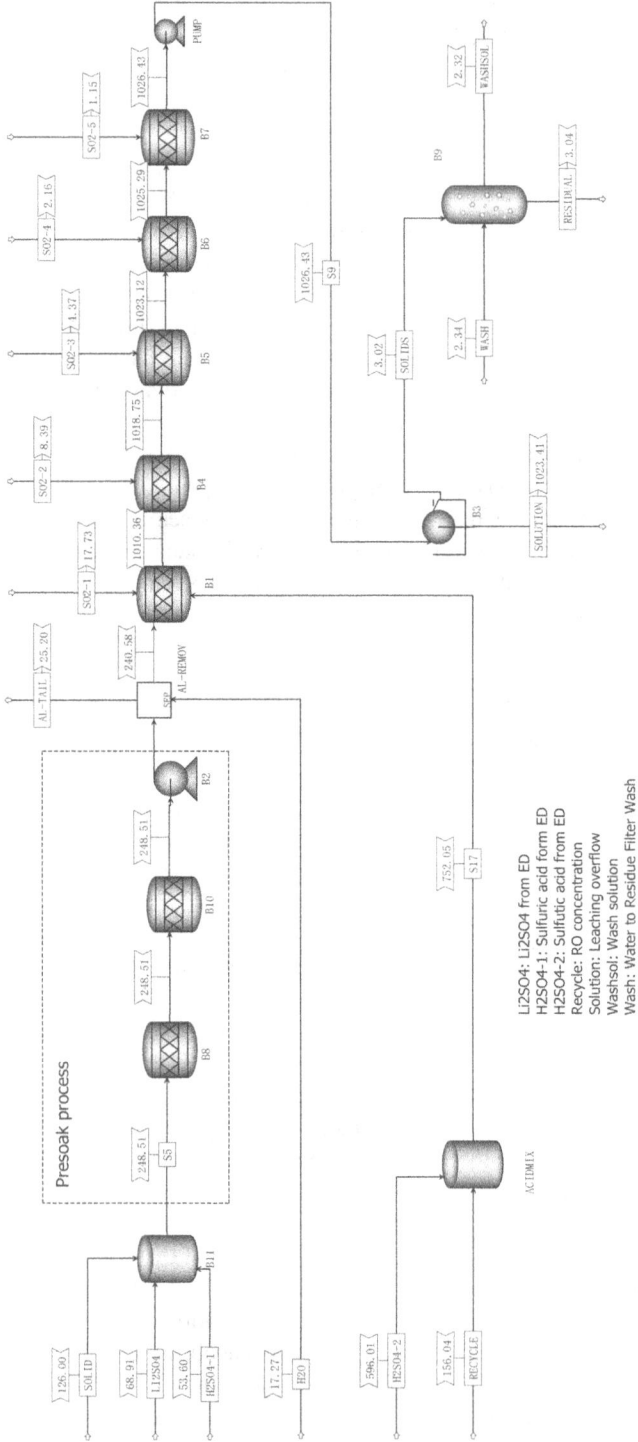

Li2SO4: Li2SO4 from ED
H2SO4-1: Sulfuric acid form ED
H2SO4-2: Sulfutic acid from ED
Recycle: RO concentration
Solution: Leaching overflow
Washsol: Wash solution
Wash: Water to Residue Filter Wash

FIGURE 6.6 Flowsheet of a leaching process implemented with Aspen Plus.

Fortunately, some simplification of the problem can be done to shorten the computational time of CFD applications in modeling recycling processes. Two standard practices to accelerate CFD simulation are the "reduced dimensionality" and "steady-state" assumptions. For instance, general reactors such as the CSTR have cylindrical geometry where the reacting flow follows a one-dimensional manner from the inlet to the reactor outflow. A thumb rule for reducing dimensionality is eliminating the dimensions where a low gradient is expected, e.g., when the rotor speed is high enough, radial variation of concentration is negligible, the problem can be simplified to be a one-dimensional (1D) problem from the reactor inlet to its outlet. If the flow in the reactor is well mixed, the problem can be even further reduced to a zero-dimension (0D) configuration, i.e., the governing equations become a set of point equations, which is similar to the approach of system software such as METSIM and Aspen Plus.

In this chapter, we use the ED as an example to demonstrate the use of CFD for recycling processes. The ED technology has seen applications in desalination, separation, food industries, pesticide sorption, etc. A 2D CFD model of the ED process and a simplified 1D model are developed to show spatial variations of Li_2SO_4, LiOH, and H_2SO_4 along the ED channels. Experimental validation of the model is also discussed.

To close the recycling loop, electrodialysis devices can be included to perform salt-splitting of the Li-rich solution after all precipitation processes are completed. A schematic diagram of a typical ED device is shown in Figure 6.7. The primary function of an electrodialysis device is similar to water electrolysis, where electricity splits water into hydrogen and hydroxyl ions on the cathode and oxygen and proton on the anode. Lithium-ion and sulfate ions move through the membranes to the cathode and anode separately, forming LiOH and H_2SO_4, respectively. Part of the LiOH can be reused to precipitate other metals, and excessive LiOH can be crystallized and used to make new LIBs. The sulfuric acid is delivered back to the leaching process.

6.4.1 MATHEMATICAL MODEL OF ELECTRODIALYSIS PROCESSES

This section presents a mathematical model to simulate a three-channel ED cell that is employed to regenerate sulfuric acid and LiOH from the Li_2SO_4 solution after the precipitation stages in the recycling process. Figure 6.7 shows a 2-dimensional schematic diagram of the ED setup. The salt solution of interest is supplied to the *dilute* channel in a typical three-channel ED process. The adjacent channels with the acid and base solutions are the *concentrate* channels. The dilute channel is separated from the concentrate channels by ion exchange membranes (IEMs). The terminals are connected to a power supply, which splits water to produce H^+ and OH^-:

Cathodic reaction:

$$2H_2O + 2e^- \rightarrow H_2 + 2OH^- \tag{I}$$

Anodic reaction:

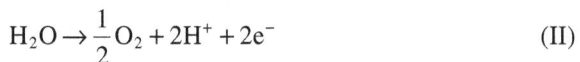

$$H_2O \rightarrow \frac{1}{2}O_2 + 2H^+ + 2e^- \tag{II}$$

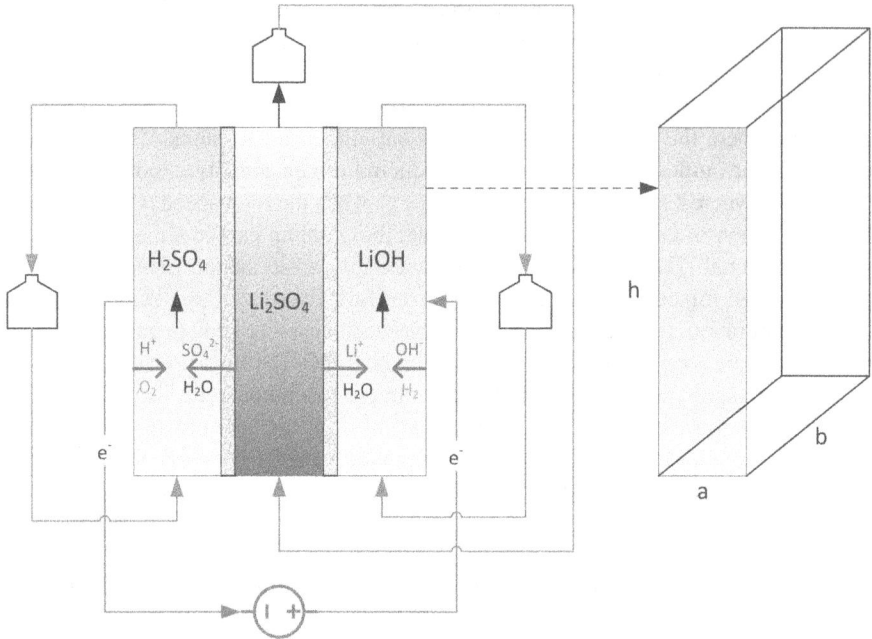

FIGURE 6.7   Schematic diagram of an electrodialysis in recirculating configuration.

In scaling up an ED system, a stack configuration is desirable, where these terminals are replaced by bipolar membranes situated between two consecutive acidic and basic concentrate channels.

The Li^+ and SO_4^{2-} ions that travel from the dilute channel to the concentrate channels will form acid and base with the H^+ and OH^- following these reactions:

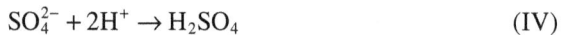

$$Li^+ + OH^- \rightarrow LiOH \tag{III}$$

$$SO_4^{2-} + 2H^+ \rightarrow H_2SO_4 \tag{IV}$$

6.4.1.1   Governing Equations
The following assumptions are made to simplify the ED model:

1. Steady-state operation and laminar flow for all channels. This assumption is generally valid for most channels except the concentrate channels with terminal plates where bubbly flows develop downstream due to electrolysis. Nevertheless, misty flows are observed from flow visualization experiments, and the flow field appears reasonably stable during operation.
2. The applied current is kept constant during the ED operation, which can be achieved by setting the power supply to constant current mode. In this mode, the voltage supplied to the ED cell would depend on the concentrations and temperature in the channels.

3. The under-limiting current density condition in the entire cell is assumed; thus, no water dissociation occurs in the dilute channel.
4. Uniform temperature distribution is considered. This assumption can be relaxed for comprehensive CFD calculation when solving the energy conservation equation.
5. The effect of pressure on the ion transport equations, water velocity, and Donnan equilibrium approach is neglected.

The governing equations for a typical ED process include the conservation of mass, ion species, gaseous species, momentum, and energy. We are primarily interested in the transfer of ions through the channels via migration, which often is accompanied by water transfer due to the movement of ions, i.e., the so-called electro-osmosis. When an ED process is added to the LIB recycling process, one can manipulate the inlet flow rates and ion concentrations to adjust the amount of acid and base needed to close the recycling loop.

Based on the assumptions above, a simplified model is developed as follows. The steady-state Navier-Stokes equations are solved to determine the velocity field in the channels:

$$\nabla \cdot V = 0 \tag{6.11}$$

$$\rho V \cdot \nabla V = -\nabla P + \mu \nabla^2 V + f \tag{6.12}$$

where V is the velocity vector, P is pressure, f is a body force term, μ is dynamic viscosity, and ρ is solution density. The conservation of species can be expressed as follows:

$$\nabla \cdot N_i = R_{i,tot} \tag{6.13}$$

where $R_{i,tot}$ is the total molar reaction rate, and N_i is the flux of species i. The electrochemical reaction rate of species i is written as:

$$R_i = -a_v \frac{v_i \cdot I_{loc}}{n_e \cdot F} \tag{6.14}$$

where n_e is the number of participating electrons in the electrochemical reaction; I_{loc} is the local CD, v_i is the stoichiometric coefficients of the species i, and a_v is the specific surface area.

Ions move through channels and membranes due to diffusion, migration, and convection fluxes which are the first, second, and third terms on the right-hand side of the Nernst-Planck (NP) equation [15, 16]:

$$N_i^j = -D_i^j \nabla c_i^j - F z_i M_i^j c_i^j \nabla \phi^j + u^j c_i^j \tag{6.15}$$

where superscript j refers to the membrane (m) or channel (c) parts of the ED cell; F is the Faraday's constant, which is equal to 96485 A·s·mol^{-1}; ϕ is the electrolyte potential; u is the velocity of electrolyte solution; and c_i, D_i, M_i, and z_i are the concentration, diffusion coefficient, mobility, and valence of species i, respectively.

6.4.1.2 Model Closure

There are a few parameters in the governing equations discussed in Section 6.4.1.1. Correlations of these parameters are needed to close the model. The mobility of species i is calculated using the Nernst-Einstein equation as [17, 18]:

$$M_i = \frac{D_i}{RT} \tag{6.16}$$

where T is temperature and R is the universal gas constant, 8.314 J·mol^{-1}·K^{-1}.

The diffusion coefficient of ions in the flow channel, D_i^c, can be calculated based on the limiting (zero concentration) ionic conductance using the following relation [19, 20]:

$$D_i^c = \frac{RT^2}{334\eta z_i^2 F^2} \lambda_i^0 \tag{6.17}$$

where η is the water viscosity at the desired temperature in centipoises, and λ_i^0 is the limiting ionic conductance of species i at 298 K, available in the literature for various ions [21, 22].

The Mackie and Meares correlation is employed to estimate the effective diffusion coefficient of species i in membranes, D_i^m, as follows [23, 24]:

$$D_i^m = \left(\frac{\omega}{2-\omega}\right)^2 D_i^c \tag{6.18}$$

in which D_i^c is the diffusion coefficient of ions in channels, and ω is membrane water volume fraction calculated as follows:

$$\omega = \frac{\lambda}{\lambda + \dfrac{\rho_w}{\rho_p}} \tag{6.19}$$

where ρ_p and ρ_w are dry membrane and water densities, and λ is water uptake.

$$\lambda = \frac{m_w - m_d}{m_d} \times 100 \tag{6.20}$$

where m_d and m_w are the dry and wet weight of the membrane, respectively. The total current is computed using Faraday's law:

$$I = F \sum_k z_i N_i \tag{6.21}$$

In an electrolyte solution consisting of K species, the flux and concentration of all species ($2K$ unknowns) are calculated by simultaneously solving Eqs. (6.13) and (6.15). Current density is evaluated from Eq. (6.21), while electrolyte potential is computed based on electroneutrality conditions, i.e., Eqs. (6.22) and (6.23) for the channels and the membrane, respectively:

$$\sum z_i c_i^c = 0 \tag{6.22}$$

$$z_{fix} c_{fix} + \sum z_i c_i^m = 0 \tag{6.23}$$

where z_{fix} and c_{fix} are the valence and concentration of fixed ions of IEMs, respectively.

It is well known that water molecules may penetrate the IEM via both the osmotic and electro-osmotic mechanisms. The electro-osmotic mechanism, i.e., dragging of water molecules into the IEM due to the transport of hydrated ions under an electrical potential gradient, is the dominant mechanism causing transmembrane water flow in the ED process [16, 25, 26]. The electro-osmotic water velocity is specified on the membrane boundary based on the driving and friction forces that the ions encounter as:

$$u_m = \frac{\varepsilon \zeta}{4\pi\mu} \frac{\partial \phi}{\partial x} \tag{6.24}$$

where u_m is the transmembrane water velocity, ζ is zeta potential representing the Helmholtz double-layer potential, and ε is the solvent permittivity, which is defined as a product of the relative permittivity (dielectric constant), ε_r, and the vacuum permittivity, ε_0:

$$\varepsilon = \varepsilon_r \varepsilon_0 \tag{6.25}$$

where ε_0 is equal to 8.854×10^{-12} F·m^{-1} and ε_r for dilute solutions is calculated as [27–29]:

$$\varepsilon_r = 185.765 - 0.35963T \tag{6.26}$$

The following relation is used to calculate zeta potential based on the surface charge density:

$$\zeta = 4\pi z_i d_i F c_{i,ave}^m \frac{r_i}{\varepsilon} \tag{6.27}$$

where d_i and r_i are the diameter and radius of the counter-ion, i transferring through the membrane, respectively, and $c_{i,ave}^m$ is the average concentration of species i in a membrane.

6.4.1.3 Computational Domain and Boundary Conditions

The BCs for the governing equations are shown in Figure 6.8. The velocity and normal flux of the outer walls is zero, and isolation BC, i.e., zero current density, is assumed at the inlets and outlets of channels. The BCs set at the IEM-solution, and electrode-electrolyte interfaces are explained in detail.

The continuities of CD and species flux are assumed in the IEM-solution interfaces as:

$$n \cdot I^c = n \cdot I^m \tag{6.28}$$

$$n \cdot N_i^c = n \cdot N_i^m \tag{6.29}$$

However, electrolyte potential variation at these boundaries is due to considerable concentration differences occurring at the IEM-solution interfaces [15]. This potential difference is estimated based on the Donnan potential theory. With the assumption of the neglected effect of pressure difference, the Donnan potential equation is written as [30–32]:

$$\phi^m - \phi^s = \frac{-RT}{z_i F} \ln\left(\frac{c_i^s}{c_i^m}\right) \tag{6.30}$$

where superscripts m and s refer to the membrane and solution sides of the IEM-solution interface. To consider the effect of transmembrane water velocity, which is

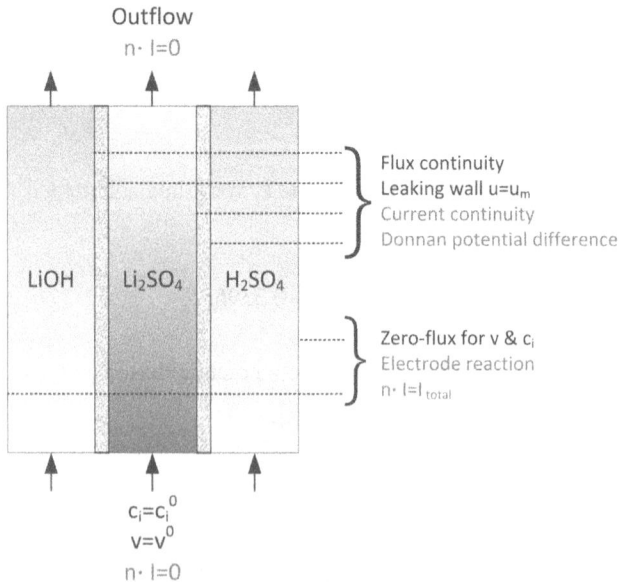

FIGURE 6.8 Boundary conditions for ED.

estimated using Eq. (6.24), the leaking wall BC ($u = u_m$) is used instead of no-slip BC ($u = 0$) at the IEM-solution BCs.

At the surfaces of anode and cathode electrodes, the no-flux conditions for Li^+ and SO_4^{2-} species and no-slip BC are imposed. Also, in these boundaries, the applied CD is set as:

$$n \cdot I = I_{total} \tag{6.31}$$

where I_{total} is the normal CD supplied by an external power supply. Although the amount of electrode potential can also be prescribed instead of CD, owing to the possibility of a water layer formation at the surfaces of electrodes and subsequently resistance and voltage enhancement, setting CD as a BC on these electrode surfaces provides more accurate results.

Considering electrochemical reactions occurring at the cathode and anode surfaces is crucial to obtaining a converged solution and fulfilling the electroneutrality condition in the simulation of large-scale EDs operated in practical CDs. The cathodic and anodic reactions in the ED are shown in (I) and (II). The molar fluxes of ionic species in an electrochemical reaction occurring at an electrode-electrolyte interface are the summation of all fluxes caused by the electrochemical reactions, as [13]:

$$N_i = \sum_m \frac{v_{i,m} \cdot I_{loc,m}}{n_{e,m} \cdot F} \tag{6.32}$$

where m refers to the number of reactions occurring at the electrode surface, $n_{e,m}$ is the number of participating electrons in the electrode reaction of index m and is always a positive number; $I_{loc,m}$ is the local CD generated from the electrochemical reaction m; and $v_{i,m}$ is the stoichiometric coefficients of the species i participating in the reaction m. It is noted that stoichiometric coefficients $v_{i,m}$ is a positive or negative value for reduced and oxidized species, respectively, e.g., the amount of v_H and v_{OH} in the anodic and cathodic reactions are -4 and $+2$.

The sum of all currents generated due to the charge transfer reaction occurring at an electrode-electrolyte interface gives rise to total normal CD as follows [13]:

$$I \cdot n = \sum_m I_{loc,m} \tag{6.33}$$

Considering the number of reactions occurring at the surface of each electrode during the ED process, $m = 1$, Eqs. (6.32) and (6.33) are written as follows:

$$N_i = \frac{v_i \cdot I_{loc}}{n_e \cdot F} \tag{6.34}$$

$$I \cdot n = I_{loc} \tag{6.35}$$

The above problem is implemented using Multiphysics COMSOL to demonstrate what one can achieve with CFD software to investigate the coupling among flow, ion species, and electrical field. The setup of this software is straightforward, i.e., problem setup, geometry creation and meshing, computation, and post-processing. One advantage of COMSOL is its versatility in dealing with many transport phenomena and flexibility in adjusting the coupling level within the solution procedure. Similar CFD software products such as Ansys Fluent [10], Siemens Star-CCM+ [11], etc., are available commercially. There have also been open-source codes, such as OpenFOAM [12], capable of parallel, and high-performance computing.

6.4.1.4   Simulation Results

The concentrations of the OH^-, H^+, Li^+, and SO_4^{2-} species are presented in Figures 6.9(a), 6.9(b), 6.9(c), and 6.9(d), respectively. As shown in these figures, the concentrations of all species decrease at the dilute interface of IEMs, while the concentrations at the concentrate side of the IEMs all increase downstream. Considering electrochemical reactions, which lead to the production of hydroxide and protons at the anode and cathode surfaces, the concentration of OH^- and H^+ at the electrode surfaces is higher than their bulk concentrations at the cathode and anode channels, as depicted in Figure 6.9(a) and 6.9(b). Also, as shown in Figures 6.8(c) and 6.9(b), due to the electroneutrality assumption, the concentration of Li^+ and SO_4^{2-} reach their maximum in the CEM and AEM, respectively; high Li^+ and SO_4^{2-} concentration regions appear at the cathode and anode, respectively.

6.4.2   SIMPLIFIED MODEL

Comprehensive CFD simulations can provide spatial and temporal information on transport and chemical phenomena inside physical devices. These simulation tools, however, require substantial computational resources, and sophisticated simulations are often time-consuming. It is impractical to embed these simulation procedures into system-level simulations such as METSIM or Aspen Plus discussed in previous sections. Models with modification are thus desired to allow efficient and rapid system applications.

Standard practices of simplification are seeking analytical solutions and dimension reduction. Analytical solutions can be found generally for simple mathematical problems that one can solve for solutions without resorting to heavy numerical calculation. Cases that can be solved analytically are rare, especially for problems that involve complex, coupled transport processes. Dimension reduction, on the other hand, is more common and feasible.

A simplified ED process model introduced previously is developed. The model reduces the 2D problem to a 1D problem. A rule of thumb for dimension reduction is keeping the dimensions with significant gradients. Only the dimension between the two external terminals is considered in this case. To develop this model, it is assumed that co-ions' flux through IEMs is zero, i.e., only counter-ions can penetrate through the IEMs. Derivation of the model is described as follows.

The model development starts with the variations along the flow channel direction. Ion concentration and electrolyte velocity vary along the channels due to the flux of

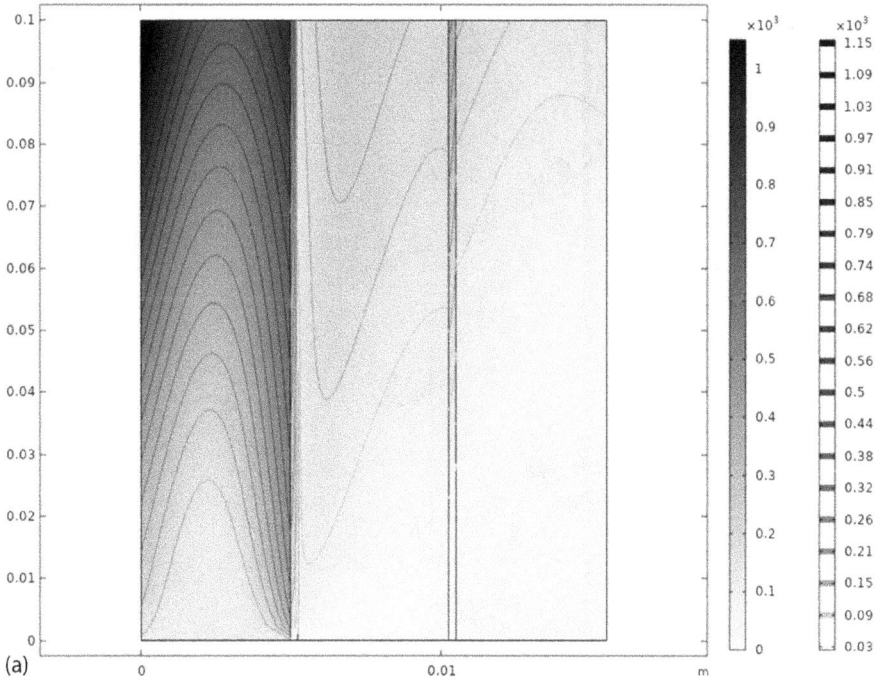

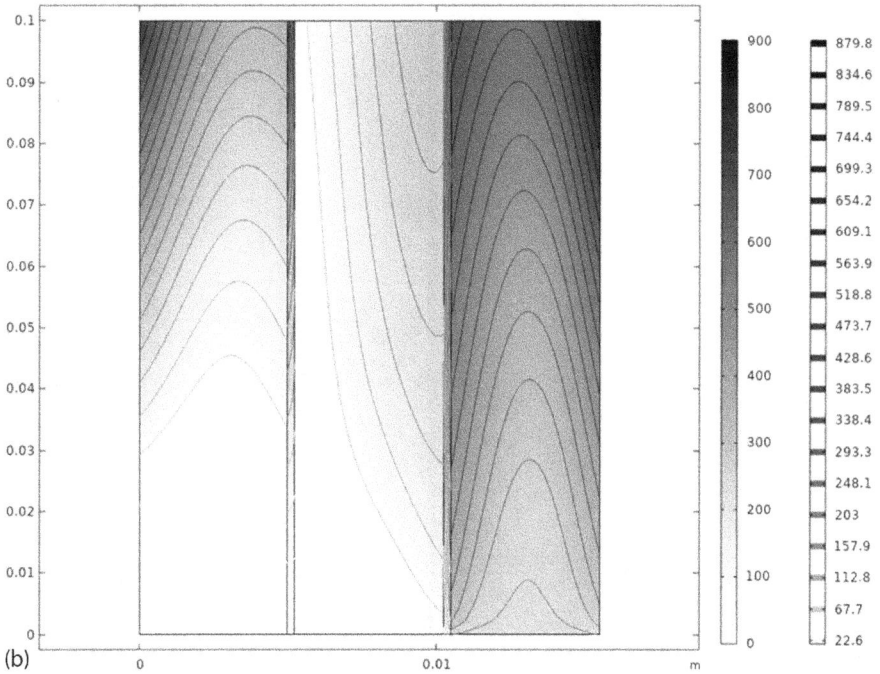

FIGURE 6.9 Concentration distribution in all channels and IEMs of ED system: (a) OH⁻ ions; (b) H⁺ ions; (c) Li⁺ ions; and (d) SO_4^{2-} ions. *(Continued)*

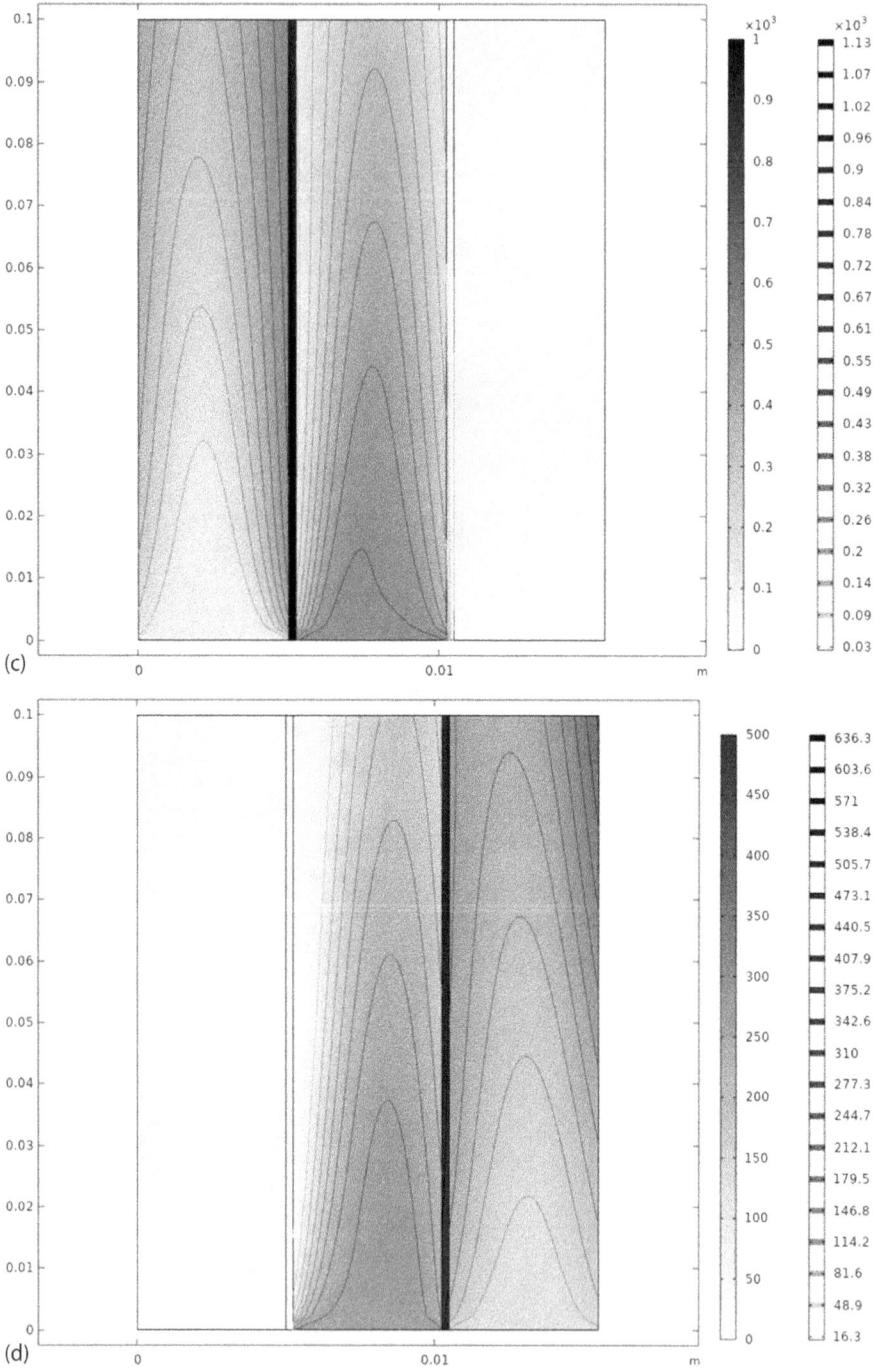

FIGURE 6.9 *(Continued)*

ions and water through the IEMs. These variations can be determined using material balance relations. The flow rate of the liquid solution is related to the flux by:

$$|J_w| = ab|u_{out} - u_{in}| \tag{6.36}$$

where a, b, and h, see Figure 6.7, are the width, length, and height of channels, respectively, and u_{out}, u_{in}, and J_w are the outlet, inlet velocities, and the transmembrane flux of water through IEMs, respectively. At the dilute channel, J_w consists of the flux of water through both anion and cation exchange membranes, i.e., $J_w = J_w^{CEM} + J_w^{AEM}$, where J_w^{AEM} and J_w^{CEM} are the transmembrane water flux at the anode and cathode through the CEM and AEM, respectively. Similarly, the species flow rate is related to flux by:

$$S_m|J_i| = ab|c_{i,out}u_{out} - c_{i,in}u_{in}| \tag{6.37}$$

where $S_m = bh$ is the membrane surface area, J_i is the flux of species i through an IEM, and $c_{i,in}$ and $c_{i,out}$ are the concentration of specie i at the inlet and outlet of channels, respectively.

According to Eqs. (6.36) and (6.37), the flux of water and ions transferring through IEMs must be determined first to find the outlet velocities and outlet concentration of lithium and sulfate ions. Assuming constant current operation and zero flux of co-ions through IEMs, the fluxes of Li^+ and SO_4^{2-} ions across the IEMs can be determined using Faraday's law:

$$I = z_i F J_i^m \tag{6.38}$$

where superscript m refers to the membrane, z_i is the valence of ion i, and I is the current density. The transmembrane water velocity remains the same as Eq. (6.24). The flux of water is:

$$S_m u_m = J_w \tag{6.39}$$

6.4.2.1 Concentration Polarization at Membrane-Solution Interfaces

In order to estimate the concentration of ions at the solution side of an IEM-solution interface, Tanaka [33] firstly developed Nernst-Planck equations for co- and counterions transferring at the boundary layer adjacent to the IEM and then proposed Eq. (6.40), which relates the concentration of species at the solution adjacent to an IEM to the bulk concentration, transmembrane water velocity, and diffusion coefficient of mobile ions in a single-salt solution:

$$c_i^s = \left[\frac{\beta I}{\alpha u_m} + \left(c_i^{bulk} - \frac{\beta I}{\alpha u_m} \right) e^{-\alpha \delta |u_m|} \right] \tag{6.40}$$

where δ is the thickness of the diffusion layer, c_i^s is the concentration of species i in the solution side of a solution-membrane interface, and superscript *bulk* refers to the bulk solution. Tanaka introduced the following relations to define α and β coefficients for monovalent salts, e.g., NaCl:

$$\alpha = \frac{D_+^s + D_-^s}{2D_+^s D_-^s} \tag{6.41}$$

$$\beta = \frac{t_+ D_-^s - t_- D_+^s}{2FD_+^s D_-^s} \tag{6.42}$$

where subscripts $+$ and $-$ refer to cations and anions, t is the transport numbers of ions in the membrane, and D^s is the diffusion coefficient of ions in the solution. The formulations listed in Table 6.5 are proposed to calculate the α and β values for multivalent salts such as Li_2SO_4 and H_2SO_4 based on the development of Nernst-Plank equations at the boundary layers of IEMs vicinity.

The Donnan potential theory is used to predict the ions concentration in the membrane side of a solution-membrane interface. According to Eq. (6.30), the potential difference between an electrolyte solution and an IEM is a function of pressure difference across the IEM, neglected in this study, and the ion concentration at both sides of the IEM-solution interface.

On the other hand, the potential difference occurring at the IEMs-solution interfaces, Eq. (6.43), is developed based on the Helmholtz double-layer model:

$$\phi^m - \phi^s = \pm \left(\frac{l_i^d}{\varepsilon} F d_i c_i^s \right) \tag{6.43}$$

where positive and negative signs allocate to CEM and AEM, respectively, and l_i^d is the thickness of the double layer and is considered equal to the hydrated ion radius

---

TABLE 6.5
α and β Coefficients for Mono- and Multivalent Ions for All Solution-Membrane Interfaces

Solution-Membrane Interface	Solution	α	β
Cathode channel-CEM	$LiOH$	$\dfrac{D_{Li}^s + D_{OH}^s}{2D_{Li}^s D_{OH}^s}$	$\beta = \dfrac{t_{Li}D_{OH}^s - t_{OH}D_{Li}^s}{2FD_{Li}^s D_{OH}^s}$
Dilute channel-CEM	Li_2SO_4	$\dfrac{D_{Li}^s + 2D_{So4}^s}{3D_{Li}^s D_{So4}^s}$	$\beta = \dfrac{2t_{Li}D_{So4}^s - t_{So4}D_{Li}^s}{6FD_{Li}^s D_{So4}^s}$
Dilute channel-AEM	Li_2SO_4	$\dfrac{D_{Li}^s + 2D_{So4}^s}{3D_{Li}^s D_{So4}^s}$	$\beta = \dfrac{2t_{Li}D_{So4}^s - t_{So4}D_{Li}}{6FD_{Li}^s D_{So4}^s}$
Anode channel-AEM	H_2SO_4	$\dfrac{D_H^s + 2D_{So4}^s}{3D_H^s D_{So4}^s}$	$\beta = \dfrac{2t_H D_{So4}^s - t_{So4}D_H^s}{6FD_H^s D_{So4}^s}$

(r_i^H) or Debye length (r_D) based on the Helmholtz and Gouy-Chapman double layer models, respectively. The Debye length is given as [34]:

$$r_D = \left(\frac{\varepsilon RT}{\sum_{i=1}^{N} c_i z_i^2 F^2} \right)^{0.5}. \tag{6.44}$$

Therefore, c_i^m is evaluated by substituting Eq. (6.43) into Eq. (6.30).

To determine the potential difference across a membrane, the Nernst-Planck equation for cation- and anion-exchange membranes can be approximated as:

$$J_i = \frac{z_i FD_i c_i^m}{RT} \frac{\Delta \varphi_m}{l_m} + D_i^m \left(\frac{dc_i^m}{l_m} \right) + u_m c_i^m \tag{6.45}$$

where dc_i^m indicates the concentration difference of species i across the IEM, and c_i^m is the average concentration of ion i at the IEM.

6.4.2.2 Computation and Solver

At the first step of a continuous ED modeling, the initial concentration, $c_{i,tank}^0$, and volume, V_{tank}^0, for all tanks containing LiOH, H_2SO_4, and Li_2SO_4 solutions, and the inlet velocity of solutions, which enter the ED, are inputted to calculate u_{out} and $c_{i,out}$ for all channels. u_{out} and $c_{i,out}$ parameters are determined using the simultaneously solving of mass conservation relations, as well as the equations proposed to find the flux of ions, potential difference, electro-osmosis water velocity, and the concentration of ions at both electrolyte solution and membrane phase of solution-IEM interfaces, by employing the Equation Engineering Solver (EES) [35]. Then, V_{tank}^1 and $c_{i,tank}^1$ are estimated for all tanks based on the mass conservation equations. Finally, the output parameters obtained at each step are inputted as the initial values of the next time step, and this process goes on until $n = n_{max}$. More details about the modified 1D model can be found in [36].

6.4.2.3 Experiments and Model Validation

To verify the simplified numerical model, experiments of ED were carried out, and the results were compared with the simulation results. The experimental setup of the ED device is shown in Figure 6.10. There are three channels in the test cell: the dilute channel filled with Li_2SO_4 solution, sandwiched by two concentrate channels filled with H_2SO_4 (anode) and LiOH solutions (cathode), respectively. The width of each channel is 22 mm. A cation-exchange membrane (Nafion NRE-212) and an anion-exchange membrane (Astom ACM) are installed to separate the dilute channel with the anode and the cathode channels, respectively. The cell has an active area of 336 cm². Dilute H_2SO_4 (0.1M) and LiOH (0.1M), and high concentration Li_2SO_4 (0.55 M) solutions are pumped into the cell at 20 L/h in a co-flow fashion. All the fluids from the test cell are collected with individual tanks to cool down, and the fluids are pumped back into the test cell continuously. Concentration measurements

FIGURE 6.10   Experimental setup of ED.

of the effluent from the test cell are taken periodically using an inductively coupled plasma-optical (ICP) emission spectrometer (Agilent 5100 ICP-OES, USA). The temperature of the test cell is maintained at 45°C.

To investigate the present model's accuracy, the present model's obtained results are compared with the experimental data of the ED system. All the input parameters required for the ED modeling, including applied current, inlet concentrations, the initial volume of tanks, model dimensions, and experimental conditions, are listed in Table 6.6. In addition, the characterization of the membranes used in the under-study ED systems is shown in Table 6.7. More details about the experiments can be found in [37].

Table 6.8 shows the experimental results and the results obtained by the present model to investigate the accuracy of the present model. As indicated in this table, this comparison was made after 2, 4, and 6 hours of the ED operation.

Figure 6.11 shows the outlet concentrations of lithium and sulfate ions, which were evaluated by the present model and also measured experimentally. As indicated in Figure 6.11, the lithium and sulfate concentrations increase at con-centrate channels while decreasing at the dilute channel, and the mean error between modeling and experimental data enhances as the operation time of the ED increases.

TABLE 6.6
Experimental Condition and Input Parameters for ED Modeling

Parameter	Value
Temperature, K	343.25
Current, A	28
Inlet flow rate, $m^3 min^{-1}$	1.8×10^{-4}
Channel width, m	0.022
Channel length, m	0.14
Channel height, m	0.22
Inlet concentration of $Li+$, dilute channel, $mol\ m^{-3}$	1100
Inlet concentration of $So_4 2-$, dilute channel, $mol\ m^{-3}$	550
Inlet concentration of $H_2 So_4$, anode channel, $mol\ m^{-3}$	100
Inlet concentration of $LiOH$, cathode channel, $mol\ m^{-3}$	100
$LiOH$ tank initial volume, m^{-3}	1.8×10^{-3}
$Li_2 SO_4$ tank initial volume, m^{-3}	1.8×10^{-3}
$H_2 SO_4$ tank initial volume, m^{-3}	1.0×10^{-2}

CEM = Nafion NRE-212, AEM = Neosepta ACM.

TABLE 6.7
Properties of IEMs

Membrane Type		Thickness, μm	Water Uptake, %	Membrane Dry Weight, g m⁻²	Ion-Exchange Capacity, meq g⁻¹
CEM	Nafion NRE-212	50.8	25–30	100	0.95–1.01
AEM	Neosepta ACM	110	15	118	1–1.5

6.4.3 HYBRID SIMULATION METHODOLOGY

As discussed in previous sections, there have been a few simulation tools available that can be employed to understand the multi-physics problem and optimize the design and operation of hydrometallurgical recycling for spent LIB materials. These simulation tools have their pros and cons; each has its limitations in terms of length and time scales. For large system design, the use of system software like METSIM suffices. The model parameters involved in system-level simulation can be either measured by experiments or obtained by detailed simulation at the unit operation level. A wise strategy to fully utilize these tools to achieve efficient and economical recycling is to construct a hybrid simulation.

The notion of hybrid simulation is similar to the multiscale simulation currently under joint development in materials science and thermo-physics engineering. The

TABLE 6.8

Comparison Between Experimental Data and Present Model Results

Time (hr)		Concentration, mol·m^{-3}				Mean Error, %
		Li at Base Channel	Li at Dilute Channel	SO$_4$ at Dilute Channel	SO$_4$ at Acid Channel	
2	Experimental value	877	1002	518	473	
	Present modeling results	935	1017	509	495	
	Error (%)	6.6	1.4	1.7	4.6	3.5
4	Experimental value	1335	874	465	830	
	Present modeling results	1346	891	445	802	
	Error (%)	0.8	1.9	4.3	3.3	2.5
6	Experimental value	1622	742	405	1137	
	Present modeling results	1553	713	357	1078	
	Error (%)	4.2	3.9	11.8	5.1	6.3

basic idea of multiscale modeling is to homogenize the low-scale simulation results and use them in the higher-scale simulation. For instance, the conversion efficiency from a CSTR in the METSIM simulation can be determined from empirical correlations, which are often obtained experimentally. Alternatively, one can perform CFD simulation for the reactor and extract useful information from the results suitable for the high-level METSIM simulation. On the other hand, for macroscopic CFD simulations, model parameters may exist, e.g., parameters regarding the kinetics and thermodynamics involved or the transport properties such as diffusivity and ion conductivities. These parameters can be obtained experimentally or numerically by carrying out pore-scale or even ab-initio simulations. The ED case demonstrated in the previous section is a good example. There remain quite a few unknowns about the phenomena inside the device. This situation arises because many possible processes, e.g., migration, water dissociation, electro-osmotic drag, etc., occur within the device and the results observed are strongly coupled. A better approach to understanding these coupled processes is to de-couple them and investigate each process individually and independently. Such a divide-and-conquer strategy has been practiced in topics like fuel cells and ED. With a fundamental understanding of the processes, one can then upgrade the simulation procedure step by step, from microscale, device, unit operation, and finally up to the system level of simulation.

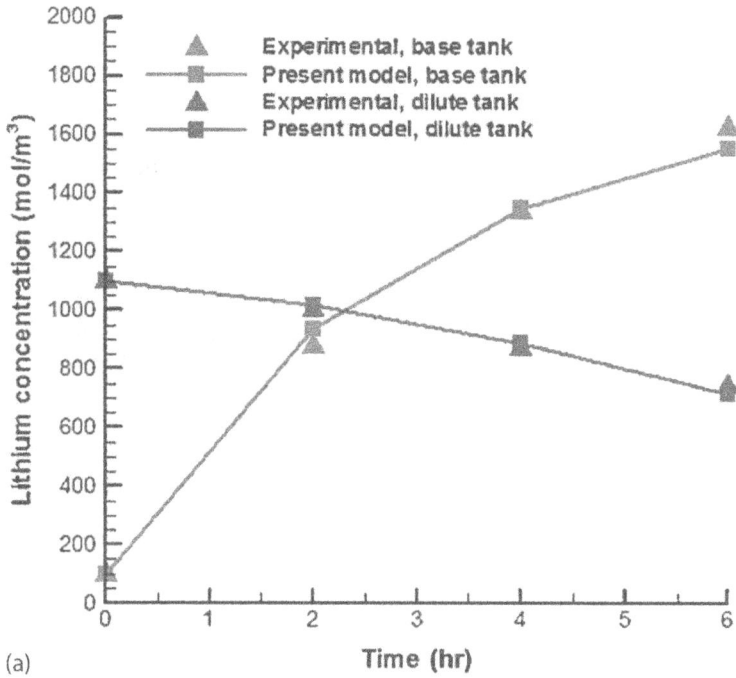

(a)

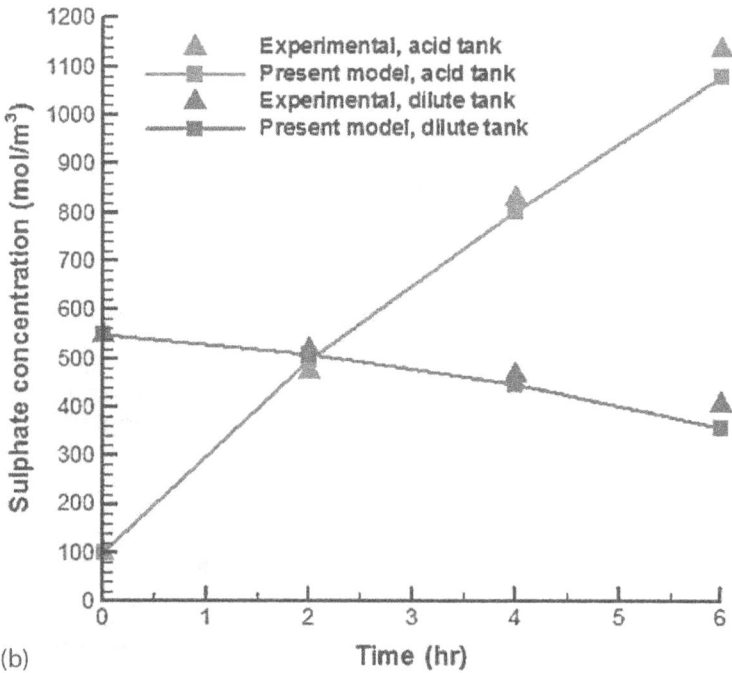

(b)

FIGURE 6.11 A comparison between the experimental data and present model results for different ions: (a) lithium and (b) sulfate ions.

6.5 CONCLUDING REMARKS

Modeling and various simulation methods suitable for spent LIB material recycling are introduced in this chapter. With the help of these simulation tools, process engineers can attempt to virtually adjust design parameters and operating conditions to optimize overall recycling performance. Expedient, simplistic models are excellent for a first-cut estimate of system design, whereas sophisticated, comprehensive CFD software can provide detailed information and accurate predictions to pinpoint places in the components or reactors where performance can be improved. System-level software such as METSIM is an industry-standard and powerful platform for system integration and optimal operation of recycling procedures. It should be pointed out that experimental validation is key to making these simulation tools truly useful for the actual design, research, and development of LIB recycling. Aside from the simulation methods discussed in this chapter, there are new possibilities for future modeling work, e.g., investigation of the transport phenomena at the particle level or the nanoscale [38]. Such methods have been developed and successfully applied in studying porous electrodes [39] and have the potential to bridge the gaps between material microstructure and recycling processes.

ACKNOWLEDGMENTS

PCS is grateful for the indispensable assistance from Ms. Jinxing Peng (analysis of METSIM results, Aspen Plus software), Mrs. Anahitha Asadi and Dr. Hesam Bazargan Harandi (COMSOL and simplified models for ED), and Mr. Bolin Kang (ED validation experiment), all from Wuhan University of Technology, Wuhan, China.

REFERENCES

1. Marsden J.O., Botz M.M., Heap leach modeling – A review of approaches to metal production forecasting, Minerals & Metallurgical Processing, 34 (2017). https://doi.org/10.19150/mmp.7505
2. Bard A., Faulkner L., Bard A.J., Faulkner L.R., Electrochemical Methods: Fundamentals and Applications (New York: Wiley, 2001).
3. Li Y. Jiao, Li L., Su Q. Ye, Lu W. Sheng, Han Q., Li L., Chen Y. Xiang, Deng S. Yi, Lei T. Xing, Thermodynamic analysis of Li-Ni-Co-Mn-H2O system and synthesis of LiNi0.5Co0.2Mn0.3O2 composite oxide via aqueous process, Journal of Central South University, 26 (2019). https://doi.org/10.1007/s11771-019-4204-6
4. White F.M., Fluid Mechanics Seventh Edition, (2011).
5. METSIM | The Premier Steady-State & Dynamic Process Simulator, (n.d.). https://metsim.com/
6. Aspen Plus | Leading Process Simulation Software | AspenTech, (n.d.). https://www.aspentech.com/en/products/engineering/aspen-plus
7. Bartlett J., Holtzapple A., Rempel C., A brief overview of the process modeling/simulation and design capabilities of METSIM, in: COM 2014 – Conf. Metall. Proc., (2014).
8. Ferziger J.H., Perić M., Street R.L., Computational Methods for Fluid Dynamics (Switzerland: Springer Nature Switzerland AG, 2020). https://doi.org/10.1007/978-3-319-99693-6

9. McBride D., Gebhardt J., Croft N., Cross M., Heap leaching: Modelling and forecasting using CFD technology, Minerals, 8 (2018). https://doi.org/10.3390/min8010009
10. Ansys Fluent | Fluid Simulation Software, (n.d.). https://www.ansys.com/products/fluids/ansys-fluent
11. Multiphysics computational fluid dynamics (CFD) simulation software | Siemens Software, (n.d.). https://www.plm.automation.siemens.com/global/en/products/simcenter/STAR-CCM.html
12. OpenFOAM | Free CFD Software | The OpenFOAM Foundation, (n.d.). https://openfoam.org/
13. COMSOL Multiphysics® Software - Understand, Predict, and Optimize, (n.d.). https://www.comsol.com/comsol-multiphysics
14. Home – Palabos – UNIGE, (n.d.). https://palabos.unige.ch/
15. Strathmann H., *Ion-Exchange Membrane Separation Processes* (Amsterdam: Elsevier, 2004).
16. Jiang C., Wang Y., Zhang Z., Xu T., Electrodialysis of concentrated brine from RO plant to produce coarse salt and freshwater, Journal of Membrane Science, 450 (2014). https://doi.org/10.1016/j.memsci.2013.09.020
17. Honarparvar S., Reible D., Modeling multicomponent ion transport to investigate selective ion removal in electrodialysis, Environmental Science & Ecotechnology, 1 (2020). https://doi.org/10.1016/j.ese.2019.100007
18. Koryta J., Dvorak W., Kavan L., *Principles of Electrochemistry Second Edition* (Chichester: John Wiley & Sons, 1987).
19. Leaist D.G., Diffusion in aqueous solutions of sulfuric acid, Canadian Journal of Chemistry, 62, 1692–1697, (1984).
20. Leaist D.G., Goldik J., Diffusion and ion association in concentrated solutions of aqueous lithium, sodium, and potassium sulfates, Journal. Solution Chemistry, 30 (2001). https://doi.org/10.1023/A:1005296425604
21. Harned H.S., Owen B.B., Physical chemistry of electrolytic solutions, 2nd edition (New York: Reinhold Publishing Company, 1950).
22. Poling B.E., Prausnitz J.M., O'connell J.P., Properties of gases and liquids (McGraw-Hill Education, 2001).
23. Kamcev J., Paul D.R., Manning G.S., Freeman B.D., Predicting salt permeability coefficients in highly swollen, highly charged ion exchange membranes, ACS Applied Materials & Interfaces, 9 (2017). https://doi.org/10.1021/acsami.6b14902
24. Harandi H.B., Asadi A., Fathi H., Sui P.-C., Combined macroscopic and pore scale modeling of direct contact membrane distillation with micro-porous hydrophobic membranes, Desalination, 514 (2021).
25. Strathmann H., Ion-exchange membrane processes in water treatment, Sustainability Science and Engineering, 2, 141–199 (2010).
26. Jiang C., Wang Q., Li Y., Wang Y., Xu T., Water electro-transport with hydrated cations in electrodialysis, Desalination, 365 (2015). https://doi.org/10.1016/j.desal.2015.03.007
27. Mohammadi T., Moheb A., Sadrzadeh M., Razmi A., Modeling of metal ion removal from wastewater by electrodialysis, Separation and Purification Technology, 41 (2005). https://doi.org/10.1016/j.seppur.2004.04.007
28. Dean J.A., *Lange's Handbook Of Chemistry*. New York: McGrawHill (1999).
29. Weast R.C., *CRC Handbook of Chemistry and Physics* (Ohio: CRC Press, 1977).
30. Generous M.M., Qasem N.A.A., Zubair S.M., The significance of modeling electrodialysis desalination using multi-component saline water, Desalination, 496 (2020). https://doi.org/10.1016/j.desal.2020.114347
31. Zourmand Z., Faridirad F., Kasiri N., Mohammadi T., Mass transfer modeling of desalination through an electrodialysis cell, Desalination, 359 (2015). https://doi.org/10.1016/j.desal.2014.12.008

32. Donnan F.G., Guggenheim E.A., Die genaue thermodynamik der membrangleichgewichte, Zeitschrift Für Phys. Chemie, 162A (1932). https://doi.org/10.1515/zpch-1932-16225

33. Tanaka Y., *Ion Exchange Membranes: Fundamentals and Applications*, 2nd edition (Amsterdam: Elsevier, 2015). https://doi.org/10.1016/C2013-0-12870-X

34. Chein R., Chen H., Liao C., Investigation of ion concentration and electric potential distributions in charged membrane/electrolyte systems, Journal of Membrane Science, 342, 121–130 (2009). https://doi.org/10.1016/j.memsci.2009.06.033

35. Engineering Equation Solver, (n.d.). https://www.fchartsoftware.com/ees/

36. Asadi A., Kang B., Harandi H.B., Jung J.C.-Y., Shen Z., Sui P.-C., Electrodialysis of lithium sulphate solution: Model development and validation, Journal of the Electrochemical Society, 169, 53508–53521 (2022).

37. Kang B., Kang D., Jung J.C.-Y., Asadi A., Shen Z., Sui P.-C., Electrodialysis of a lithium sulphate solution: An experimental investigation, Journal of the Electrochemical Society, 169(6), 63515–63526 (2022).

38. Raabe D., Overview of the lattice Boltzmann method for nano- and microscale fluid dynamics in materials science and engineering, Modelling and Simulation in Materials Science and Engineering, 12, R13–R46 (2004).

39. Lange K., Sui P.C., Djilali N., Pore-Scale simulation of transport and electrochemical reactions in the catalyst layer of a PEM fuel cell, Journal of the Electrochemical Society, 157, 600 (2010).

7 Prospective Life Cycle Assessment Study of Hydrometallurgical Recycling Process with RecycLiCo™ Process and Electrodialysis

Phoebe Whattoff and Jordan Lindsey

CONTENTS

DOI: 10.1201/9781003269205-7

7.1 OUR STATEMENT

Information contained in this report has been compiled and computed from sources believed to be credible. Application of the data is strictly at the discretion and responsibility of the reader. Minviro is not liable for any loss or damage arising from the use of the information in this document.

Life-cycle assessment is an environmental accounting method with an inherent level of uncertainty, and it should not be seen as having the same level of precision as financial accounting. Life-cycle assessment requires a very large amount of data, particularly to calculate all the inputs and outputs for every step.

Primary data is collected from RecycLiCo™ Battery Materials Inc., 2021 engineering studies. Databases are often used for secondary data since it is impractical to collect all the necessary data from the original sources.

The report does not claim to be exhaustive, nor does it claim to cover all relevant products. While steps have been taken to ensure accuracy, the listing or featuring of a particular product or company does not constitute an endorsement by Minviro.

7.2 EXECUTIVE SUMMARY

Minviro was appointed by RecycLiCo™ Battery Materials Inc. (RecycLiCo™) to conduct a gate-to-gate life cycle assessment (LCA) on the production of the high purity cathode precursor material and LiOH·H$_2$O, using the RecycLiCo™ with electrodialysis process at RecycLiCo™'s pilot plant in British Columbia, Canada, for the reference year 2021. Mass and energy balances were developed as part of RecycLiCo™'s detailed engineering studies. The RecycLiCo™ process will co-produce aluminum cake and sodium sulfate. The environmental impacts are allocated between the primary product (NCM) and the two co-products (aluminum and sodium sulfate) using mass allocation.

The results of the LCA on the production of 1 kg of NCM and 1 kg of LiOH·H$_2$O are presented in Table 7.1. Five impact categories were evaluated: global warming potential (GWP), water scarcity footprint, acidification potential, minerals and metal depletion, and fossil fuel depletion. International acceptance exists for these impact

TABLE 7.1
Results Summary of Life Cycle Assessment Study

Impact Category	NCM Impact Value	LiOH·H$_2$O Impact Value	Units (per kg of functional unit)
Global warming potential	7.1	3.3	kg CO$_2$ eq.
Water scarcity footprint	33	16	kg water eq.
Acidification potential	2.7E-2	0.6	mol H$^+$ eq.
Minerals and metal depletion	6.3E-5	5.7E-6	Kg SB. Eq.
Fossil fuel depletion	102	48	MJ

categories. Water scarcity footprint includes regional water scarcity as applied by the AWARE methodology.

The total GWP for RecycLiCo™'s NCM and LiOH·H$_2$O products are 7.1 kg CO$_2$ eq. per kg NCM, and 3.3 kg CO$_2$ eq. per kg LiOH·H$_2$O, respectively. The main drivers of both products are the consumption of natural gas for thermal energy and total electricity use. In addition to the base case scenarios, a number of benchmarks were evaluated to compare RecycLiCo™ with electrodialysis process to other global production routes.

Comparison scenarios were complete to evaluate RecycLiCo™ process, with primary raw material cathode precursor production and conventional hydrometallurgical recycling of lithium-ion battery (LIB) waste. Overall, RecycLiCo™'s recycling process with electrodialysis to produce NCM precursor and LiOH·H$_2$O indicate a lower environmental impact than the equivalent products produced via raw material production and conventional recycling. This study does not intend to support direct comparative assertions.

7.3 INTRODUCTION

RecycLiCo™ Battery Materials Inc. (RecycLiCo™) has retained Minviro to model the environmental performance of manufacturing their high purity cathode precursor material: NCM and LiOH·H$_2$O, using RecycLiCo™ process with electrodialysis at their pilot plant, located in British Columbia, Canada, using LCA. The results will be used to improve production processes and for long-term strategic planning. This chapter is a summary of the methodology applied by Minviro in the LCA.

7.3.1 PROJECT DESCRIPTION

The RecycLiCo™ process together with electrodialysis is a proposed LIB recycling process which takes consumer battery waste and reprocesses it to produce NCM and LiOH·H$_2$O to reuse in cathode manufacturing of LIBs. RecycLiCo™'s aim is to have this process implemented at major gigafactories worldwide, however for this LCA, the impact is calculated for producing NCM and LiOH·H$_2$O located at the pilot plant in British Columbia, Canada. The engineering studies are based on detailed test work of samples produced from the recycling process of NMC-622 batteries. Despite the engineering studies being based on recycling NMC-622 batteries, the aim is to use the RecycLiCo™ process with electrodialysis on all types of NMC battery chemistries [2].

TABLE 7.2

RecycLiCo™ Process Production Overview

Production Parameter	Value	Units
Pilot plant location	British Columbia	
NCM 622 LIB process feed	21	kg/h
NCM production	16.2	kg/h
LiOH·H₂O production	2.1	kg/h
Aluminum co-product production	1.0	kg/h
Sodium sulfate co-product production	1E-6	kg/h

The recycling process follows a closed-loop hydrometallurgical process, which involves the reprocessing of NCM 622 LIBs by leaching, bulk precipitation, and precursor precipitation for NCM production. For LiOH·H₂O production, the NCM 622 LIB feed undergoes leaching and bulk precipitation. Post-bulk precipitation, the LiOH·H₂O stream is then separated from the NCM stream, where it undergoes electrodialysis and evaporation to produce the LiOH·H₂O end product. Within the leaching stage of the process, two co-products are produced: aluminum and sodium sulfate. The technical estimates for the process are presented in Table 7.2. The RecycLiCo™ process with electrodialysis produces no landfill waste, with a recycling process in place for water and reagents to be recycled. This includes the remaining Li₂SO₄, which is produced within the bulk precipitation process, and not consumed within the LiOH·H₂O stream. The additional small amount of mass loss from the feed material entering the process and the product outputs, is as a result of the extensive venting system. The vents allow for evaporation, of which all outputs are saturated with water. It is assumed that the outputs are non-toxic.

7.3.2 SCOPE OF ASSESSMENT

The goal of this LCA is to determine the major project and process parameters contributing to the environmental life cycle impact of the production of NCM and LiOH·H₂O produced by the RecycLiCo™ process with electrodialysis at RecycLiCo™'s pilot plant in British Columbia, Canada. International acceptance exists for the selected category indicators. Additionally, comparison scenarios were completed to compare the environmental impact of:

1. RecycLiCo™ process and electrodialysis with NCM and LiOH·H₂O produced from conventional primary raw material extraction.
2. RecycLiCo™ process with NCM produced from conventional LIB's hydrometallurgical recycling methods.

Efforts are made to consider the system boundaries of the process benchmarked are comparable to the system boundary defined for the RecycLiCo™ process with electrodialysis. This is described further in Section 7.7.

LCA is a method to assess the environmental impacts associated with all stages of a product, process, or activity [3]. Importantly, LCA makes it possible to evaluate indirect impacts that occur in the development of a product or process system over its entire life cycle, providing information that otherwise may not be considered. A wide range of environmental impacts can be captured both scientifically and quantitatively. The holistic approach generates results on how decisions made at one stage of the life cycle might have consequences elsewhere, ensuring that the balance of potential trade-offs can be made and avoiding shifting of the environmental burden [4, 5]. It must be noted that LCA is a suitable method for determining impacts on a global scale, however, the methodology is less suitable for determining local impacts.

7.4   LIFE CYCLE ASSESSMENT METHODOLOGY

This LCA study was conducted according to the requirements of the ISO-14040:2006 and ISO-14044:2006 standards [6, 7]. LCA has four fundamental steps: (i) goal and scope definition, (ii) inventory analysis, (iii) impact assessment, and (iv) interpretation, as presented in Figure 7.1.

The life cycle impact assessment (LCIA) was carried out with a combination of data provided by RecycLiCo™ and Ecoinvent databases of characterization factors. Ecoinvent version 3.7.1 provides a well-documented process for products supporting the understanding of their environmental impacts [8]. The Ecoinvent database comprises inventory data for most economic activities. The consistency and cohesion of this background life cycle inventory (LCI) dataset increases the credibility and acceptance of the LCA. The baselines of this database are LCI datasets that consider human activities and their interactions with the environment.

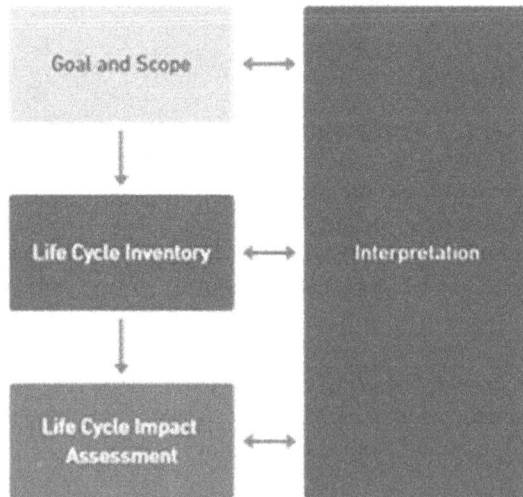

FIGURE 7.1   General phases of a life cycle assessment as described by ISO 14040, extracted from ISO 14040 (2006).

7.4.1 GOAL AND SCOPE

This study assesses the life cycle impact of the production of 1 kg of NCM and 1 kg of LiOH·H₂O produced from recycling NMC 622 LIBs, using the RecycLiCo™ process with electrodialysis, at RecycLiCo™'s pilot plant in British Columbia, Canada. The production chain includes waste NCM cathode material production waste being fed into the RecycLiCo™ process, where the battery undergoes acid leaching. Aluminum co-product is separated from the rest of the metals as a result of the acid leach, where it is precipitated and sold.

Post leaching and aluminum removal, bulk precipitation occurs where a mixed hydroxide precipitate (MHP) and lithium sulfate (Li₂SO₄) are produced. MHP is fed into the precursor production stage, where the mixed hydroxide precipitate is redissolved and co-precipitated to form the final NCM product. Li₂SO₄ undergoes electrodialysis and evaporation, forming the final LiOH·H₂O product.

This LCA is a gate-to-gate study, meaning the product life cycle impact is being assessed from the waste LIB feed going through the RecycLiCo™ process with electrodialysis, with the preceding stages of production and use of the NCM batteries being excluded from the LCA. The end-gate (a set point at the end of processing) has been set to NCM and LiOH·H₂O produced at the RecycLiCo™ pilot plant facility. The use of the product in cathode manufacturing and end-of-life (the disposal of the battery) is outside the scope of this LCA study. Understanding the full life cycle impact of NCM and LiOH·H₂O from cradle-to-grave or cradle-to-cradle requires the extension of the system boundary into the use-phase and the end-of-life phases.

The next objective of this study is to understand and quantify the environmental impact of RecycLiCo™'s NCM product compared with the production of cathode precursor (NCM) via conventional primary raw material extraction. This is a cradle-to-gate study, with the cradle being the point of raw material extraction and the end-gate being set at the point of NCM and LiOH·H₂O production. For this comparison, the RecycLiCo™ with electrodialysis LCA system boundary was extended to cradle-to-gate. The final objective is to quantify the environmental impact of RecycLiCo™'s NCM product compared with the production of NCM produced via conventional hydrometallurgical recycling processes. This is a gate-to-gate study, with the start-gate being the plant feed entering the recycling process and the end-gate route being set at the point of production of NCM.

The scope boundaries for each process result in suitable coverage of comparable sets of processes, which is discussed further in Section 7.8.

The data generated during these studies provide estimates on the technical parameters for NCM production and precursor production associated with the production of NCM and LiOH·H₂O. The aim of carrying out this study is to quantify the environmental impacts of the proposed recycling process and to identify the environmental hotspots for the production of NCM and LiOH·H₂O. The secondary aim of this study is to assist in project development and improvement, with a third motivation being to assist with strategic planning.

This study has been conducted according to the requirements of the ISO-14040:2006 and ISO-14044:2006, including a third-party review from two independent LCA experts to ensure that it is scientifically robust. The intended audience for this study includes parties that are interested in LIB value chain, ranging from investors to customers of NCM and end-users. This study does not intend to support comparative assertions. It is recognized that the data provided by this LCA study may be used by others for comparative assertions in separate future studies. These comparisons should be made on a product system basis only and carried out in accordance with ISO 14040 and 144044 standards.

7.4.2 Functional Unit

LCA uses a functional unit as a reference to evaluate the components within a single system or among multiple systems on a common basis. The functional unit is the quantitative reference used for all inventory calculations and impact evaluations.

7.4.2.1 RecycLiCo™ Process with Electrodialysis

The functional unit for this study is defined as: 1 (one) kilogram of high purity cathode precursor (NCM), produced from the recycling of NMC 622 LIBs, via RecycLiCo™ process at RecycLiCo™'s pilot plant facility in British Columbia, Canada.

Additionally, the impact of one kilogram of battery grade LiOH·H$_2$O, produced from the recycling of NMC 622 LIBs, via electrodialysis is also assessed.

7.4.2.2 Conventional NCM Precursor Production via Primary Raw Material Extraction

For the scenario of producing NCM precursor and LiOH·H$_2$O from primary raw material extraction, the functional unit is defined as: 1 (one) kilogram of NCM precursor, produced from primary raw material extraction.

Additionally, the impact of one kilogram of battery grade LiOH·H$_2$O, produced from 70% spodumene ore and 30% brine, is investigated.

7.4.2.3 NCM Precursor Production via Conventional LIB Recycling

For the scenario of producing NCM precursor from conventional hydrometallurgical recycling, the functional unit is defined as: 1 (one) kilogram of NCM precursor, produced from hydrometallurgical recycling of spent NMC 811 batteries at recycling centers in Wuhan and Wixi, China.

7.4.3 System Boundary

7.4.3.1 RecycLiCo™ with Electrodialysis Production Route Gate-to-Gate

This LCA models the production of NCM and LiOH·H$_2$O from the recycling of NMC 622 LIBs, via RecycLiCo™ process with electrodialysis. This LCA is conducted using regional-specific data, assuming the process is conducted at RecycLiCo™'s pilot plant facility in British Columbia, Canada. It should be noted that this process is designed to be implemented in gigafactories worldwide. As a result of this, the impact is likely to differ depending on the environmental impact of each country that provides the electricity consumed within the process.

The system boundary for the LCA study covering these stages is presented in Figure 7.2(a). The life cycle impact of the RecycLiCo™ process with electrodialysis is

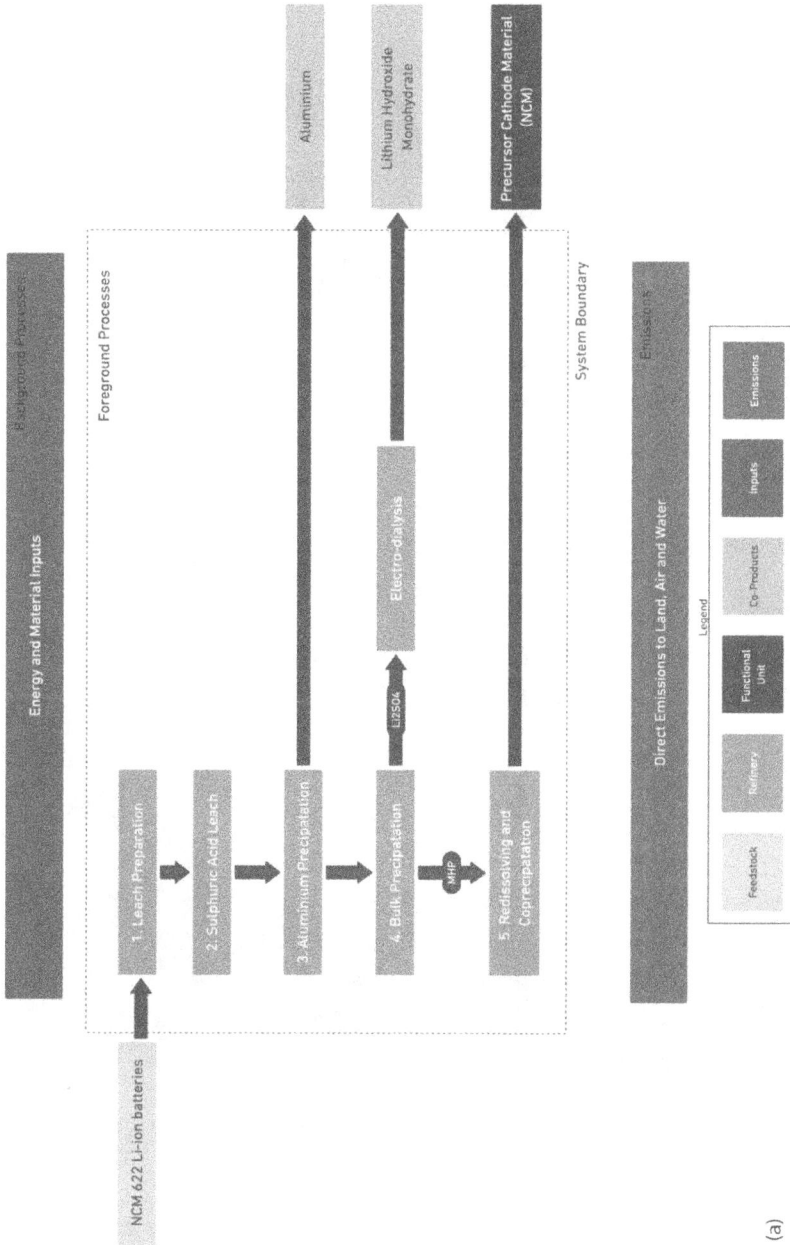

FIGURE 7.2 (a) Gate-to-Gate system boundary applied to the life cycle assessment study and (b) Cradle-to-gate system boundary applied to the life cycle assessment study. (*Continued*)

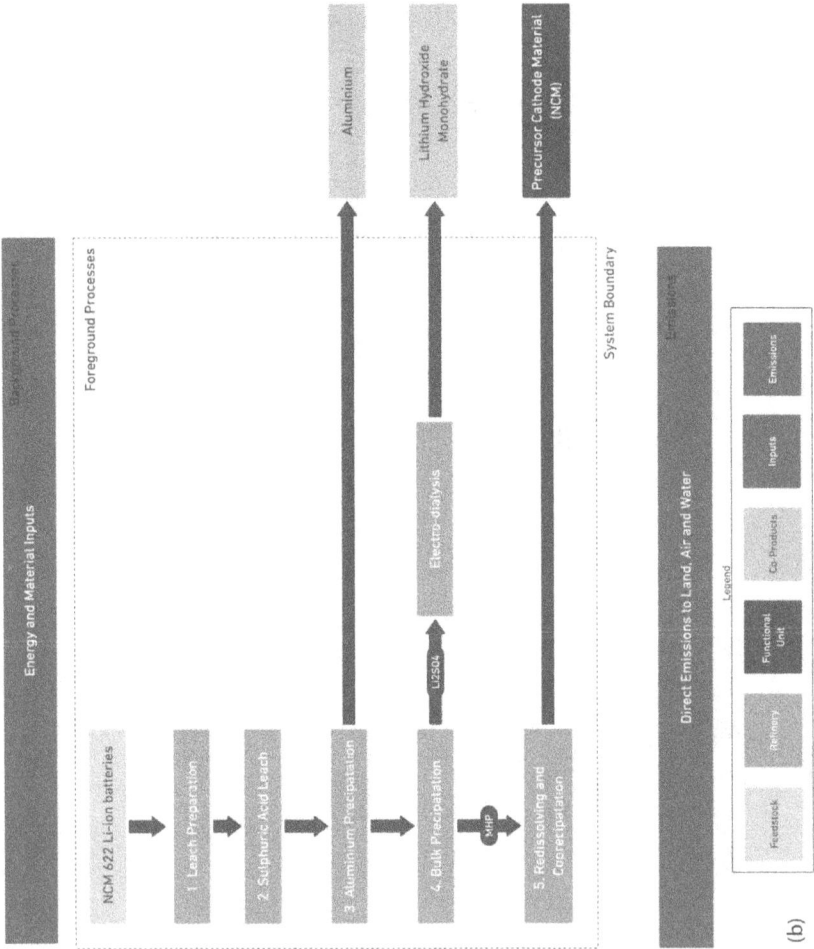

FIGURE 7.2 (*Continued*)

modeled, considering the different flows relating to the specific co-products. The upstream production of the NMC 622 LIB is not included in this study.

Waste NMC batteries are fed into an acid leach, using concentrated sulfuric acid and sulfur dioxide (SO_2), where aluminum is separated and precipitated out of the solution. This is sold as a co-product. The metals remaining in the solution undergo bulk precipitation, where MHP is precipitated while Li_2SO_4 remained in the solution. Lithium hydroxide is the reagent added within this stage.

Following bulk precipitation, the MHP and Li_2SO_4 intermediate products are separated into two streams. The MHP undergoes a redissolving and co-precipitation stage to produce the high-purity precursor cathode material, NCM. Sulfuric acid, ammonium hydroxide, sodium hydroxide, and hydrogen peroxide are consumed within this stage.

The Li_2SO_4 undergoes electrodialysis, evaporation, and crystallization to produce LiOH solution and LiOH monohydrate crystal. Low concentration of sulfuric acid and lithium hydroxide solutions are fed to the electrodialysis and upgraded. The remaining low concentration Li_2SO_4 is recycled back into the process. The final stage is the evaporation of the solution, and crystallization to produce $LiOH \cdot H_2O$.

Electricity consumed by the process is sourced from the British Columbia electricity grid. Thermal energy is supplied by natural gas from a small scale < 100kW condensing boiler.

7.4.3.2 RecycLiCo™ with Electrodialysis Production Route Cradle-to-Gate

Within this LCA study, the system boundary of the RecycLiCo™ process with electrodialysis was extended from a gate-to-gate process to a cradle-to-gate process to allow for the comparison in impact of RecycLiCo™'s process compared to conventional primary raw material extraction (Sections 7.4.3.3 and 7.7.1). The extended system boundary is shown in Figure 7.2(b).

7.4.3.3 Conventional NCM Precursor and LiOH·H₂O Production via Primary Raw Material Extraction

As part of this LCA study, the environmental impact of producing NCM precursor from primary raw materials, was calculated. The refinery and cathode manufacturing plant is based in Vancouver, British Columbia, Canada. The system boundary of the route is shown in Figure 7.3. Electricity is sourced from the British Columbia electricity grid. The transport of raw materials is accounted for within this LCA. The transport routes for the primary raw materials are:

- Nickel sulfate hexahydrate from Ontario to British Columbia.
- Manganese sulfate monohydrate from China to British Columbia.
- Cobalt sulfate heptahydrate from China to British Columbia.
- Lithium hydroxide monohydrate from China to British Columbia.

7.4.3.4 NCM Precursor Production via Conventional LIB Recycling

The environmental impact of producing NCM precursor via conventional hydrometallurgical recycling of NMC 811 was calculated and compared to the environmental

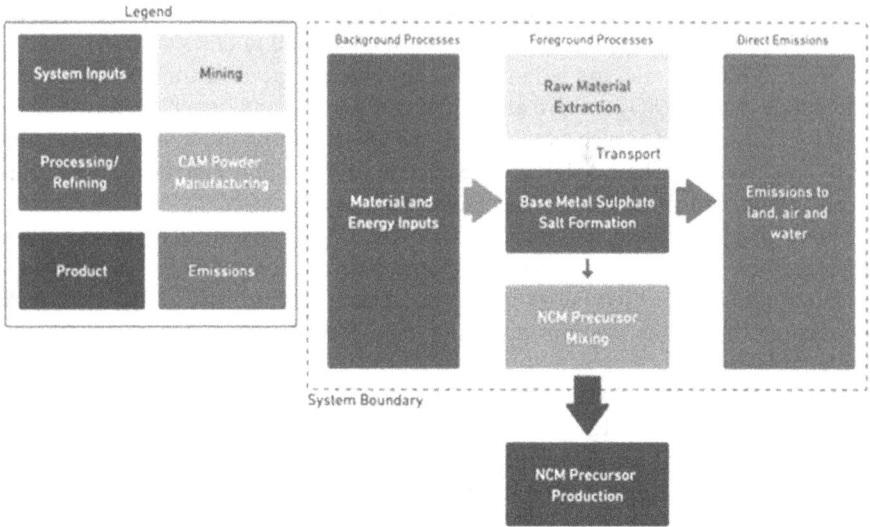

FIGURE 7.3 System boundary applied to the life cycle assessment study for primary raw materials extraction comparison.

impact of RecycLiCo™ producing NCM precursor using their RecycLiCo™ process. The system boundary is shown in Figure 7.4. Research has identified that regardless of the NMC chemistry (e.g., NMC 811, NMC 622), there is a negligible difference in GWP, when same energy density is assumed (see Section 7.4.5.2) [9, 10].

This scenario was chosen as it is well-defined process, and representative of hydrometallurgical battery recycling. The transportation was highway transportation, with

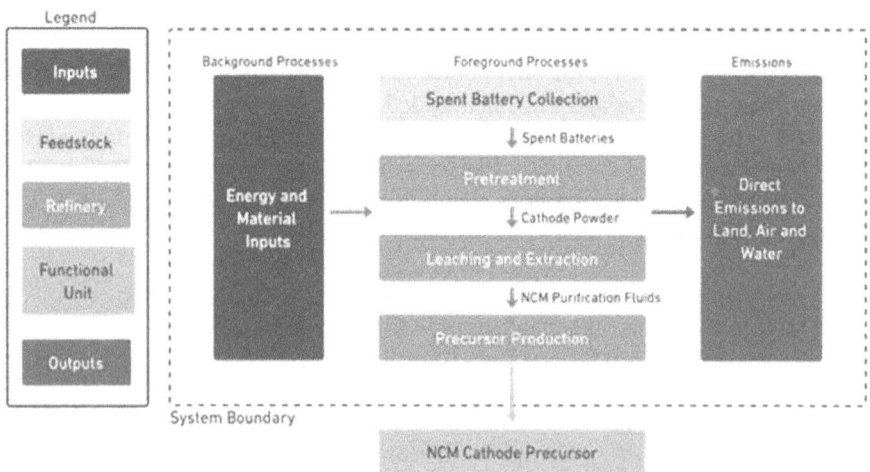

FIGURE 7.4 System boundary applied to the life cycle assessment study for conventional hydrometallurgical recycling comparison.

an estimated average distance of 888.52 km [11]. The energy mix was based on China's estimated 2022 electricity generation mix, based on the China Electricity Council's power supply installed capacity plans [12].

7.4.4 MULTI-OUTPUT ALLOCATION

In LCA, it is critical to ensure that environmental impacts are divided among different products of a process operation in a way that is scientifically valid and best practice. Following the guidance provided in ISO-14044:2006 standards, it is recommended to avoid allocation as much as possible [3].

Avoiding allocation can be achieved by allocating specific impacts to specific process streams which produce only one product, or by using a method that is known as system expansion. It is recommended to allocate the impacts following the physical relationships between the products, such as mass-based allocation, before any alternative allocation methodologies using non-physical relationships. An alternative is economic allocation, in which the economic value of different products is used to allocate the impacts among them. For the mining and metals industry, it is recommended that mass-based allocation is used for base metal production and economic allocation for precious and platinum metal production [5].

The co-products are produced from the same process feed using shared processes. In this study, mass allocation was used. Sodium sulfate product is excluded because it contributes < 1% of the total products produced. This means sodium sulfate has very minimal impact on the overall impact results. The mass allocation breakdown is shown in Table 7.3.

7.4.5 LIFE CYCLE INVENTORY

7.4.5.1 RecycLiCo™ with Electrodialysis Production Route

This study was desk-based, meaning that all data was either provided by RecycLiCo™, collected from public sources, or assembled from public and private databases. Background data was used from Ecoinvent 3.7.1. Assumptions and limitations for this study are discussed in Section 7.4.8. An LCI summary is included in Appendix A. An analysis of the material and energy flows within the system boundary were made and all major material and energy flows related to the recycling of NMC 622 batteries to produce NCM and LiOH·H_2O has been included in the LCI and included in the LCIA. This includes materials and energy consumed within the process.

TABLE 7.3
Mass Allocation Breakdown for the LCA of RecycLiCo™ Process

Product	Production (t/h)	Mass Allocation	Economic Allocation
NCM	16.2	83.9%	79.1%
LiOH·H_2O	2.1	11.1%	20.9%
Aluminum Co-product	1.0	5.0%	

7.4.5.2   Comparison Studies

For the benchmarking exercises, the LCIs were collected internally from Ecoinvent 3.7.1., and from public databases [11, 13]. The system boundaries can be obtained from Section 7.4.3 of this report. The studies were desk-based, meaning that all data was either collected from public sources or assembled from public and private databases. Foreground and background data were used from Ecoinvent 3.7.1.

7.4.5.2.1   *Conventional NCM Precursor and LiOH·H₂O Production via Primary Raw Material Extraction*

To evaluate the life cycle impact of producing NCM precursor and LiOH·H$_2$O from primary raw material extraction, the following assumptions within the LCI were made:

- The bill of materials to complete the NCM precursor of an NCM-811 battery was taken from Argonne National Labs GREET model [13].
- It is assumed that the NMC-811 battery and NMC-622 battery have the same energy density of ~240 Wh/kg [13].
- Electricity inputs for each stage of precursor manufacturing are based on 2019 study by Thomitzek et al. [14] and reviewed by Erakca et al. in 2021 [15], exploring energy demand in cell manufacturing.
- Electricity inputs for refining and powder manufacturing are calculated using region-specific information for assumed production in Vancouver, British Columbia, Canada, in all instances.
- The assumptions for transport of the raw materials to RecycLiCo™'s facility include road transport by 7.5–16 metric ton EURO3 freight lorry, and shipping via freight container ship. Empty returns are not included.

Given that supply chain impacts for individual battery raw materials can be highly variable, the model assumes the following routes for major components taken from major global industry studies or Minviro's internal database:

- Nickel sulfate from the Nickel Institute global average [16].
- Cobalt sulfate from primary data from a Chinese facility.
- Manganese sulfate from the global average for primary production.
- Lithium hydroxide from a calculated representative global average (70% from Australian spodumene processed in China and 30% from brines extracted and processed in Chile).

The metal sulfate quantities assumed within each NCM-811 battery are shown in Table 7.4.

7.4.5.2.2   *NCM Precursor Production via Conventional LIB Recycling*

To evaluate the life cycle impact of producing NCM precursor from hydrometallurgical LIB recycling, the data originates from investigations of established enterprises in Hubei and Jiangxi Provinces [11]. The spent batteries collected are sourced from major cities including Beijing, Guangzhou, Shanghai, Tianjin, Chengdu, and Wuhan.

TABLE 7.4
NCM Precursor Metal Salts Component Breakdown

Component	Value (GREET)	Unit
Nickel Sulfate Hexahydrate	1.34	kg per kg cathode
Cobalt Sulfate Heptahydrate	0.17	kg per kg cathode
Manganese Sulfate Monohydrate	0.16	kg per kg cathode

Energy and transport data was derived from CNMLCA Sino-Centre database [11]. It is assumed in this scenario the spent batteries are composed of NMC-811, however, as long as the battery chemistries have the same energy density (assumed to be ~240 Wh/kg in this instance), there is a negligible difference in the environmental impacts [9, 10, 13].

7.4.6 CUT-OFF CRITERIA

Cut-off criteria are used in LCA to decide which inputs should be included in the assessment based on mass, energy, or environmental significance. As this study is based on data from detailed engineering studies, cut-off criteria have been applied. All flows provided by RecycLiCo™ were considered in the LCA study, except for the sodium sulfate co-product, which was excluded because it contributes < 1% to the total products produced. It is possible that cut-off effects have been applied to the background flows from Ecoinvent 3.7.1 due to missing flows in the background dataset. LCI related to the manufacturing of equipment, maintenance, packaging, and infrastructure have been excluded in this LCA. The reason for excluding these flows is that they are often very small compared to flows of reagents or energy consumed in the process over decades of operation. Upstream transport of reagents, and other consumable flows such as pipes, reagent packing materials, and staff-related impacts have not been included due to the lack of data.

7.4.7 LIFE CYCLE IMPACT ASSESSMENT

The LCIA categories selected for this study include GWP, water scarcity footprint, acidification potential, minerals and metal depletion, and fossil fuel depletion. These impact categories were chosen for this project to understand the environmental value proposition of NCM and LiOH·H₂O vs alternative processing routes. The LCIA results are relative expressions and do not predict impacts on category endpoints, the exceeding of thresholds, safety margins, or risks.

7.4.7.1 Global Warming Potential

7.4.7.1.1 Baseline Model of 100 Years Based on IPCC 2013 [17]
Climate change can be defined as the change in global temperature caused by the greenhouse effect of "greenhouse gases" (GHG) released by human activity. There is

now scientific consensus that the increase in these emissions have a noticeable effect on climate. Climate change is one of the major environmental effects of economic activity, and one of the most difficult to control because of its global scale [18]. The environmental profiles characterization model is based on factors developed by the UN's Intergovernmental Panel on Climate Change. Factors are expressed as GWP over the time horizon of different years, the most common historically being 100 years, measured in the reference unit, kg CO_2 eq.

The Greenhouse Gas Protocol identifies three "scopes" of GHG emissions which have been included in this study, however, it should be noted that scopes of emissions are not a framework inherent to LCA. The GHG Protocol defines scopes of emissions as:

Scope 1: direct GHG emissions (e.g., furnace off-gas, combustion of fuels).

Scope 2: indirect GHG emissions from consumption of purchased electricity, heat, or steam (e.g., emissions embodied in grid power or embodied in steam at an industrial park).

Scope 3: other indirect emissions such as the extraction and production of purchased materials and fuels, transport-related activities in vehicles not owned or controlled by the reporting entity, electricity-related activities (e.g., transmission & distribution losses) not covered in scope 2, outsourced activities, and waste disposal. Scope 3 emissions can be either "upstream" or "downstream." In a cradle-to-gate LCA, "upstream" scope 3 must be included.

7.4.7.2 Water Scarcity Footprint (AWARE Methodology)

7.4.7.2.1 *"Available WAter REmaining" (AWARE)*

The AWARE method is applied to quantify the environmental performance of products and operations regarding freshwater. This method is developed by Water Use in Life Cycle Assessment (WULCA), a working group of the UNEP-SETAC LCI, on a water scarcity midpoint method for use in LCA and for water scarcity footprint assessments. This approach is based on the available water remaining per unit of surface area in a given watershed after human and ecosystem demands have been met relative to the world average. The resulting characterization factor ranges between 0.1 and 100 and can be used to calculate water scarcity footprints as defined by the ISO standard [19]. A value close to 0.1 means that plenty of water is available in that region, while a water scarcity of a 100 means that water in that region is extremely scarce. The water scarcity factor is defined as the water scarcity index (WSI). The world average non-agricultural water scarcity factor is 1.0. Units of the characterization factor are dimensionless, expressed in m^3 world eq./m^3. Figure 7.5 provides the AWARE overlay map for the region of RecycLiCo™'s pilot plant facility, British Columbia, Canada. It is important to note that this impact relates to the potential of water deprivation to humans or ecosystems, rather than direct water use by the project. Another way to think about this is that it is a LCIA value, not an inventory flow. This value also includes embodied water impacts of reagents and energy.

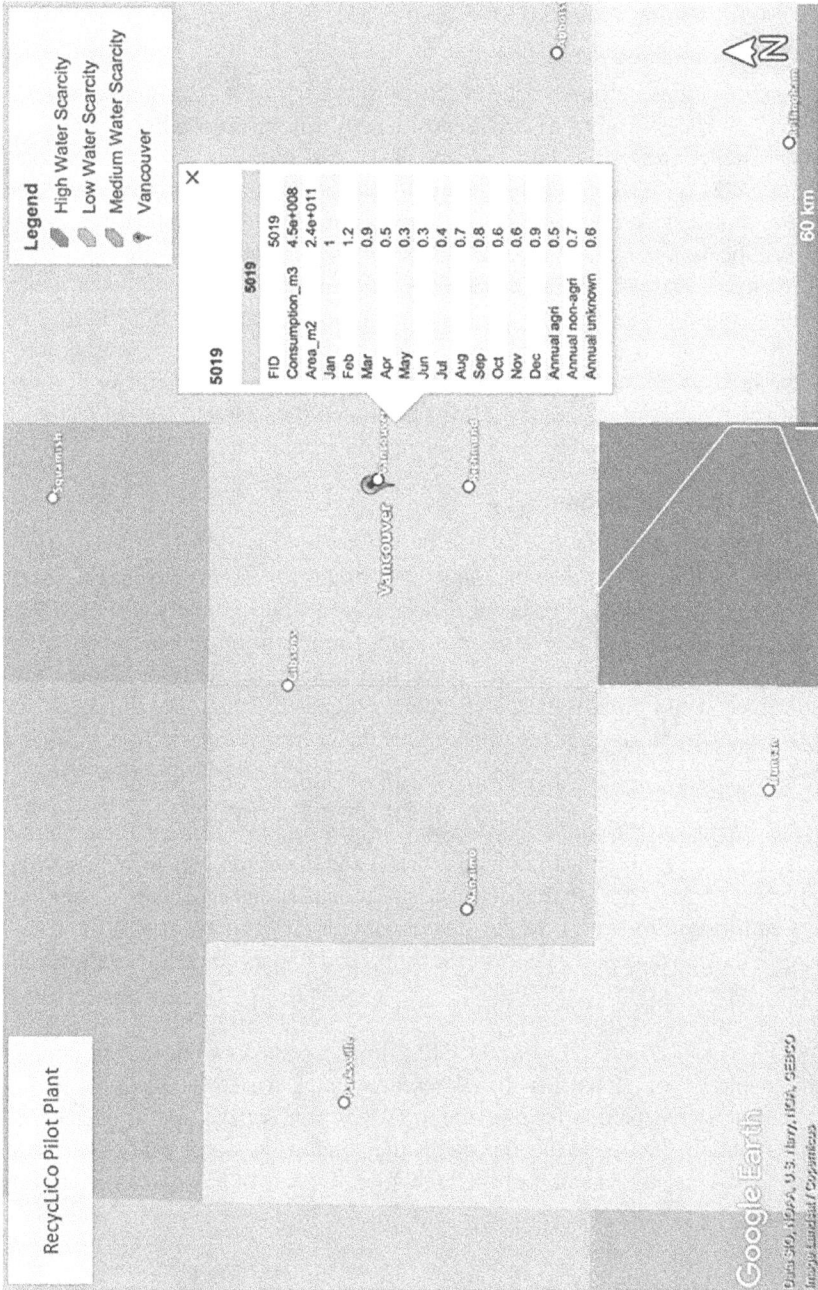

FIGURE 7.5   Regionalized water scarcity impact for RECYCLICO's pilot plant, located in British Columbia, Canada.

The annual non-agricultural water scarcity factor used in the AWARE analysis for this LCA is 0.7.

7.4.7.3 Acidification Potential

7.4.7.3.1 Accumulated Exceedance

Acidic gases such as sulfur dioxide reacts with water in the atmosphere to form "acid rain," a process known as acid deposition. When this rain falls, often a considerable distance from the original source of the gas, it causes ecosystem damage to varying degrees depending upon the nature of the ecosystems. Gases that cause acid deposition include ammonia (NH_3), nitrogen oxides (NO_x), and sulfur oxides (SO_x). Acidification potential is expressed using the reference unit, mol H^+ equivalent. The model does not take an account of regional differences in terms of which areas are more or less susceptible to acidification. It accounts only for acidification caused by SO_x and NO_x. This includes acidification due to fertilizer use, according to the method developed by the Intergovernmental Panel on Climate Change. Centrum voor Milieukunde Leiden (CML) has based the characterization factor on the RAINS model developed by the University of Amsterdam [20, 21].

7.4.7.4 Resource Depletion

7.4.7.4.1 Scarcity with Relation to Ultimate Reserves

This impact category quantifies the depletion of natural resources from the earth, based on: the concentration of reserves; deaccumulation rate; quantity of fuels; ratios of annual production to available reserves; and damage to resources based on the increased cost of extraction. Characterization factors are given for minerals and metals collectively, and for fossil fuels as a separate impact category – in this LCA, we use both categories to describe the impacts that the project in question has on global reserves [16].

7.4.7.4.1.1 Mineral and Metal Depletion This impact category describes abiotic resource depletion from global ultimate mineral and metal reserves using the CML 2002 methodology [16]. Resource depletion for this midpoint indicator is measured in kg of antimony (Sb) equivalent per functional unit, and this represents the mass-based resource consumption as part of the embodied impacts of any given material or energy input.

7.4.7.4.1.2 Fossil Fuel Depletion This impact category describes abiotic resource depletion of global fossil fuel reserves using the CML 2002 methodology [16]. Resource depletion for this midpoint indicator is measured in megajoules (MJ) per functional unit, and this represents the potential common energy output of all fossil fuels consumed as part of the embodied impacts of any given material or energy input.

7.4.8 Assumptions and Limitations

The primary limitation of this study is the uncertainty associated with the level of definition of a project at the detailed engineering stage compared to an operational facility. This has been addressed by assigning 15% uncertainty to the data [22]. Additional limitations and assumptions within this study are:

- Natural gas produced in different countries are likely to have varying impacts. The characterization factor used in this study is not regional, it is based on a global characterization factor.
- Transport of reagents to the site has been excluded from this study due to a lack of data. Transport of reagents has been excluded from the benchmarking studies as well to ensure like-for-like comparison.
- For electricity, location-based grid mixes have also been used, not residual market grid mixes.
- Background data was sourced from Ecoinvent 3.7.1, which was updated in 2020. However, some of the highest impact contributors to each route were sampled to check the time relevance for the unit processes. While most datasets were updated for a time period up to 2020, some flows had a reference year of more than 4 years old.
- Multifunctionality is dealt with using mass allocation. However, as the use-phase and end-of-life-phase of the products are not included, this study only covers a part of the product life cycle.
- As the location of production of the reagents is unknown, a WSI of the world, average of one is used (see Section 7.5.2 for definition).
- The assumptions associated with the life cycle inventories collected for the comparison scenarios are addressed in Section 7.4.5.2.
- RecycLiCo™ assumes the RecycLiCo™ process with electrodialysis is a closed loop recycling process, meaning that it is implemented at gigafactories located worldwide, where the manufacturers at the gigafactories take back their own LIB product at the end-of-life and upcycle it to create a similar product to the first use phase - in this case, cathode precursor material [23].
- The waste air emissions, as a result of venting within the RecycLiCo™ process with electrodialysis, are assumed to be non-toxic.
- The NCM precursor production from primary raw material extraction comparison scenario uses a characterization factor for nickel sulfate hexahydrate from the Nickel Institute [16]. This is a global average, however depending on where the nickel sulfate hexahydrate is produced globally, has a significant impact on the environmental impact, thus impacting the overall impact of NCM precursor production. To address this, sensitivity analysis (Section 7.9.2) has been carried out to show the range in impact of NCM precursor from primary raw material extraction, sourcing nickel sulfate hexahydrate from different locations globally.

7.4.9 INTERPRETATION

The results were interpreted with reference to the goal and scope, comparing the impacts associated with the identified process routes, geographic regions, and technology implemented. Contribution analysis, sensitivity analysis, and uncertainty analysis were carried out to support the interpretation of the LCA.

7.4.9.1 Data Quality Review

The key data criteria used to evaluate the quality of the LCI used for this LCA study were:

- Technological, time, and geographical representativeness: data is representative if it matches geographical, temporal, and technological aspects of the goal and scope of the study. By utilizing representative data for all foreground processes, the study can be made as representative as possible. When primary data are not available, best-available proxy data is used, ideally from databases or academic LCA literature.
- Completeness: a dataset is judged based on the completeness of inputs and outputs per unit processes and the completeness of the unit processes. The goal is to capture all relevant data in terms of unit processes.
- Precision: measured primary data is considered to be of the highest precision, followed by calculated data, data from the literature, and estimated data. This study is carried out using mass and energy stream tables (measured and/or calculated) from RecycLiCo™'s 2021 detailed engineering studies. It must be noted that measured data can be precise but inaccurate. Accuracy can be obtained by cross-validation of measured data.
- Methodological appropriateness and consistency: data is considered appropriate and consistent if the differences between data reflect actual differences between distinct product systems and are not due to inconsistencies in data collection or modeling.

Table 7.5 presents the grading system of data quality indicators [1]. An evaluation of the data quality for this LCA on RecycLiCo™'s NCM product can be found in later chapters of this report [1].

Minviro assigned a data quality assessment of 1–5 for every life cycle inventory item provided by RecycLiCo™ and data quality assessment of each area of the LCA. An assignment of 1 corresponds to a "very good" data quality, while an assignment of 5 corresponds to a "very poor" data quality. Typically, Minviro assigns data qualities to life cycle inventory items with the levels of process and project definition shown in Table 7.6, however, the exact assignment depends on the specific situation and expert judgement.

The data quality rating of the life cycle inventory item was calculated using the following equation [1].

$$DQR = (TeR + GR + TiR + C + P + M + 4 * X_w) / (i + 4)$$

Where:

"DQR" is the data quality rating.
"TeR, GR, TiR, C, P, and M" are acronyms for the data quality indicators in Table 7.7.
"X_w" is the lowest data quality assignment in the set of data quality assignments for the life cycle inventory's different data quality indicators.
"i" is the number of data quality indicators that were assigned values (in this LCA, six data quality indicators were used for assessing data quality).

TABLE 7.5

Grading Guidelines for Data Quality Assessment as Environmental Footprint 2.0 Pedigree Matrix [1] (PEF = Product Environmental Footprint) [1]

Data Quality Indicator	Very Poor	Poor	Fair	Good	Very Good
Technological representativeness	Old to dissimilar technology used	Technology dissimilar to what is used	Generic technology average	From technology specific to the application	All technology aspects of data have been modeled
Time representativeness	The dataset is older than 8 years	The dataset is less than 8 years old	The dataset is less than 6 years old	The dataset is less than 4 years old	The dataset is less than 2 years old
Geographical representatives	Data represented is from a distinctly dissimilar region of project location	Similar regions are represented in data	Global average is represented in data	Country of interest is represented in the data	Region of interest is fully represented in data
Completeness	Unknown coverage	Data is from small parts of the target region	Data is less than 50% from the target region	Data is more than 50% from the target region	Data is representative of the entire target region
Precision	Rough estimate with known deficits	Estimates based on calculations not checked by the reviewer	Estimates based on expert judgement	Estimates based on measured and prior values	Measured and verified values with <7% uncertainty
Methodological appropriateness and consistency	Attribution process-based approach and following none of the three method requirements of the PEF guide: dealing with multi functionality, end of life modelling, and system boundary	Attribution process-based approach and following one out of three method requirements of the PEF guide: dealing with multi functionality, end of life modelling, and system boundary	Attribution process based approach and following two out of three method requirements of the PEF guide: dealing with multi functionality, end of life modelling, and system boundary	Attribution process based approach and following three method requirements of the PEF guide: dealing with multi functionality, end of life modelling, and system boundary	Full compliance with all requirements of the PEF guide

TABLE 7.6

Data Quality Assignment According to Level of Process Definition

Very Poor	Poor	Fair	Good	Very Good
Back-of-the-envelope calculations	Scoping study or preliminary economic assessment	Pre-feasibility study or definitive feasibility study	Detailed design and engineering	Operating facility

TABLE 7.7

Environmental Impact Results of LiOH·H$_2$O Impact Value

Impact Category	LiOH·H$_2$O Impact Value	Units (per kg LiOH·H$_2$O)
Global warming potential	3.3	kg CO$_2$ eq.
Water use	16.2	kg water eq.
Acidification potential	0.58	mol H$^+$
Minerals and metal depletion	5.7E-6	kg Sb. eq
Fossil fuel depletion	48	MJ

The definition of the acronyms from the equation state above are detailed below. The definitions of the acronyms are in Table 7.7.

- TeR = Technical Representativeness
- GR = Geographical Representativeness
- TiR = Time Representativeness
- C = Completeness
- P = Precision
- M = Methdology

7.4.9.2 Critical Review

Following internal review processes, a critical review was carried out by two independent external experts, and together they cover the required competencies relevant to the critical review. The critical review was performed at the end of the LCA study. Their findings and suggestions to improve the study are included in Appendix C.

7.5 RESULTS

The results are presented by stage, followed by contribution analysis. The individual stream inputs per stage are defined within Appendix A, Table 7.16. The below figures aggregate contributors to each environmental impact category worth less than 1% of the total impact as "other" for visualization. These small contributors are still included in the overall result. If the impact of the contributors is less than the decimal places shown, "other" will be displayed as zero. Appendix B shows the non-aggregated contribution analysis.

7.5.1 GLOBAL WARMING POTENTIAL

7.5.1.1 Global Warming Potential – Total

The total GWP for the RecycLiCo™ NCM product is around 7.1 kg CO_2 eq. per kg NCM according to the LCA model produced by Minviro. The total GWP is presented, broken down by stage of the LCA, in Figure 7.6.

7.5.1.2 Global Warming Potential – Contribution Analysis

Contribution analysis of the GWP is presented in Figure 7.7. The top three most significant contributors to GWP in the production of RecycLiCo™ NCM product are:

- 2.4 kg CO_2 eq. per kg NCM associated with natural gas consumed for drying within NCM processing.
- 1.7 kg CO_2 eq. per kg NCM associated with electricity supplied to NCM stream.
- 1.2 kg CO_2 eq. per kg NCM associated with sodium hydroxide consumed within the precursor production stage.

7.5.1.3 Global Warming Potential – Total Breakdown by Scope

GWP impact is classified into scope 1, 2, and upstream scope 3 emissions (defined in Section 7.3.7.1). The total GWP impact is 7.1 kg CO_2 eq. per kg NCM. The GWP broken down by scopes (1, 2, and 3) is presented in Figure 7.8.

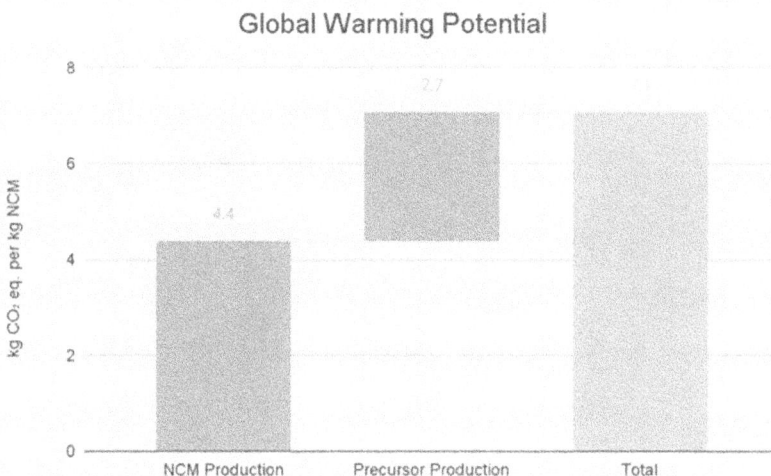

FIGURE 7.6 Total global warming potential.

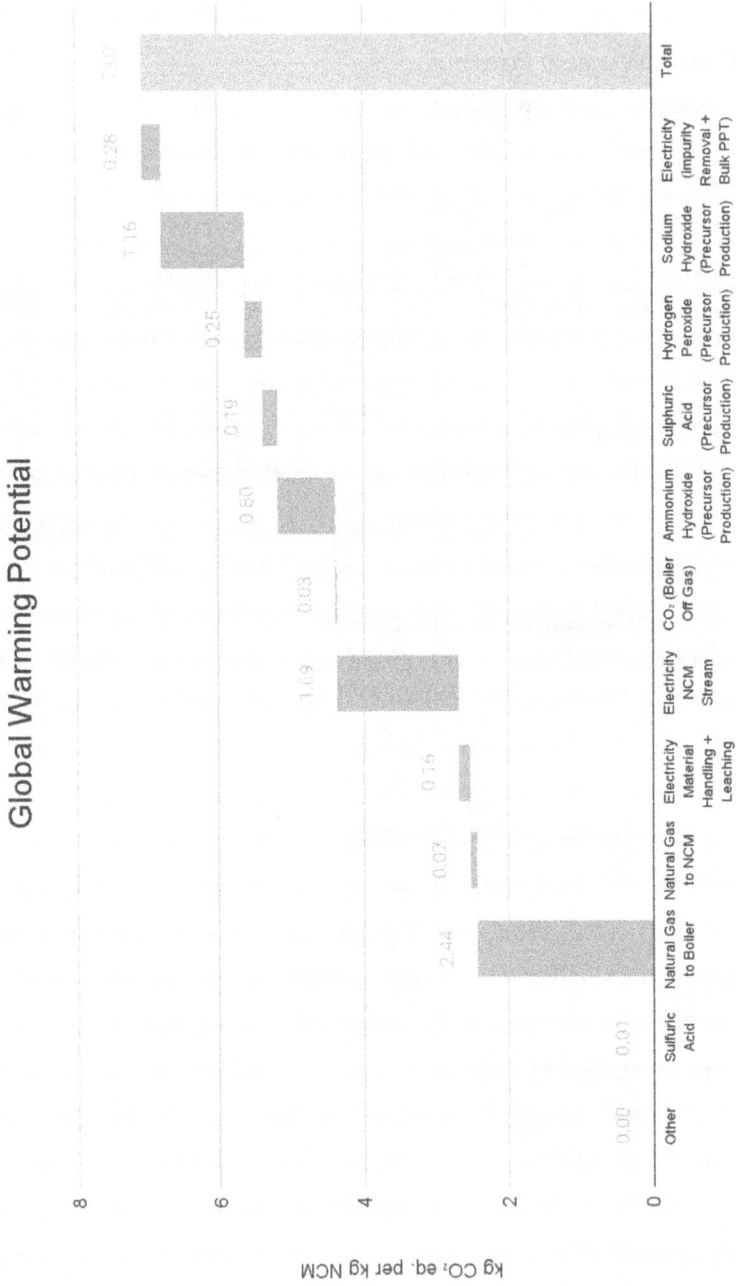

FIGURE 7.7 Global warming potential contribution analysis.

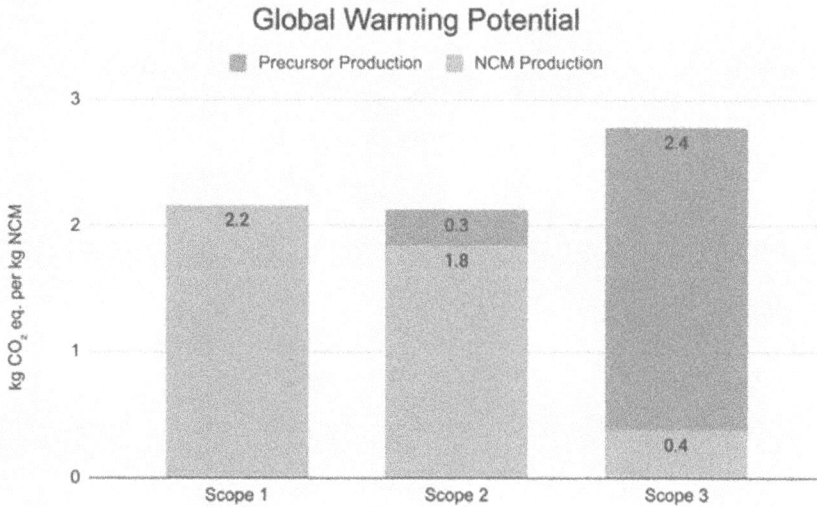

FIGURE 7.8 Global warming potential contribution analysis by scope of emissions.

Scope 1 emissions, which reflect direct emissions associated with the combustion of fossil fuels on-site (natural gas), total 2.2 kg CO_2 eq. per kg NCM. These emissions come from the combustion of natural gas and direct CO_2 emissions.

Scope 2 emissions are the embodied impact of imported electrical energy. These total 2.1 kg CO_2 eq. per kg NCM.

Upstream scope 3 emissions reflect the embodied impact of consumables and thermal energy. Total scope 3 emissions equate to 2.8 kg CO_2 eq. per kg NCM.

7.5.2 WATER SCARCITY FOOTPRINT (AWARE METHODOLOGY)

7.5.2.1 Water Scarcity Footprint – Total

The total water scarcity footprint for the RecycLiCo™ NCM product is around 33 kg water eq. per kg NCM according to the LCA model produced by Minviro. The results are presented in Figure 7.9. As the location of production of the reagents is unknown, a WSI of the world average of one is used. Only water lost within the processes is accounted for.

As discussed in the introduction of this report, the regionalized WSI for RecycLiCo™ process is 0.7 m³ word equivalent per m³. This means that for each cubic meter of water consumed by RecycLiCo™ process utilized at the pilot plant in British Columbia, Canada, the impact will be scaled up with a factor of 0.7 to draw global comparisons. This factor is included in the water scarcity footprint presented in this LCA. This is considered a low scarcity factor compared to other regions globally.

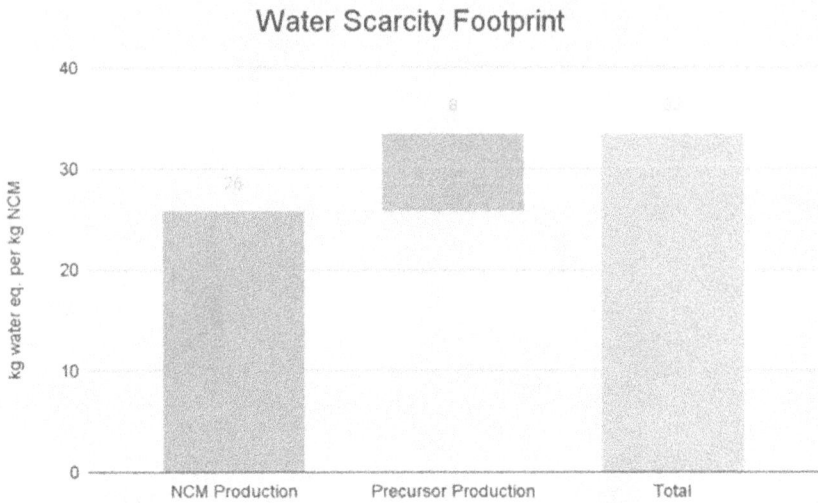

FIGURE 7.9   Total water scarcity footprint.

7.5.2.2   Water Scarcity Footprint – Contribution Analysis

Contribution analysis for water scarcity footprint is presented in Figure 7.10. The top three most significant contributors to GWP in the production of the RecycLiCo™ NCM product are:

- 15 kg water eq. per kg NCM associated with electricity for NCM stream in NCM production.
- 9 kg water eq. per kg NCM associated with fresh water consumed in the NCM production.
- 2 kg water eq. per kg NCM associated with fresh water supply in NCM precursor production.

The top contributors that impact in NCM production are electricity used for the NCM stream, freshwater consumption, and electricity for material handling and leaching.

The top contributors of impact in precursor production are electricity and freshwater consumption.

Electricity has a high embodied impact due to being sourced from the British Columbia electrical grid, which is predominantly supplied by hydroelectric power. Hydroelectric power generated in Canada is run-of-river and alpine/non-alpine reservoirs [24]. A large volume of water is lost from alpine/non-alpine reservoirs from evaporation. This leads to a high embodied impact of water lost, which is accounted for when assessing the water scarcity footprint.

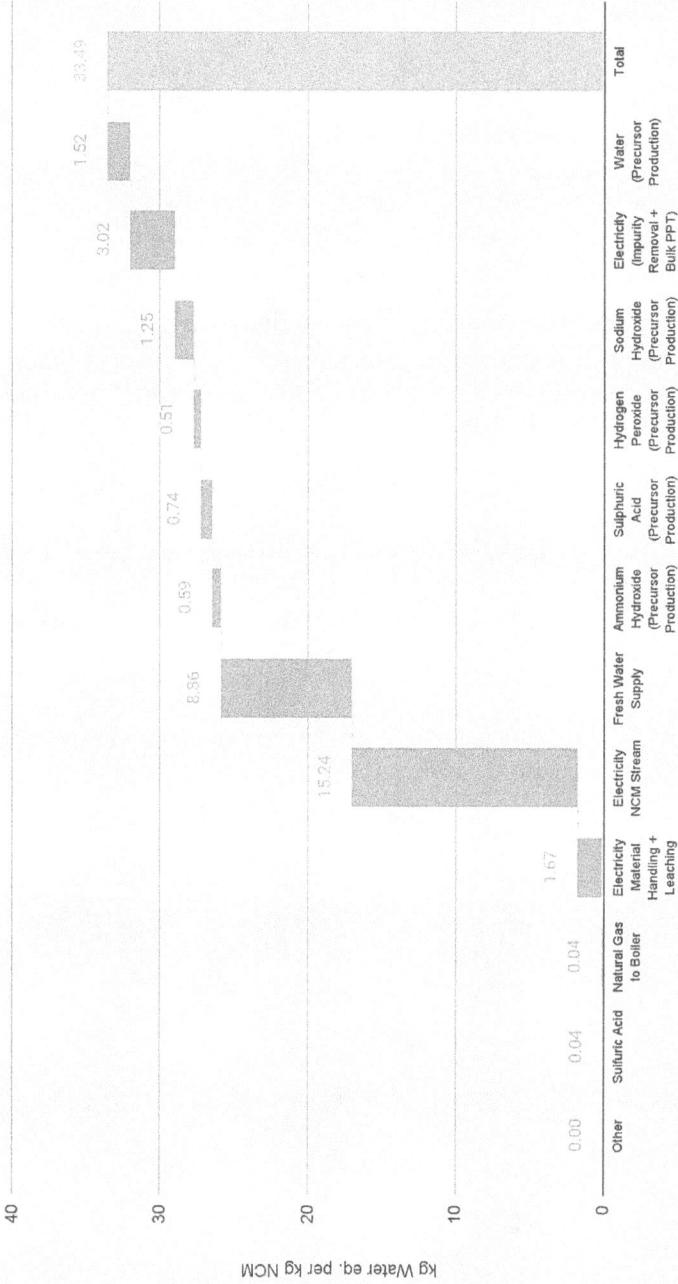

FIGURE 7.10 Water scarcity footprint contribution analysis.

Approximately 11 kg water eq. per kg NCM of the water scarcity footprint comes from the use of actual freshwater at site. The remainder of the contributions come from the embodied water impacts of reagent and energy inputs.

7.5.3 ACIDIFICATION POTENTIAL

7.5.3.1 Acidification Potential – Total

The total acidification potential for the RecycLiCo™ NCM product is around 2.7E-2 mol H^+ eq. per kg NCM according to the LCA model produced by Minviro. The total acidification potential is presented, broken down by area of the LCA, in Figure 7.11.

7.5.3.2 Acidification Potential – Contribution Analysis

Contribution analysis of the acidification potential is presented in Figure 7.12. The top three most significant contributors to acidification potential in the production of the RecycLiCo™ NCM product are:

- 9.1E-3 mol H^+ eq. per kg NCM associated with sulfuric acid in precursor production.
- 6.3E-3 mol H^+ eq. per kg NCM associated with sodium hydroxide in precursor production.
- 3.9E-3 mol H^+ eq. per kg NCM associated with electricity for NCM stream in NCM production.

Acidification Potential

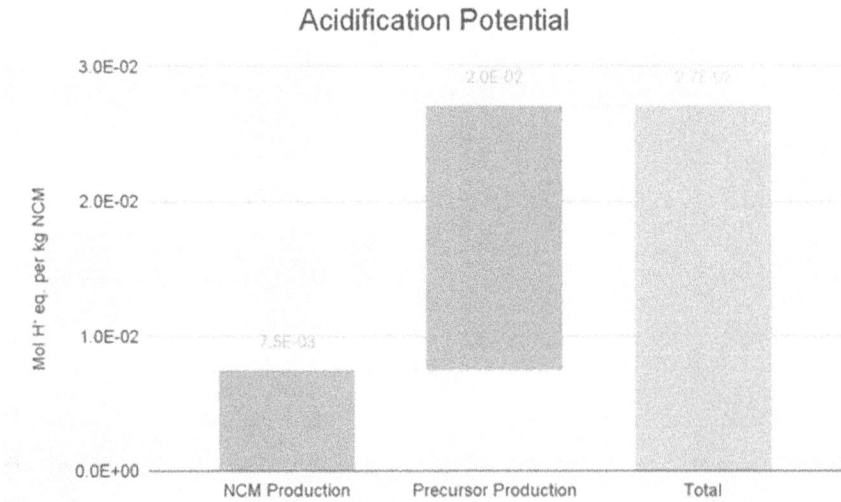

FIGURE 7.11 Total acidification potential.

Acidification Potential

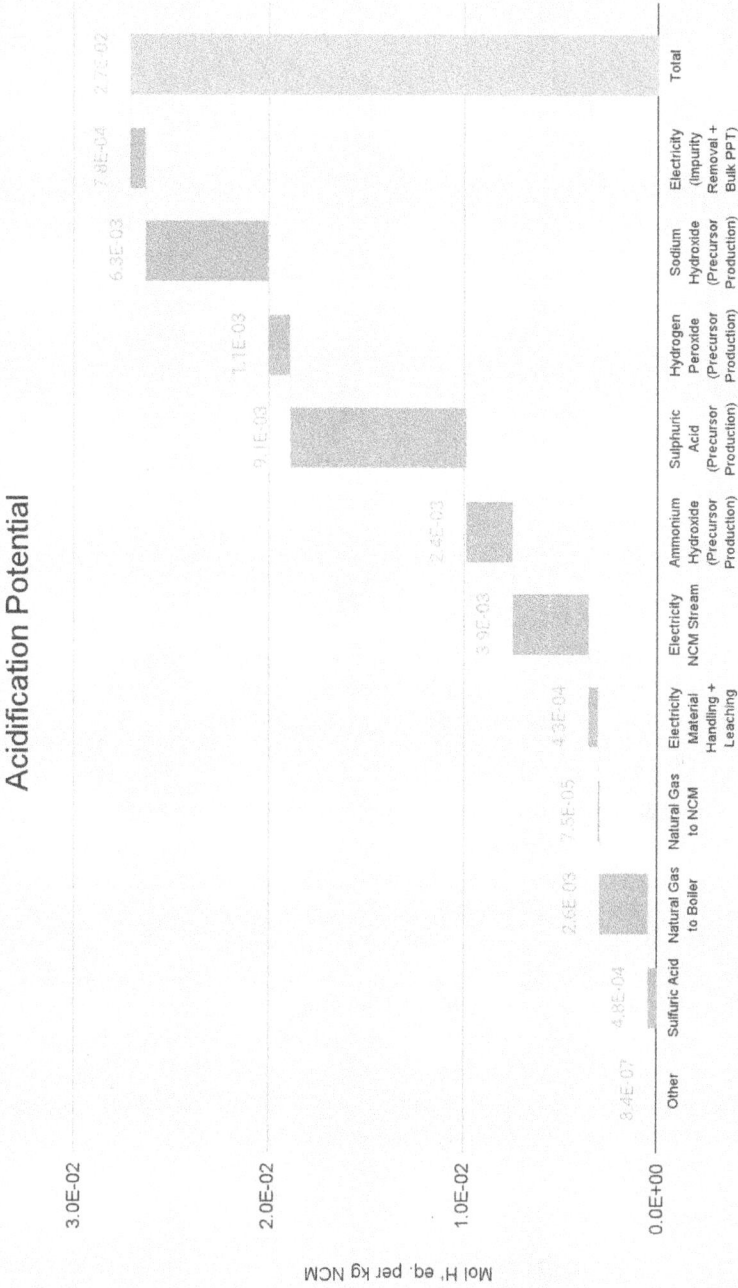

FIGURE 7.12 Acidification potential contribution analysis.

Minerals + Metals Depletion

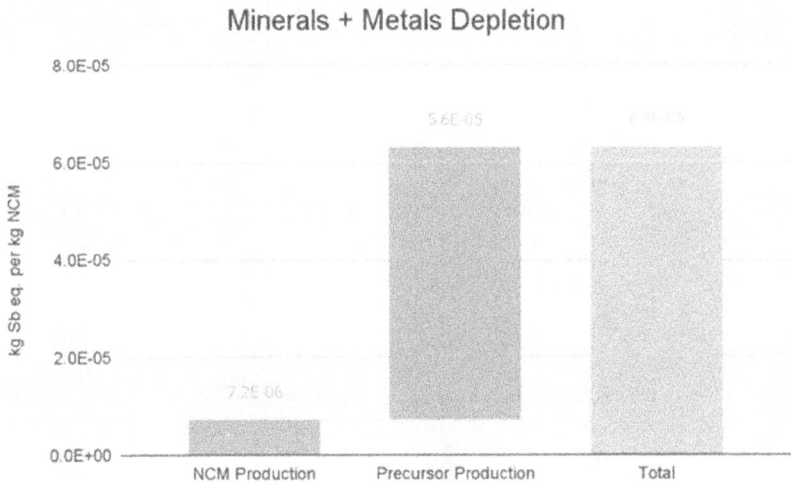

FIGURE 7.13   Total minerals and metal depletion.

7.5.4   MINERALS AND METAL DEPLETION

7.5.4.1   Minerals and Metals Depletion – Total

The total minerals and metal depletion for the RecycLiCo™ NCM product is around 6.3E-5 kg Sb eq. per kg NCM according to the LCA model produced by Minviro. The minerals and metal depletion is presented, broken down by area of the LCA, in Figure 7.13.

7.5.4.2   Minerals and Metal Depletion – Contribution Analysis

Contribution analysis of the minerals and metal depletion is presented in Figure 7.14. The top three most significant contributors to minerals and metal depletion in the production of the RecycLiCo™ NCM product are:

- 3.2E-5 kg Sb eq. per kg NCM associated with the embodied impact of sulfuric acid consumed in precursor production.
- 1.7E-5 kg Sb eq. per kg NCM associated with the embodied impact of sodium hydroxide consumed in precursor production.
- 3.8E-6 kg Sb eq. per kg NCM associated with the embodied impact of ammonium hydroxide consumed in precursor production.

7.5.5   FOSSIL FUEL DEPLETION

7.5.5.1   Fossil Fuel Depletion – Total

The total fossil fuel depletion for the RecycLiCo™ NCM product is around 102 MJ per kg NCM according to the LCA model produced by Minviro. The total fossil fuel depletion is presented, broken down by area of the LCA, in Figure 7.15.

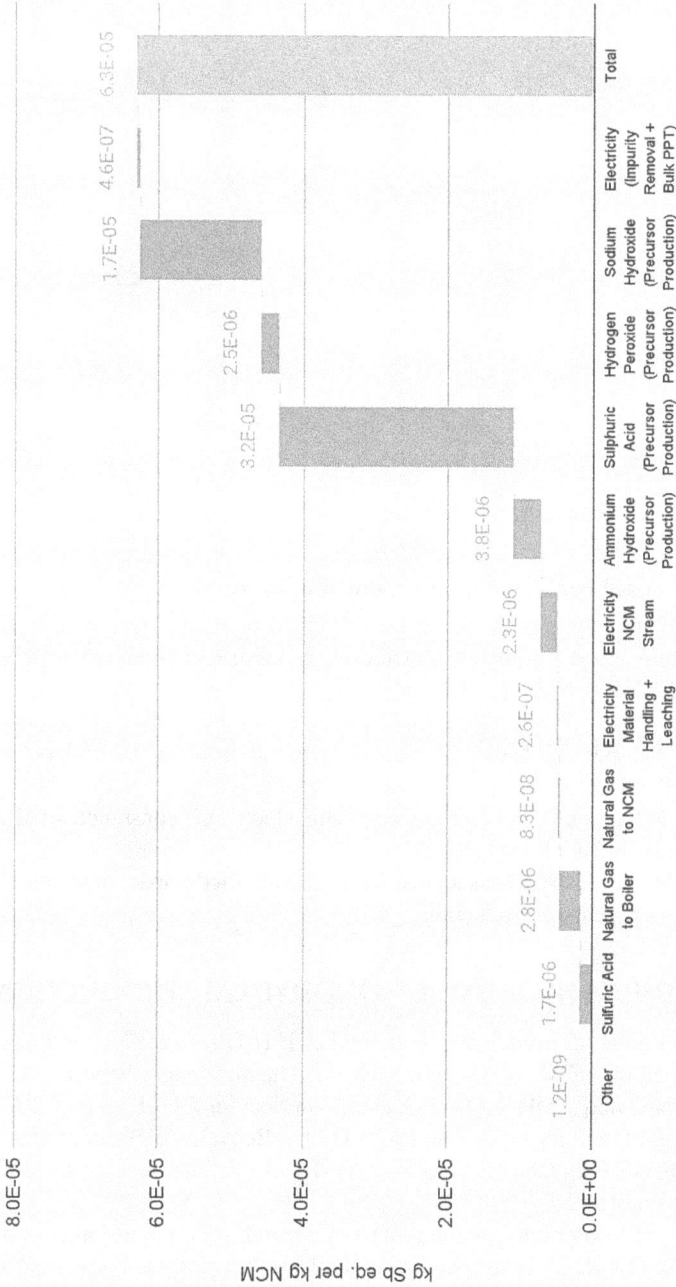

FIGURE 7.14 Minerals and metals depletion contribution analysis.

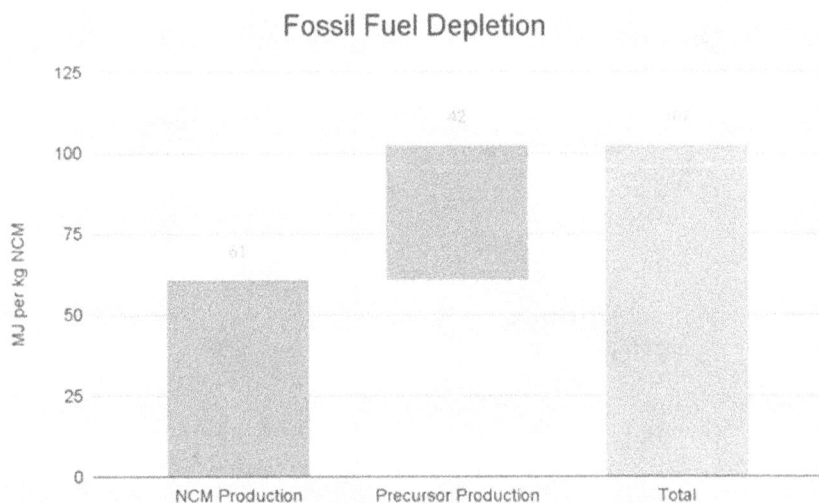

FIGURE 7.15 Total fossil fuel depletion.

7.5.5.2 Fossil Fuel Depletion – Contribution Analysis

Contribution analysis of the RecycLiCo™ NCM product is presented in Figure 7.16. The top three most significant contributors to fossil fuel depletion in the production of RecycLiCo™'s NMC product are:

- 35 MJ per kg NCM associated with natural gas to boilers in NCM production.
- 22 MJ per kg NCM associated with electricity consumed in the NCM stream in NCM production.
- 18 MJ per kg NCM associated with the embodied impact of sodium hydroxide in precursor production.

7.6 LITHIUM HYDROXIDE MONOHYDRATE PRODUCTION

The environmental impact of producing LiOH·H$_2$O for the same impact categories as evaluated for NCM, is shown in Table 7.7. The differences between the NCM production route and LiOH·H$_2$O are described in Section 7.4.3.1.

The total GWP for producing LiOH·H$_2$O at RecycLiCo™'s pilot plant in British Columbia, is 3.3 kg CO$_2$ eq. per kg LiOH·H$_2$O. Contribution analysis of the GWP is shown in Figure 7.17.

The largest contributor is the impact of natural gas combustion within the boiler for drying (2.4 kg CO$_2$ eq. per kg LiOH·H$_2$O). Natural gas consumed to dry lithium sulfate (Li$_2$SO$_4$), an intermediate product, contributes 0.4 kg CO$_2$ eq. per kg LiOH·H$_2$O. Electricity powering the LiOH·H$_2$O production stream contributes 0.3 kg CO$_2$ eq. per kg LiOH·H$_2$O.

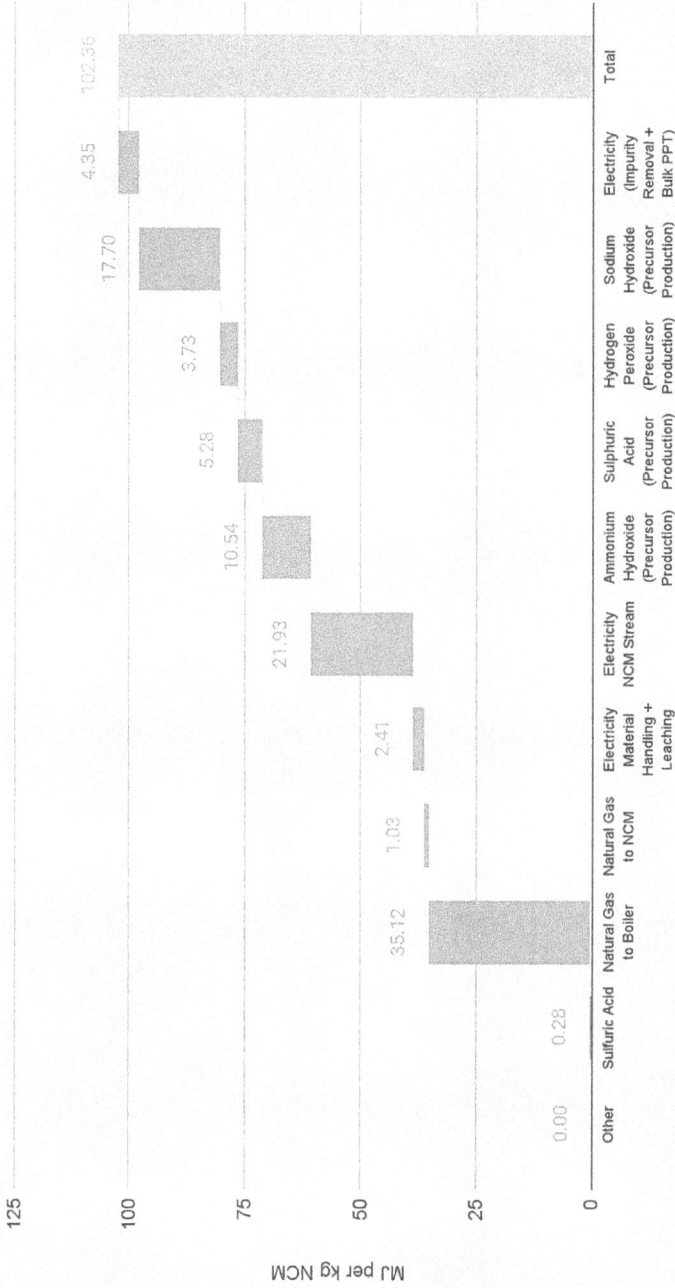

FIGURE 7.16 Fossil fuel depletion contribution analysis.

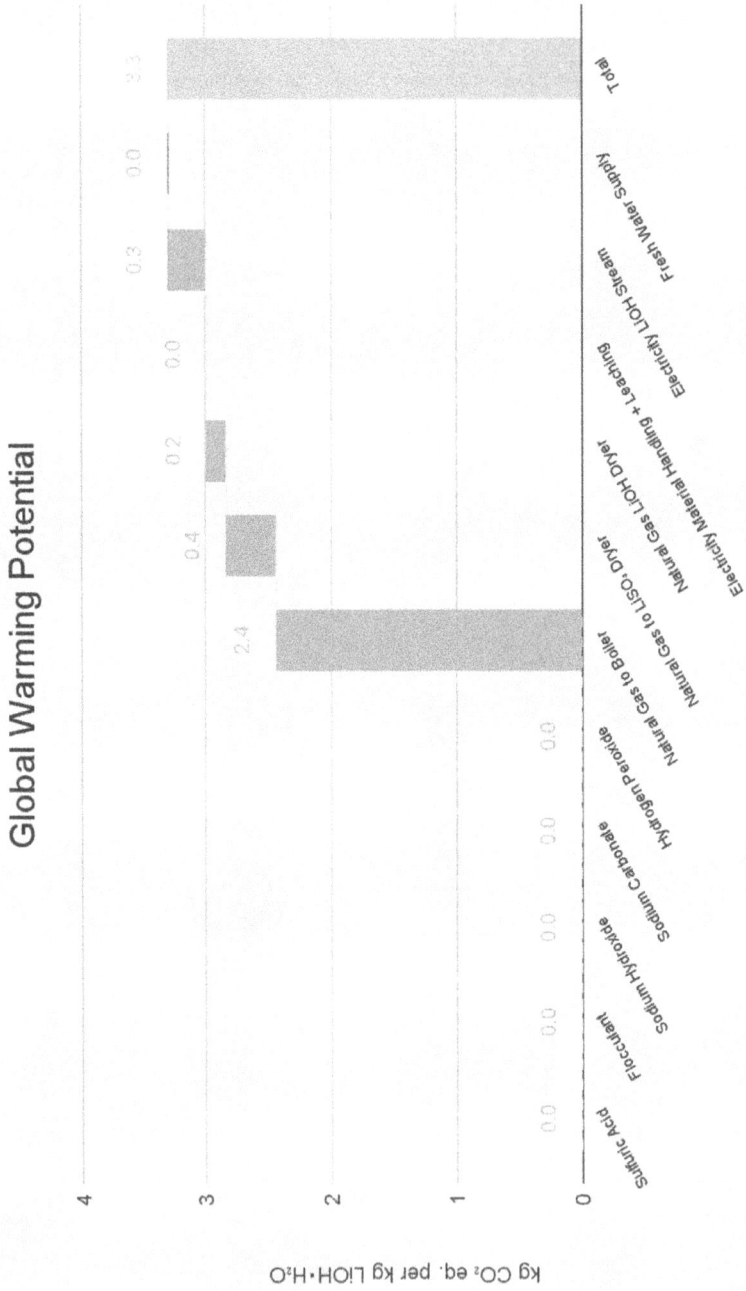

FIGURE 7.17 LiOH·H$_2$O global warming potential contribution analysis.

Fossil fuel depletion, minerals and metal depletion, and acidification potential follow the same trend as the GWP impact category. However, sulfuric acid is also a major contribution to the acidification potential and minerals and metal depletion. This is due to the high embodied impact of producing sulfuric acid, which is later consumed within RecycLiCo™'s recycling process.

Freshwater supply is the highest contribution to the water scarcity footprint, followed by electricity supplied to the LiOH·H$_2$O stream. There is a high embodied water impact when sourcing electricity from the British Columbia grid due to large volumes of water lost through evaporation when electricity is sourced from alpine/non-alpine reservoirs [24]. The embodied impact of material and thermal energy consumption has minimal impact on the overall LiOH·H$_2$O water scarcity footprint.

7.7 COMPARISON SCENARIOS

Minviro has developed generic process route data for making comparisons between existing NCM and LiOH·H$_2$O production pathways and new pathways in development. These comparison scenarios are presented in this chapter. They do not refer to specific operators and do not use the confidential information of other parties. These studies do not intend to support direct comparative assertions. NCM precursor and LiOH·H$_2$O have different environmental impacts depending on the natural resource they are produced from, and the process technology chosen in flowsheets.

The objective of this chapter is to compare the results of producing RecycLiCo™'s NCM and LiOH·H$_2$O products with (i) the production of NCM and LiOH·H$_2$O from primary material sources and (ii) NCM produced from conventional hydrometallurgical LIB recycling. The comparison scenarios assume a feed material of NMC-811, compared to RecycLiCo™'s process that is based on NMC-622 batteries. Nonetheless, there is robust evidence that supports that as long the energy density for the NMC-811 and NMC-622 batteries are the same (~240 Wh/kg) [13], there is negligible difference in the environmental impact for the recycling process itself [9, 10]. NMC-811 was used due to the better available process data for a more accurate indication in the difference in impact of RecycLiCo™'s process versus conventional NMC production from primary raw materials as well as conventional hydrometallurgical recycling process.

The LCI used for the comparison studies are collected from public data sources, and the background databases used are Ecoinvent 3.7.1 [8] and the Environmental Footprint 3.0 database [25]. A comparison could not be made for the water use impact category due to the lack of data available on the water lost and degraded for the comparison processes, to allow for a fair comparison.

Efforts have been made to ensure the system boundaries for the comparison exercises are the same as the system boundary assumed for RecycLiCo™ process with electrodialysis. When assessing different recycling processes, the total impacts are used for a direct comparison. For comparisons with primary raw material production, the system boundary of RecycLiCo™ process with electrodialysis is extended from a gate-to-gate study to a cradle-to-gate LCA study. When expanding the system boundary to account for the impact of the waste LIB's going into the RecycLiCo™ process with electrodialysis, and assuming the RecycLiCo™ process with electrodialysis is a closed loop system, a credit is given to the waste LIB feed-stock, as it is being upcycled, rather than being disposed of. This credit is calculated using system expansion (see Section 7.4.4).

7.7.1 NCM AND LiOH·H₂O FROM CONVENTIONAL PRIMARY RAW MATERIAL PRODUCTION

The aim of this comparison exercise is to compare the results of producing RecycLiCo™'s NCM precursor and LiOH·H₂O products with the production of NCM and LiOH·H₂O from primary raw material extraction.

These are cradle-to-gate studies, with the cradle defined as the point of resource extraction, and the end-gate defined as the final NCM precursor and LiOH·H₂O products. The system boundary of the route chosen is shown in Section 7.4.3.2.

The functional units chosen for this process are:

a. 1 (one) kilogram of NCM precursor, produced from primary raw material extraction.
b. 1 (one) kilogram of lithium hydroxide monohydrate from spodumene: brine (70:30) deposits.

7.7.1.1 NCM Precursor Production

The stages modeled for primary extraction include:

1. Primary Extraction/Refining: metal sulfate hydrates are produced from raw materials. In this case, nickel sulfate hexahydrate, manganese sulfate monohydrate, and cobalt sulfate heptahydrate are produced from various natural resources around the world. This LCA uses representative product impacts for each. Each metal is extracted as ore and concentrated, before being refined into their sulfate forms and hydrated for transportation to precursor manufacturing facilities.
2. Powder Manufacturing: the metal salts are combined in proprietary sequences with sodium hydroxide and ammonium hydroxide, whilst being heated using natural gas, to produce a Ni-Co-Mn hydroxide co-precipitate precursor.

7.7.1.1.1 LCA Allocation Procedure

100% allocation is given to the functional unit, 1 kilogram of NCM precursor production. There are no co-products produced within the processes modeled.

7.7.1.1.2 Global Warming Potential

The total GWP for this scenario is 11.0 kg CO₂ eq. per kg NCM, as seen broken down by contribution analysis in Figure 7.18 and Table 7.10. This GWP value is product specific LCA data, located in Vancouver, British Columbia, Canada. Similar operations with different inputs or outputs may have different GWP, depending on a variety of factors.

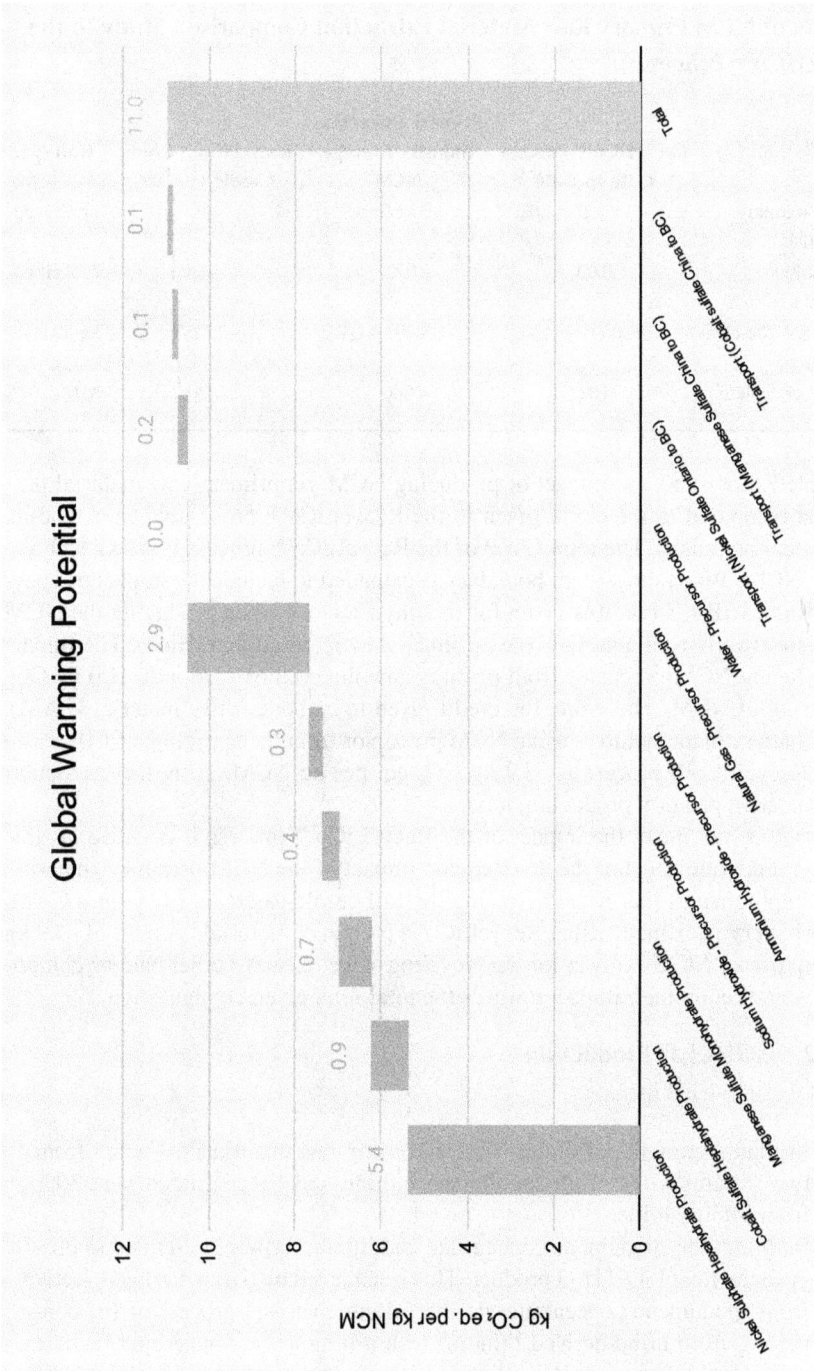

FIGURE 7.18   Global warming potential breakdown for NCM precursor production from primary raw materials.

TABLE 7.8

Results of NCM Primary Raw Material Extraction Comparison Study to the RecycLiCo™ Process

	RecycLiCo™ Process Gate-to-Gate	Primary Extraction Industry Average NCM	RecycLiCo™ Credit Cradle-to-Gate	Unit (per kg NCM)
Global warming Potential	7.1	11.0	−3.9	kg CO_2 eq.
Acidification potential	0.03	0.43	−0.4	Mol H^+ eq.
Minerals + Metal depletion	6.3E-5	8.4E-4	−7.8E-4	kg Sb eq.
Fossil fuel depletion	102	666	−564	MJ

Table 7.8 presents the impact of producing NCM via primary raw material production compared to the credit given to the RecycLiCo™ process when extending the system boundary. The total GWP of the RecycLiCo™ process is 7.1 kg CO_2 eq. per kg NCM. When the system boundary is extended to include the impact of upcycling waste LIB's, a credit is given for the impact saved when producing the NCM precursor from virgin materials (i.e., primary raw material extraction). The impact of producing NCM precursor from primary raw material extraction is 11.0 kg CO_2 eq. per kg of NCM. Therefore, the credit given to cathode active material (CAM), cell or battery manufacturers using NCM precursor from the upcycling of LIB waste using RecycLiCo™ process is −3.9 kg CO_2 eq. per kg NCM, using this particular representative primary production model.

This does not mean the impact of the RecycLiCo™ process is negative, as this does not take into account the downstream impact of the NCM precursor material. This simply means, that when assessing the full cradle-to-cradle LCA for a CAM, cell or battery pack made using RecycLiCo™ precursor material, a credit of −3.9 kg CO_2 eq. per kg NCM is given for the recycling stage only. All other battery components would retain their impact towards the total for a given configuration.

7.7.1.2　LiOH·H₂O Production

The stages modeled include:

1. Primary Extraction: lithium is created using raw materials extracted from two streams −70% from spodumene mining and concentration, and 30% from lithium brine.
2. Refining: the spodumene concentrate and lithium brines are refined to produce the final LiOH·H₂O product. This is achieved by (i) converting directly from spodumene concentrate via the sulfate chemical process or (ii) converting from brine-derived lithium, with a lithium carbonate intermediate product; the latter entails a slightly higher energy requirement due to the extra steps beyond those in the carbonate route and thus, higher project cost.

7.7.1.2.1 LCA Allocation Procedure

100% allocation is given to the functional unit, 1 kilogram of. There are no co-products produced within the process modeled. Allocation of the background data is according to the Ecoinvent 3.7.1 database.

7.7.1.2.2 Environmental Impact Results

The total GWP of this scenario is 12.7 kg CO_2 eq. per kg $LiOH \cdot H_2O$. The results of the environmental categories assessed within this LCA are shown in Table 7.11. This GWP value is product specific LCA data. Similar operations with different inputs or outputs may have a different GWP, depending on a variety of factors.

Table 7.9 presents the impact of producing $LiOH \cdot H_2O$ via primary raw material production compared to the credit given to the RecycLiCo™ process with electrodialysis when extending the system boundary. The total GWP of the RecycLiCo™ process with electrodialysis is 3.3 kg CO_2 eq. per kg $LiOH \cdot H_2O$. The impact of producing $LiOH \cdot H_2O$ from primary raw material extraction is 12.7 kg CO_2 eq. per kg NCM. Therefore, the credit given to CAM, cell or battery manufacturers using $LiOH \cdot H_2O$ from the upcycling of LIB waste using RecycLiCo™ process with electrodialysis is –9.4 kg CO_2 eq. per kg $LiOH \cdot H_2O$, using this particular representative primary production model.

7.7.1.3 Global Warming Potential Comparison

Figures 7.19 and 7.20 present the comparison between the GWP of producing NCM and $LiOH \cdot H_2O$ via the RecycLiCo™ process with electrodialysis for the gate-to-gate impact, as well as the credit given to downstream battery manufacturers when assessing the cradle-to-gate impact of NCM and $LiOH \cdot H_2O$ production via the RecycLiCo™ process with electrodialysis. The production of NCM and $LiOH \cdot H_2O$ from primary raw materials is estimated to be 11.0 kg CO_2 eq. per kg NCM and

---

TABLE 7.9

Results of the $LiOH \cdot H_2O$ from Primary Raw Material Extraction Comparison Study to the RecycLiCo™ Process with Electrodialysis

	$LiOH \cdot H_2O$ RecycLiCo™ with Electrodialysis (Gate-to-Gate)	$LiOH \cdot H_2O$ Industry Average	$LiOH \cdot H_2O$ RecycLiCo™ with Electrodialysis Credit (Cradle-to-Gate)	Units (per kg $LiOH \cdot H_2O$)
Global warming potential	3.3	12.7	–9.4	kg CO_2 eq.
Acidification potential	4.5E-3	0.1	–0.1	Mol H+ eq.
Minerals + metal depletion	5.7E-6	3.2E-5	–2.7E-5	kg Sb eq.
Fossil fuel depletion	48.1	154	–106	MJ

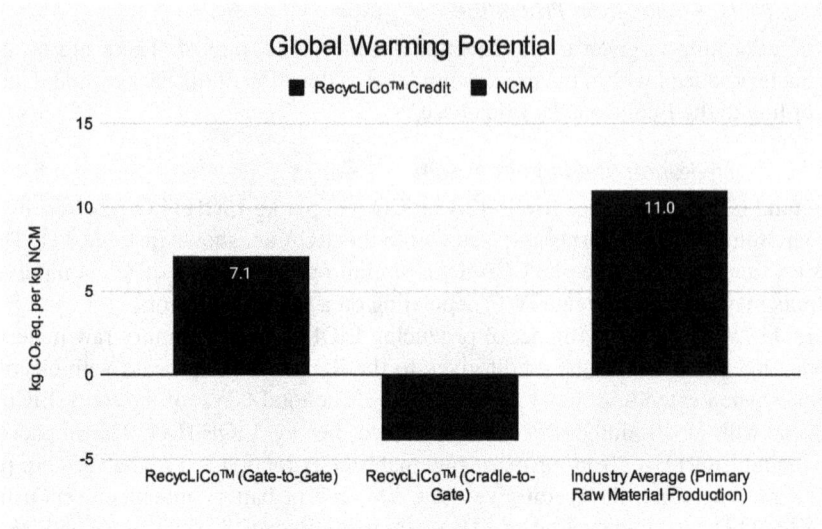

FIGURE 7.19 Comparison of global warming potential impact for producing NCM precursor from primary raw material production vs RecycLiCo™ process.

12.7 kg CO_2 eq. per kg $LiOH \cdot H_2O$, respectively, and these values are used to assign the credits for these particular scenarios. It should be noted that the credit value will vary when considering alternative primary production GWP impacts depending on the route applicable to any given downstream manufacturer, but the credit allocation methodology will be the same.

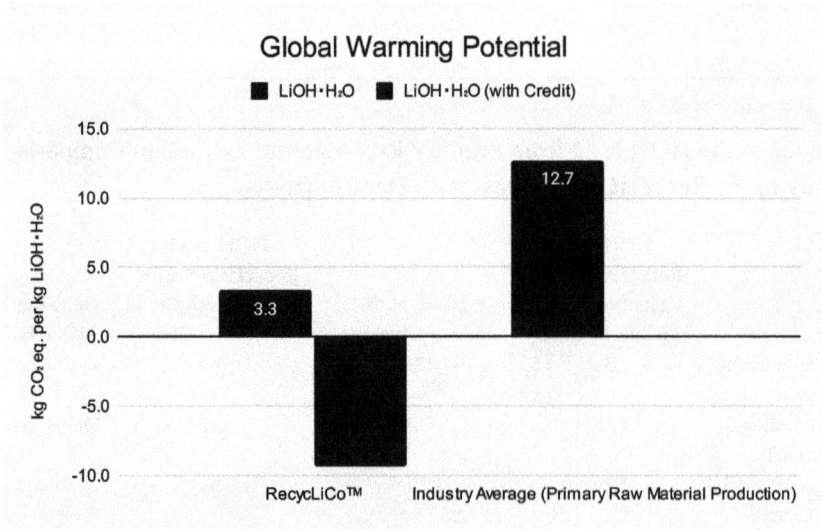

FIGURE 7.20 Comparison of global warming potential impact for producing $LiOH \cdot H_2O$ from primary raw material production vs RecycLiCo™ process with electrodialysis.

7.7.2 CONVENTIONAL LIB HYDROMETALLURGICAL RECYCLING

The aim of this comparison exercise is to compare the results of producing RecycLiCo™'s NCM precursor product with the production of NCM precursor from conventional LIB hydrometallurgical recycling. This is a gate-to-gate study, with the gate defined as the point of the spent LIB being transported to the recycling facility and the end-gate defined as the final NCM precursor product.

The functional unit chosen for this comparison is one kg of NCM precursor produced from spent LIB batteries, via hydrometallurgical recycling, in China.

The stages modeled within this study include:

1. Spent Battery Collection: the spent batteries collected and transported via highway to the recycling plant.
2. Pretreatment: the battery is crushed to separate other primary metals from the electrode containing the desired metals and sorted into separate streams.
3. Leaching and Extraction: the electrode powder and other wastes containing nickel-cobalt from the pre-treatment process are transported to the slurry drum by screw conveyor or belt conveyor for slurry treatment, and then put into the high-pressure reactor and added with sulfuric acid, water, and reducing agent (such as sodium sulfite). Nickel, cobalt, manganese, and other elements in the raw materials into the solution, while carbon, an organic binder, and other components do not enter the solution and are removed, obtaining a solution containing valuable metals. The solution after preliminary purification was first extracted to remove iron, aluminum, and other impurity ions, and the leached solution after deep impurity removal was then extracted manganese, cobalt-nickel separation, and other processes to obtain the NCM purification solution to meet the production of the NCM precursor.
4. Precursor Production: the NCM purification solution obtained from the spent purification was diluted and mixed with sodium hydroxide solution, natural gas, ammonium hydroxide to produce the NCM precursor, which is dried using steam as the heat source to produce the final product.

The LCA allocation procedure involved calculating foreground processes on a step by step basis. Each process stage has a single output. Allocation of the background data is according to the Ecoinvent 3.7.1 database.

7.7.2.1 Conventional LIB Hydrometallurgical Recycling Results

7.7.2.1.1 Global Warming Potential

The total GWP for this scenario is 18.8 kg CO_2 eq. per kg NCM, as seen broken down by contribution in Figure 7.21 and Table 7.10. This GWP value is the product of specific LCA data extracted from Du et al. 2022 [11]. Similar operations with different inputs or outputs may have different GWP, depending on a variety of factors.

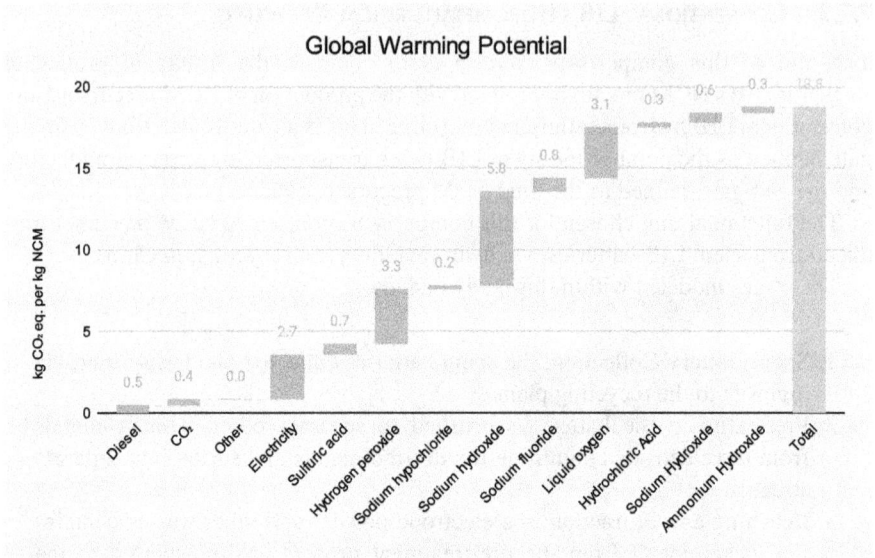

FIGURE 7.21 Global warming potential breakdown for conventional hydrometallurgical LIB recycling.

7.7.2.2 Global Warming Potential Comparison

A comparison of the GWP of conventional hydrometallurgical LIB recycling and the RecycLiCo™ process can be seen in Figure 7.22. RecycLiCo™'s impact is lower, with an estimated GWP of 7.1 kg CO_2 eq. per kg NCM. Conventional hydrometallurgical recycling has a higher impact as a result of high reagent consumption. In the context of cradle-to-gate assessments for downstream manufacturing using upcycled materials (e.g., Section 7.7.1.3), conventional hydrometallurgical recycling would produce a smaller credit than RecycLiCo™'s process as a result of their differing process-specific GWP impacts, when compared to primary production of the same product.

TABLE 7.10
Results of Conventional Hydrometallurgical Recycling Comparison Study in Terms of Percentage Variation to the RecycLiCO™ Process

	RecycLiCo™ NCM	Conventional NCM Recycling	Percentage Variation	Unit
Global warming potential	7.1	18.8	166%	kg CO_2 eq.
Acidification potential	0.03	0.6	2117%	Mol H^+ eq.
Minerals + Metal depletion	6.3E-05	7.30E-04	1058%	kg Sb eq.
Fossil fuel depletion	102	18.2	−82%	MJ

Percent Variation equation: $(\text{Impact}_{Industry} - \text{Impact}_{RecycLiCo™})/\text{Impact}_{RecycLiCo™}$

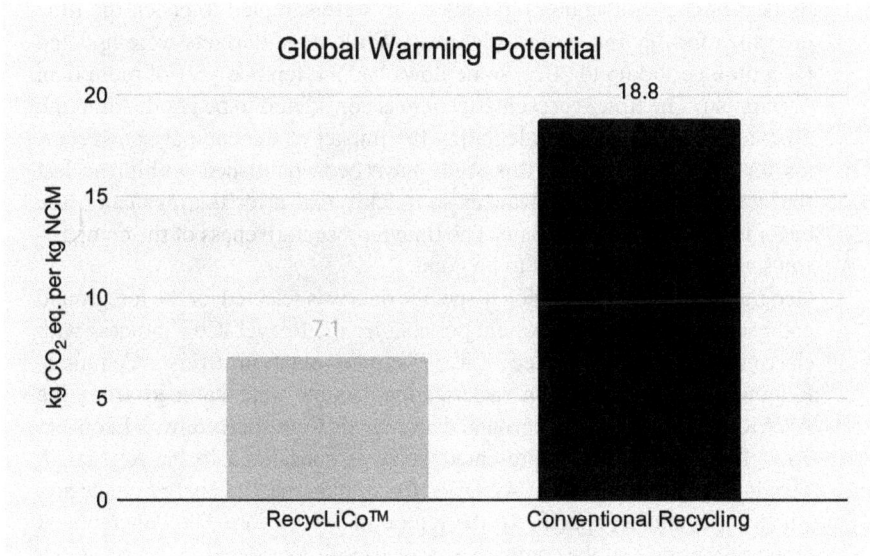

FIGURE 7.22 Comparison of global warming potential impact for producing NCM precursor from conventional hydrometallurgical recycling vs the RecycLiCo™ process.

7.8 DATA QUALITY ASSESSMENT

The foreground and background data of the LCI were judged on technological and time representativeness, geographical coverage, completeness, precision, and consistency. Foreground data used in this study was generated from data provided by RecycLiCo™. The data was derived from mass and energy balances, conducted in 2021, and prepared and reported in accordance with relevant ASX listing rules. The background data used in this LCA was from Ecoinvent 3.7.1 [8]. The RecycLiCo™ project is in the engineering stage. The limitations to the study are discussed in Section 7.4.8.

- Technological representativeness: RecycLiCo™ conducted engineering studies which produced mass and energy balances in 2021. These were used to determine the viability of the RecycLiCo™ process with electrodialysis. All background data was sourced from Ecoinvent 3.7.1. The technological representativeness is considered to be very good. For the comparison of NCM produced from primary raw materials, the data was taken from a wide range of sources. The technological representativeness is considered to be fair. For the comparison of NCM produced from hydrometallurgical recycling, the data was extracted from operating recycling plants. The technological representativeness is considered to be good.
- Time representativeness: all foreground data was based on the engineering studies provided by RecycLiCo™ in 2021. All background data was sourced from Ecoinvent 3.7.1., which was updated in 2020. However, the three

highest impact contributors to each route were sampled to check the time relevance for the unit process datasets. While most datasets were updated for a time period up to 2021, some flows had a reference year of more than 4 years old. The time representativeness is considered to be good. Although all data sources used for calculating the impact of the comparison scenarios for all calculations in this study have been published within the last four years, the research advances so quickly that prior findings can often become obsolete or out of date. The time representativeness of the comparison scenarios is considered to be good.

- Geographical Coverage: regional-specific data was selected for the foreground and background processes where possible for the RecycLiCo™ process with electrodialysis based at RecycLiCo™'s pilot plant in British Columbia. Regional water scarcity characterization factors were sourced from the AWARE database. All background data sources were sourced from Econvent 3.7.1. The geographical representativeness is considered to be very good. The data for the comparison scenarios for all material and energy inputs are, whenever available, specific to the region of production. The geographical representativeness of the comparison scenarios is considered to be very good.

- Completeness: foreground processes were checked for the completeness of the mass and emission inventory. All background data points for RecycLiCo™ process with electrodialysis were sourced from Ecoinvent 3.7.1, of which the completeness is documented. The completeness was considered to be very good. For the comparison scenario where NCM is produced from primary raw materials, the data is based off some assumptions and academic insights, the dataset is complete. The completeness for this comparison scenario was considered to be fair-good. For the comparison scenario where NCM is produced via conventional hydrometallurgical recycling, the life cycle inventory has been checked for completeness, with no data points left out. The completeness for this comparison scenario was considered to be good.

- Precision: the data for the production of NCM and LiOH·H$_2$O are based on primary information sources provided by RecycLiCo™, from their 2021 engineering studies. All background data sources were sourced from Ecoinvent 3.7.1 or LCA Commons, of which the precision was documented. The precision of the data was considered to be good. For the comparison where NCM is produced from primary raw materials, some datapoints are model upon assumptions and various unconnected studies and is the best available attempt at a unifying calculation in the current literature climate. We are confident in the results for the data we have available, but this may not reflect true impacts when more work is done to refine cathode manufacturing for the lithium-ion battery manufacturing processes. The precision of this comparison scenario was considered to be fair. For the comparison scenario where NCM is produced via conventional hydrometallurgical recycling, the data was based on published academic studies on operating hydrometallurgical recycling plants in China, in February 2022. The precision for this comparison scenario was considered to be good.

TABLE 7.11
Data Quality Assessment for the Project Life Cycle Inventory

Data Quality Indicator	NCM Production	Precursor Production
Technological representativeness	1	1
Time representativeness	1	1
Geographical representatives	2	2
Completeness	1	1
Precision	2	2
Methodological appropriateness and consistency	3	3
Data quality rating	**2.2**	**2.2**
Data quality level	**Basic quality**	**Basic quality**

- Methodological appropriateness and consistency: all primary data was pro-vided by RecycLiCo™, with the same level of detail for the NCM and pre-cursor production. Multi-functionality is dealt with, when required, using mass allocation. However, as the use-phase and end of life-phase of the products are not included, it only covers a part of the product life cycle. *The methodological appropriateness was considered to be fair.* From the LCI used, the clarity of the datasets and the complete modelling methodology used, it should be possible for an external party to reproduce this study. *The methodological appropriateness of the comparison studies was considered to be fair.*

The data quality rating and data quality level for each area are presented in Table 7.11. A data quality rating (DQR) of less than or equal to 1.6 is considered a data quality level of "high quality." A data quality rating greater than 1.6 and less than or equal to 3.0 is considered a data quality level of "basic quality." A data quality rating of greater than 3.0 is considered a data quality level of "data estimate."

7.9   SENSITIVITY ANALYSIS

Sensitivity analysis was carried out to explore the effects that variations in reagent, material, and energy consumption may have on the final product life cycle impact assessment results. The investigated variations are of the different contributing factors individually.

7.9.1   RecycLiCo™ Process with Electrodialysis – Sensitivity Analysis

The analysis studies the effect of variation of the top five main contributors to GWP in the LCA: natural gas to boiler, electricity NCM stream, ammonium hydroxide (precursor production), sodium hydroxide (precursor production), and electricity (impurity removal + bulk precipitate). The results are presented in Figure 7.23.

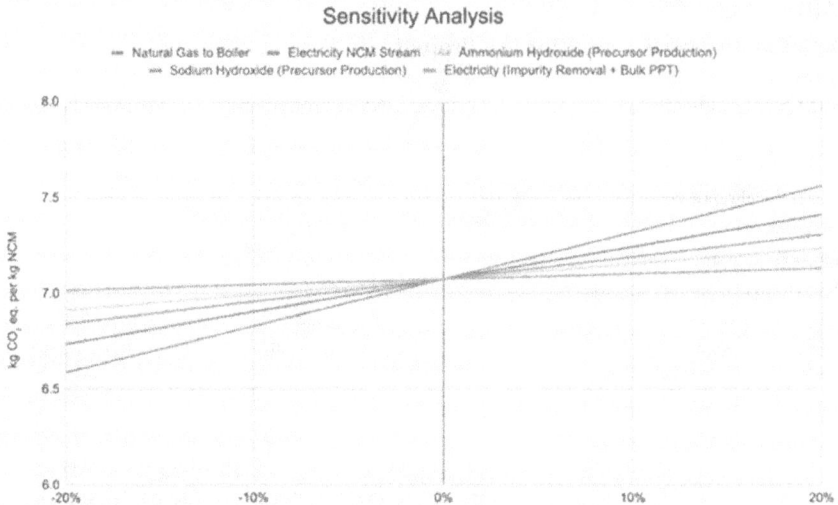

FIGURE 7.23   Sensitivity analysis of major contributors to global warming potential.

This analysis shows that the LCA model for GWP is sensitive to the GWP contributions of natural gas to the boiler, electricity to the NCM stream, and sodium hydroxide (precursor production) consumption within the process because they are some of the biggest contributors. Meanwhile, the model is less sensitive to minor inputs, such as the GWP contributions of ammonium hydroxide (precursor production) and electricity (impurity removal and bulk precipitate).

If the GWP contribution of natural gas to the boiler, the most significant contributor to GWP, increased or decreased by 20%, the total GWP would vary between 6.6 and 7.6 kg CO_2 eq. per kg NCM. Meanwhile, if the GWP contribution of electricity (impurity removal and bulk precipitate), the fifth most significant contributor, increased or decreased by 20%, the total GWP would vary between 7.0 and 7.1 kg CO_2 eq. per kg NCM. This is a clear demonstration of the high sensitivity of the LCA model to a couple major inputs to the process of manufacturing RecycLiCo™'s NCM product. These top contributors should be the main targets of environmental impact mitigation strategies.

7.9.2   PRIMARY RAW MATERIAL NCM PRODUCTION
COMPARISON – SENSITIVITY ANALYSIS

As addressed within Section 7.4.5.2.1, Table 7.6, the nickel sulfate hexahydrate component is the largest input into producing a NCM precursor material, and in turn driving the total impact. Within the comparison scenarios (Section 7.7.1) the impact of nickel sulfate hexhydrate is based off the Nickel Institute Global Average [16]. However, depending on where the nickel component is produced, the impact of NCM precursor from primary raw materials can vary considerably. Table 7.12 shows

TABLE 7.12
Sensitivity Analysis on NCM Production via Primary Raw Material Extraction

	Global Warming Potential (kg CO_2 eq. per kg NCM)				
	−20%	−10%	0%	10%	20%
Nickel Institute	10.8	10.9	11.0	11.1	11.2
Indonesian-sourced nickel	32.2	32.3	32.4	32.5	32.6
Finland-sourced nickel	7.3	7.4	7.5	7.6	7.7

the range in impact of producing NCM precursor from primary raw material extraction depending on (a) different nickel sulfate hexahydrate production routes and locations, and (b) if nickel consumed within the process was increased or decreased by 20%. The data collected on nickel sulfate hexhydrate produced in Indonesia and Finland is sourced from Minviro's internal database.

7.10 UNCERTAINTY ANALYSIS

7.10.1 UNCERTAINTY ANALYSIS FOR THE RECYCLICO™ PROCESS WITH ELECTRODIALYSIS

The uncertainty in the LCI and data quality has been explored in relation to the environmental impacts of the project using a Monte Carlo simulation, which assesses the range and likelihood of different impacts.

The Monte Carlo method, also known as a statistical simulation method, is a precise method of numerical calculation guided by probability statistical theory. The method uses random numbers to solve practical problems, whereby random variables are generated with a certain probability distribution and the numerical characteristics of the model are estimated with statistical methods. The Monte Carlo method randomly samples the values of uncertain variables and combines with the predetermined impact assessment method to simulate statistically significant environmental impact evaluation results. This can reflect the influence of uncertain factors more accurately. The use of Monte Carlo simulation in LCA can effectively present the uncertainty associated with the LCA and its inputs. The results of the Monte Carlo simulation assist with understanding the impact of risk and uncertainty in the prediction and forecasting of models.

In the Monte Carlo simulation conducted for the LCA, uncertainty was addressed by assuming a standard deviation of 15% for all items in the LCI. Emissions cannot be modeled unless the materialistic reason for the emission increases as well (e.g., direct emissions associated with the combustion of diesel). The Monte Carlo simulation does not consider the uncertainty of the background data, as it is assumed that the 15% standard deviation associated with the detailed engineering studies of the project will include the uncertainty of the background data [22]. The Monte Carlo simulation was run through 1,000 iterations using a normal probability distribution.

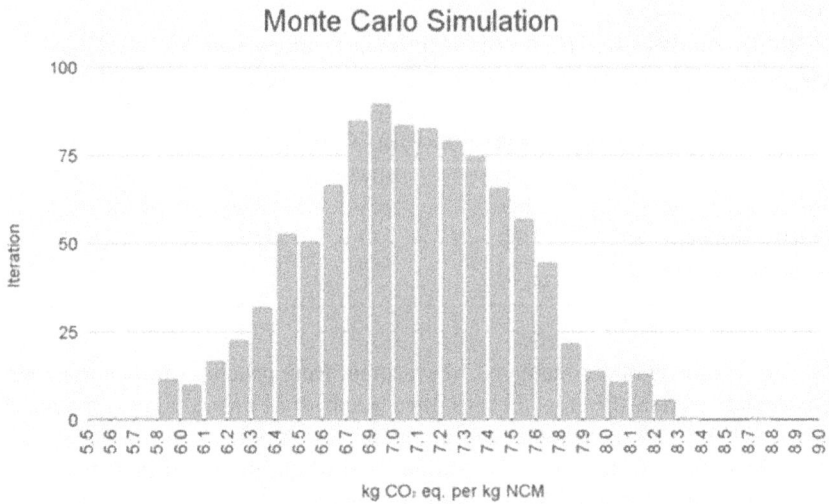

FIGURE 7.24   Monte Carlo simulation for global warming potential.

As the project moves from detailed engineering into operation, the standard deviation will decrease. The results of the Monte Carlo simulation for the GWP are presented in Figure 7.24. Values generated in the simulation range from 5.6 to 8.7 kg
CO_2 eq. per kg NCM. The mean value of the simulation is 7.0 kg CO_2 eq. per kg
NCM, which is the same as the finding of the LCA. This is expected, as an equal
uncertainty was assigned for all energy and material inputs. Statistics describing the
results of the Monte Carlo simulation are shown in Table 7.13.

TABLE 7.13
**Statistics Describing Results of the Monte
Carlo Simulation of Global Warming Potential**

Parameter	Global Warming Potential (kg CO_2 eq.)
LCA study result	7.1
Mean	7.0
Median	7.0
Minimum	5.6
Maximum	8.7
20th percentile value (P20)	6.6
80th percentile value (P80)	7.5
Standard Deviation	0.5

7.10.2   UNCERTAINTY ANALYSIS FOR THE COMPARISON SCENARIOS

7.10.2.1   Uncertainty Analysis for NCM Production
via Primary Raw Material Extraction

The results of the Monte Carlo simulation for the GWP of the comparison route for NCM production from primary raw materials is presented in Figure 7.25 and Table 7.14. Values generated in the simulation range from 6.3 to 15.6 kg CO_2 eq. per kg NCM. The mean value of the simulation is 10.6 kg CO_2 eq. per kg NCM, which is very similar to the finding to the LCA study. This is expected, as an equal uncertainty was assigned for all energy and material inputs.

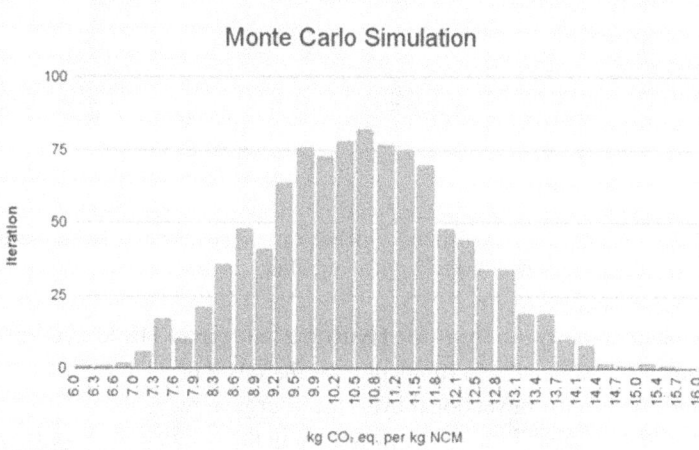

FIGURE 7.25   Monte Carlo simulation for global warming potential of NCM precursor production from primary raw material production.

TABLE 7.14

Statistics Describing Results of the Monte Carlo Simulation of the Global Warming Potential for NCM Precursor Production from Primary Raw Material Production

Parameter	Global Warming Potential (kg CO_2 eq.)
LCA Study Result	11.0
Mean	10.5
Median	10.5
Minimum	7.9
Maximum	12.8
20th Percentile Value (P20)	9.9
80th Percentile Value (P80)	11.2
Standard Deviation	0.8

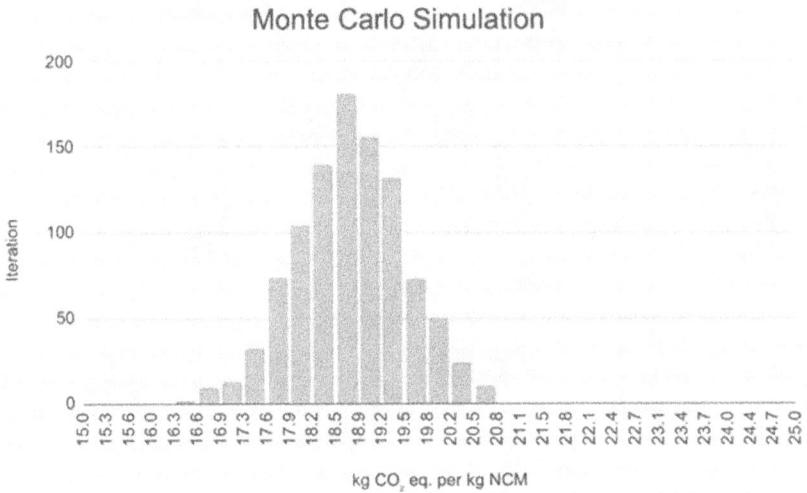

FIGURE 7.26   Monte Carlo simulation for global warming potential of NCM precursor production from conventional hydrometallurgical recycling.

7.10.2.2   Uncertainty Analysis for Hydrometallurgical LIB Recycling

The results of the Monte Carlo simulation for the GWP of the comparison route for NCM production from conventional hydrometallurgical recycling of LIB's is shown in Figure 7.26 and Table 7.15. Values generated in the simulation range from 16.3 to 20.8 kg CO_2 eq. per kg NCM. The mean value of the simulation is 18.8 kg CO_2 eq. per kg NCM, which is very similar to the finding to the LCA study. This is expected, as an equal uncertainty was assigned for all energy and material inputs.

TABLE 7.15
Statistics Describing Results of the Monte Carlo Simulation of Global Warming Potential for NCM Precursor Production from Conventional Hydrometallurgical Recycling

Parameter	Global Warming Potential (kg CO_2 eq.)
LCA Study Result	18.8
Mean	18.8
Median	18.8
Minimum	16.3
Maximum	20.8
20th Percentile Value (P20)	18.1
80th Percentile Value (P80)	19.4
Standard Deviation	0.8

7.11 CONCLUSIONS

Using RecycLiCo™ with electrodialysis data is defined in the detailed engineering studies produced in 2021, the expected environmental impacts of RecycLiCo™'s NCM and LiOH·H₂O products have been quantified using LCA. The impact categories investigated in this LCA were GWP, water scarcity footprint, acidification potential, minerals and metal depletion, and fossil fuel depletion.

It was found that the GWP of producing 1 kg of NCM is 7.1 kg CO_2 eq. per kg NCM. The main drivers of GWP impact are the consumption of natural gas in the boiler, electricity supplied to the NCM stream, and sodium hydroxide consumed in precursor production. The fossil fuel depletion impact category follows the same trend as the GWP.

The acidification potential and minerals and metal depletion follow the same trend, with the embodied impact of sulfuric acid and sodium hydroxide consumed within precursor production being the main impact drivers. For the water scarcity footprint, the embodied impact of electricity consumed within the process is the largest impact driver due to the large volume of water lost from evaporation in hydroelectric power generation from alpine/non-alpine reservoirs.

The environmental impact of LiOH·H₂O co-product was also assessed. It was found that the GWP of producing 1 kg of LiOH·H₂O is 3.3 kg CO_2 eq. per kg LiOH·H₂O. The main impact drivers for each impact category follow the same trend as NCM production.

Comparison scenarios were conducted to understand the difference in impact when using the RecycLiCo™ process with electrodialysis to produce NCM and LiOH·H₂O compared with using primary raw material extraction methods or conventional hydrometallurgical recycling. For both comparisons, the LCA results indicate RecycLiCo™ process with electrodialysis has a lower environmental impact when producing NCM and LiOH·H₂O compared to primary raw material extraction and conventional hydrometallurgical recycling methods.

APPENDIX A: LIFE CYCLE INVENTORY SUMMARY

The full LCI used in this study is presented in Table 7.16.

TABLE 7.16

Summary of the Project's Life Cycle Inventory

Inventory Item	Process Stage	Ecoinvent 3.7.1 Process Flow	Country Code	Input/Output	Data Quality Rating	Value	Unit	Reference Unit
Process feed	NCM precursor/ LiOH stage			Input	Basic quality	21	kg	per hour
Sulfuric acid	NCM precursor/ LiOH Stage	Sulfuric acid// market for sulfuric acid	[RoW]	Input	Basic quality	1.0	kg	per hour
Flocculant	NCM Precursor/ LiOH stage	Polyacrylamide// market for polyacrylamide	[GLO]	Input	Basic Quality	8.3E-4	kg	per hour
Sodium hydroxide	NCM precursor/ LiOH stage	Sodium hydroxide, without water, in 40% solution state// market for sodium hydroxide, without water, in 50% solution state	[GLO]	Input	Basic quality	1.0E-6	kg	per hour
Sodium carbonate	NCM precursor/ LiOH stage	Soda ash, dense// market for soda ash, dense	[GLO]	Input	Basic quality	1.0E-6	kg	per hour
Hydrogen peroxide	NCM precursor/ LiOH stage	Hydrogen peroxide, without water, in 35% solution state// market for hydrogen peroxide, without water, in 50% solution state	[RoW]	Input	Basic quality	1.0E-6	kg	per hour
Natural gas to boiler	NCM precursor/ LiOH stage	Heat, central or small-scale, natural gas// heat production, natural gas, at boiler condensing modulating <100kW	[RoW]	Input	Basic quality	13.1	kg	per hour

(Continued)

TABLE 7.16 (Continued)
Summary of the Project's Life Cycle Inventory

Inventory Item	Process Stage	Ecoinvent 3.7.1 Process Flow	Country Code	Input/Output	Data Quality Rating	Value	Unit	Reference Unit
Natural gas to NCM	NCM precursor	Heat, central or small-scale, natural gas// heat production, natural gas, at boiler condensing modulating <100kW	[RoW]	Input	Basic quality	0.3	kg	per hour
Natural gas to LiSO₄ dryer	LiOH stage	Heat, central or small-scale, natural gas// heat production, natural gas, at boiler condensing modulating <100kW	[RoW]	Input	Basic quality	0.2	kg	per hour
Natural gas LiOH dryer	LiOH stage	Heat, central or small-scale, natural gas// heat production, natural gas, at boiler condensing modulating <100kW	[RoW]	Input	Basic quality	0.09	kg	per hour
Electricity material handling + leaching	NCM precursor/ LiOH stage	Electricity, high voltage// market for electricity, high voltage	[CA-BC]	Input	Basic quality	1,953	kWh	per ton product
Electricity NCM stream	NCM precursor	Electricity, high voltage// market for electricity, high voltage	[CA-BC]	Input	Basic quality	17,800	kWh	per ton product
Electricity LiOH stream	LiOH stage	Electricity, high voltage// market for electricity, high voltage	[CA-BC]	Input	Basic quality	6,983	kWh	per ton product
Fresh water supply	NCM precursor/ LiOH stage			Input	Basic quality	244	kg	per hour
SO₂ (Li₂SO₄ evaporator vent)	LiOH stage	Sulfur Dioxide/sulfur dioxide direct emissions	GLO	Direct emissions	Basic quality	1	kg	per hour
CO₂ (boiler off gas)	NCM precursor/ LiOH stage	Carbon dioxide/carbon dioxide direct emissions	GLO	Direct emissions	Basic quality	1	kg	per hour

(Continued)

TABLE 7.16 (Continued)
Summary of the Project's Life Cycle Inventory

Inventory Item	Process Stage	Ecoinvent 3.7.1 Process Flow	Country Code	Input/ Output	Data Quality Rating	Value	Unit	Reference Unit
Bulk precipitate	0			Intermediate		1	kg	per kg product (NCM)
Ammonium hydroxide (precursor production)	Precursor production	Ammonia, anhydrous, liquid// market for ammonia, anhydrous, liquid	[RoW]	Input	Basic quality	0.2	kg	per kg product (NCM)
Sulfuric acid (precursor production)	Precursor production	Sulfuric acid// market for sulfuric acid	[RoW]	Input	Basic quality	1	kg	per kg product (NCM)
Hydrogen peroxide (precursor production)	Precursor production	Hydrogen peroxide, without water; in 35% solution state// market for hydrogen peroxide, without water, in 50% solution state	[RoW]	Input	Basic quality	0.2	kg	per kg product (NCM)
Sodium hydroxide (precursor production)	Precursor production	Sodium hydroxide, without water; in 40% solution state// market for sodium hydroxide, without water, in 50% solution state	[GLO]	Input	Basic quality	1	kg	per kg product (NCM)
Electricity (Impurity removal + bulk PPT)	Precursor production	Electricity, high voltage// market for electricity, high voltage	[CA-BC]	Input	Basic quality	2,963	kWh	per ton product (NCM)
Water (precursor production)	Precursor production			Input		3	kg	per kg product (NCM)
NCM product				**Product**		16	kg	per hour
LiOH product				**Product**		2	kg	per hour
Li₂SO₄ (lithium sulfate) intermediate				Intermediate		8	kg	per hour
Aluminum coproduct				Coproduct		1	kg	per hour
Sodium sulfate product				Coproduct		0	kg	per hour

APPENDIX B: NON-AGGREGATED CONTRIBUTION ANALYSIS

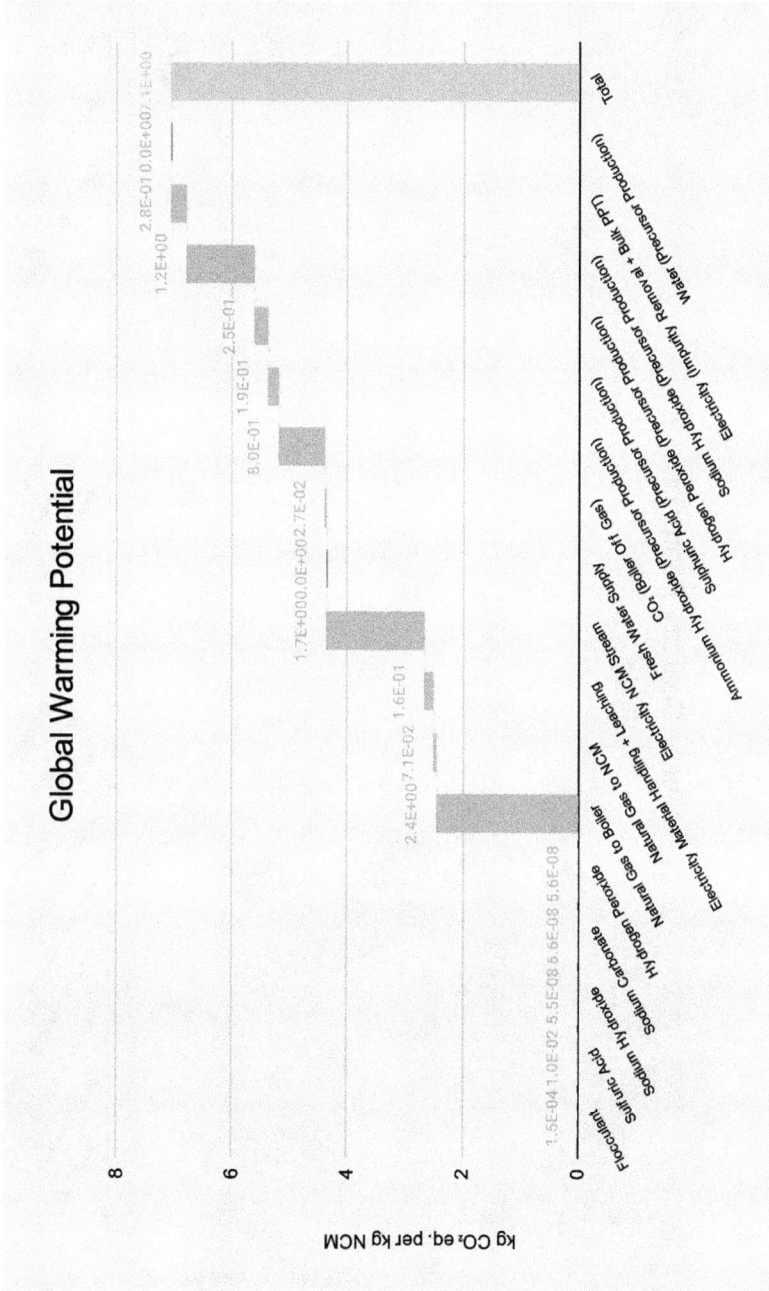

Global Warming Potential

Water Scarcity Footprint

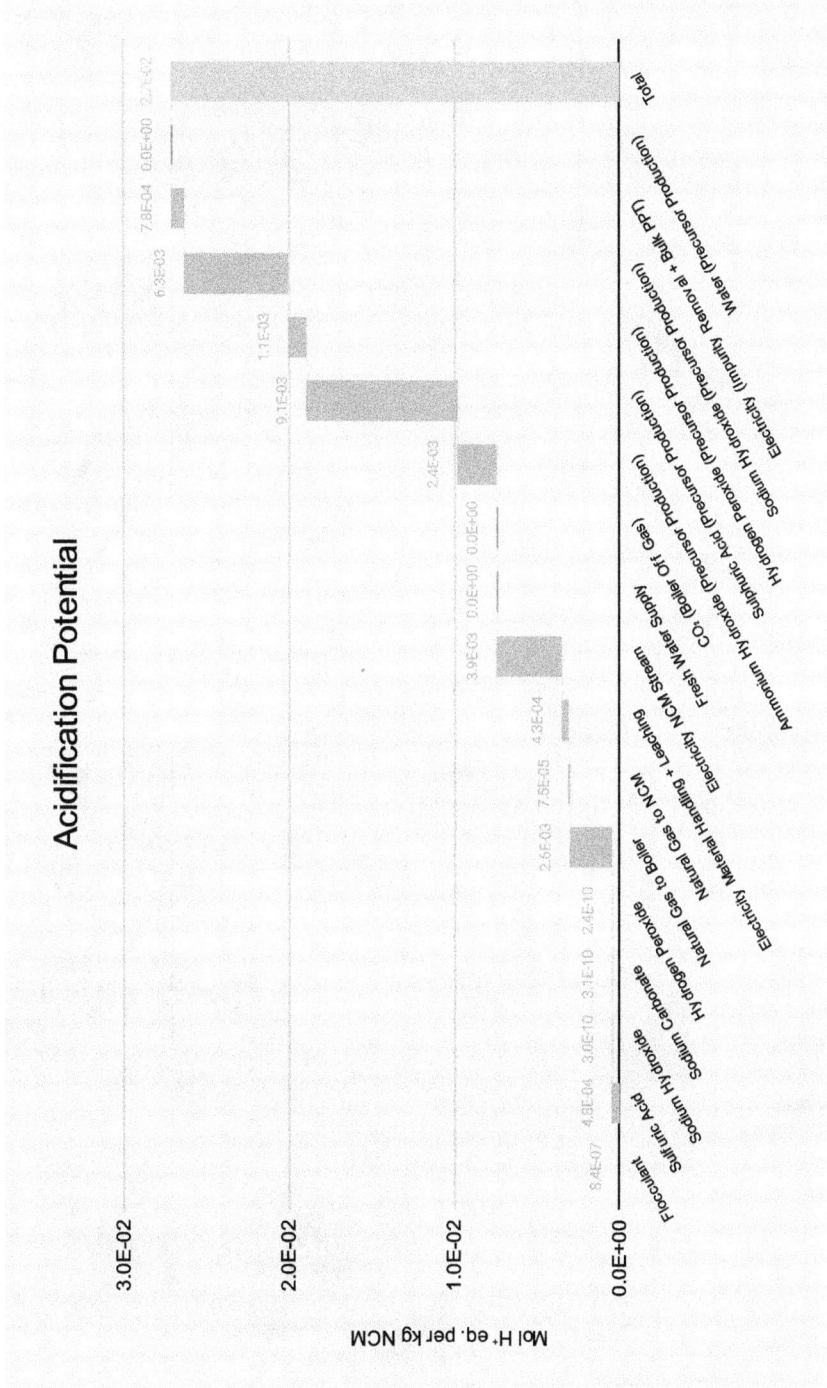

Acidification Potential

Mol H⁺ eq. per kg NCM

Minerals + Metal Depletion

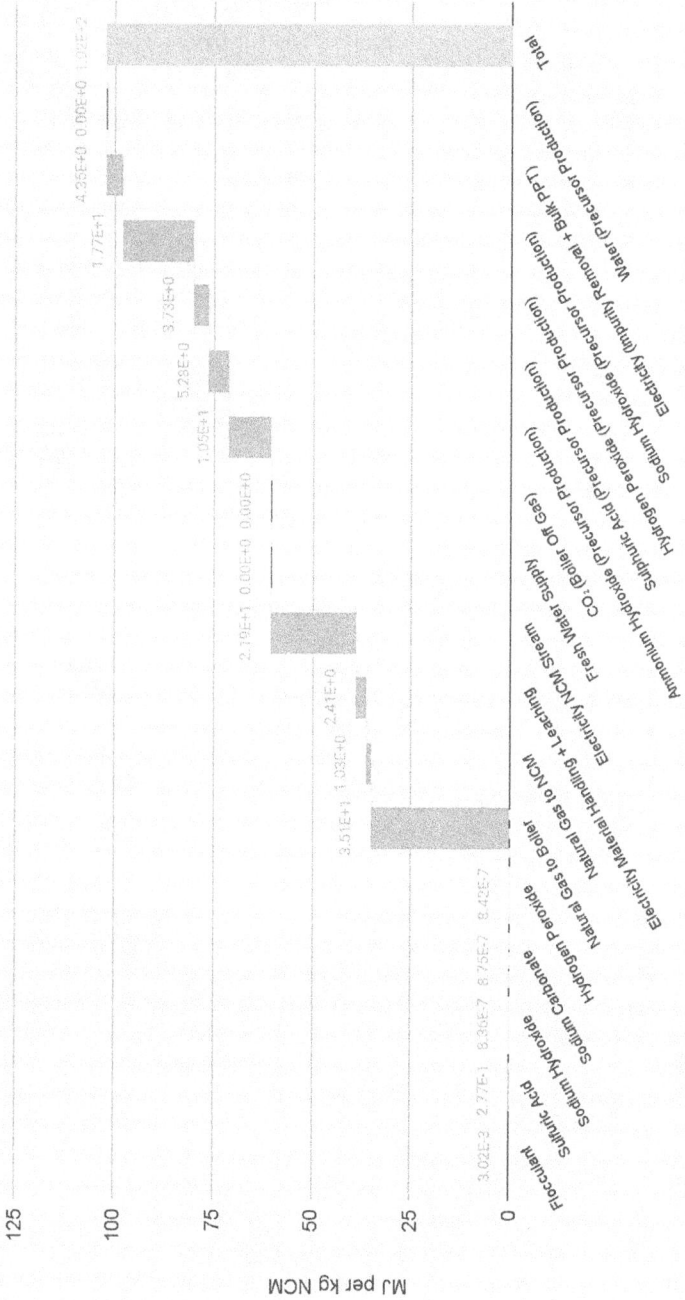

Fossil Fuel Depletion

REFERENCES

1. Corporate-Body. IES:Institute for Environment & Sustainability. *International Reference Life Cycle Data System (ILCD) Handbook: General Guide for Life Cycle Assessment: Detailed Guidance.* Publications Office of the European Union, (2011).
2. Sustainable lithium-ion battery recycling. *American Manganese Inc* https://american-manganeseinc.com/
3. Finkbeiner M., Tan R., Reginald M. Life cycle assessment (ISO 14040/44) as basis for environmental declarations and carbon footprint of products. in *ISO Technical Committee 207 Workshop, Norway* (2011).
4. Horne R.E., Grant T., Verghese K., *Life Cycle Assessment: Principles, Practice and Prospects.* Csiro Publishing, (2009).
5. Santero N., Hendry J., Harmonization of LCA methodologies for the metal and mining industry, International Journal of Life Cycle Assessment, 21, 1543–1553, (2016).
6. ISO. *ISO 14040:2006 – Environmental Management – Life Cycle Assessment – Principles and Framework.* (2006).
7. International Standard Organization (ISO). *ISO 14044: Environmental Management — Life Cycle Assessment — Requirements and Guidelines.* (2006).
8. Wernet G. *et al.* The ecoinvent database version 3 (part I): Overview and methodology, International Journal of Life Cycle Assessment, 21, 1218–1230, (2016).
9. Mohr M., Peters J.F., Baumann M., Weil M., Toward a cell-chemistry specific life cycle assessment of lithium-ion battery recycling processes, Journal of Industrial Ecology, 24, 1310–1322, (2020).
10. Alipanah M. *et al.* Value recovery from spent lithium-ion batteries: A review on technologies, environmental impacts, economics, and supply chain, Clean Technologies and Recycling, 1, 152–184, (2021).
11. Du S. *et al.* Life cycle assessment of recycled NiCoMn ternary cathode materials prepared by hydrometallurgical technology for power batteries in China, Journal of Cleaner Production, 340, 130798, (2022).
12. Wu Y., Yang L., Tian X., Li Y., Zuo T., Temporal and spatial analysis for end-of-life power batteries from electric vehicles in China, Resources, Conservation and Recycling, 155, 104651, (2020).
13. Argonne GREET Publication: Update of Bill-of-Materials and Cathode chemistry addition for Lithium-ion Batteries in the GREET® Model. https://greet.es.anl.gov/publication-bom_lib_2020
14. Thomitzek M., von Drachenfels N., Cerdas F., Herrmann C., Thiede S., Simulation-based assessment of the energy demand in battery cell manufacturing, Procedia CIRP, 80, 126–131, (2019).
15. Erakca M. et al. Energy flow analysis of laboratory scale lithium-ion battery cell production, iScience, 24, 102437, (2021).
16. Nickel life cycle assessments. https://nickelinstitute.org/policy/nickel-life-cycle-management/life-cycle-assessments/
17. Stocker T.F., *Climate Change 2013: The Physical Science Basis: Part of the Working Group I Contribution to the Fifth Assessment Report of the Intergovernmental Panel on Climate Change.* Cambridge University Press, (2013).
18. Intergovernmental Panel on Climate Change. *Climate Change 2013: The Physical Science Basis: Working Group I Contribution to the Fifth Assessment Report of the Intergovernmental Panel on Climate Change.* Cambridge University Press, (2014).
19. Boulay A.-M. *et al.* The WULCA consensus characterization model for water scarcity footprints: Assessing impacts of water consumption based on available water remaining (AWARE), International Journal of Life Cycle Assessment, 23, 368–378, (2018).

20. Alcamo J., Shaw R.W., Hordijk L., The RAINS model of acidification, Science and Strategies in Europe, 30, 1–23, (1991).
21. Seppälä J., Posch M., Johansson M., Hettelingh J.-P., Country-dependent characterisation factors for acidification and terrestrial eutrophication based on accumulated exceedance as an impact category indicator (14 pp), International Journal of Life Cycle Assessment, 11, 403–416, (2006).
22. Nethery B., The role of feasibility studies in mining ventures. in *Conference Board of Canada, Structuring More Effective Mining Ventures, AMEC Mining and Metals, Vancouver* (2003).
23. EoL and recycling. *Sustainability Impact Metrics* https://www.ecocostsvalue.com/lca/eol-and-recycling/(2020).
24. Wikipedia contributors. BC Hydro. *Wikipedia, The Free Encyclopedia.* https://en.wikipedia.org/w/index.php?title=BC_Hydro&oldid=1078121258 (last edited on 20 March 2022).
25. European Commission. *European Platform on Life Cycle Assessment.* https://eplca.jrc.ec.europa.eu/

Index

Note: Locators in *italics* represent figures and **bold** indicate tables in the text.

For Product Safety Concerns and Information please contact our EU
representative GPSR@taylorandfrancis.com
Taylor & Francis Verlag GmbH, Kaufingerstraße 24, 80331 München, Germany